AF861825

Nadia Sebbar

Thermochemistry and Kinetics for the Oxidative Degradation of Dibenzofuran and Precursors

Thermochemistry and Kinetics for the Oxidative Degradation of Dibenzofuran and Precursors

by
Nadia Sebbar

universitätsverlag karlsruhe

Dissertation, Universität Karlsruhe (TH)
Fakultät für Chemieingenieurwesen und Verfahrenstechnik, 2006

Impressum

Universitätsverlag Karlsruhe
c/o Universitätsbibliothek
Straße am Forum 2
D-76131 Karlsruhe
www.uvka.de

SOME RIGHTS RESERVED
Dieses Werk ist unter folgender Creative Commons-Lizenz
lizenziert: http://creativecommons.org/licenses/by-nc-nd/2.0/de/

Universitätsverlag Karlsruhe 2006
Print on Demand

ISBN-13: 978-3-86644-085-2
ISBN-10: 3-86644-085-5

Thermochemistry and kinetics for the Oxidative Degradation of Dibenzofuran and Precursors

Zur Erlangung des akademischen Grades

eines

DOKTORS DER INGENIEURWISSENSCHAFTEN (DR. –ING.)

Von der

Fakultät für Chemieingenieurwesen der
Universität Fridericiana Karlsruhe (Technische Hochschule)

genehmigte

DISSERTATION

von

Magister –Ing. Nadia Sebbar
aus Oran

Tag des Kolloquiums	31. Juli 2006
Referent:	Prof. Dr.-Ing H. Bockhorn
Korreferent:	Prof. Dr. J. W. Bozzelli

A papa
A maman

I am exceptionally appreciative to my advisors, Prof. Dr.-Ing. Henning Bockhorn and Distinguished Prof. Joseph W. Bozzelli, for their guidance, support, and patience.

I am deeply thankful to Dr. Hans-Heinrich Carstensen for the time he spent reading this thesis and for his valuable advices and assistance.

I would like to thank, Prof. Dr.-Ing Matthias Kind, Prof. Dr. Karlheinz Schaber, and Prof. Dr.-Ing. Clemens Posten for accepting to be in the dissertation committee.

My special thanks are for Prof. Dr. rer. nat. Karl Griesbaum who is the "initiator" of this accomplishment.

I would like to thank Dr. habil. Rainer Suntz for his "intensive courses" and for helping me understand different physical problems.

To Leonhard Rutz I address my warm thanks for the constructive discussions and for his permanent help.

I thank my colleagues, Dr.-Ing. habil Jochen Fröhlich, Dr. Olivier Roussel, Dr. Mickael Lecanu, Dr. Jordan Denev, Dr. Matthias Hettel, and Walter Pfeffinger for their friendship and encouragments.

To Dr. Astrid Schön, a special thank for the special relationship I have with Lilith.

Je remercier infiniment Yazid, Hinda et Yasmina « d'avoir été là » tout simplement.

A mes parents, qui m'ont tout donné, je ne puis qu'exprimer l'immense gratitude que j'ai pour eux.

Zusammenfassung

Mit dem Beginn der Industrialisierung und dem damit einhergehenden Anstieg des Lebensstandards hat der Bedarf an Primärenergie, Chemikalien und Transport eine enorme Expansion erfahren. Diese Zunahme ist durch technische Errungenschaften im Bereich der technischen Chemie, der chemischen Prozess- und Anlagentechnik möglich gewesen.

Trotz des enormen Fortschrittes führt die unkontrollierte Produktion unvermeidbar zu Umweltproblemen, wie Klimaveränderungen oder zur Emission toxischer Produkte. Es werden beispielsweise Dioxine und Furane von den verschiedenen Quellen wie offenen Müll-verbrennungsanlagen, bei thermischen Prozessen in der metallurgischen Industrie, Hausfeuerungen, Kraftfahrzeugen, industriellen Dampfkesseln, Verbrennungsanlagen für Abfallöle unbeabsichtigt gebildet und in die Umwelt freigegeben.

Die unkontrollierte Freisetzung von Dioxinen und Furanen in die Umwelt führt dazu, dass diese Substanzen sich in sehr niedrigen Konzentrationen in vielen Nahrungsquellen wieder finden. Als Folge davon können diese je nach Expositionsgrad toxisch sein. Oberhalb des „Tolerable Daily Intake“ (TDI) können Dioxine oder Furane Hautveränderungen hervorrufen oder die Leberfunktion beeinträchtigen. Langfristige Exposition mit diesen Substanzen kann zur Schädigung des Immunsystems, des Nervensystems, des Drüsensystems und zu genetischen Veränderungen führen. Dauerhafte Exposition von Tieren mit Dioxinen hat bei einigen Arten Krebs hervorgerufen.

Das Problem mit Dioxinen und Furanen ist, dass sie sehr stabile chemische Substanzen sind. Sie sind daher in der Umwelt überall vorhanden und beide chemische Substanzgruppen bauen sich sehr langsam ab, so dass der Vermeidung der Emissionen besondere Bedeutung zukommt. Um dieses Ziel zu erreichen, stehen zwei Möglichkeiten zur Verfügung:

1. die Bildung dieser Dioxine und Furane a priori zu verhindern
2. diese Dioxine und Furane zu zerstören, bevor sie in die Umwelt gelangen.

Evidenterweise sollte die beste Strategie auf der Vermeidung der Bildung von Dioxinen und Furane beruhen, jedoch ist dieses häufig aus ökonomischen Gründen nicht realisierbar. Da unter Verbrennungsbedingungen sowohl die Bildung als auch der Abbau von Dibenzofuran ablaufen, konzentriert sich die vorliegende Arbeit auf beide Prozesse. Die Beschränkung in der vorliegenden Arbeit auf Dibenzofuran ist gerechtfertigt, da es als eine geeignete Modellsubstanz für Dioxine angesehen werden kann.

Soweit es der numerische Aufwand erlaubt wird in der vorliegenden Arbeit versucht ein besseres Verständnis des Abbaumechanismus des Dibenzofuranyl + O_2- System sowie dessen wichtige Reaktionspfade der sauerstofffreien Aufspaltung zu gewinnen. Um dieses Ziel zu erreichen, muss ein Haupthindernis überwunden werden, das in der Größe der aromatischen Spezies des Dibenzofuranyl + O_2- System liegt. Für große Spezies sind nur ungenaue quantenmechanische Berechnungen realisierbar, die aber nicht imstande sind genaue thermochemische Daten zur Verfügung zu stellen. Um dieses Problem zu umgehen, wurde bei der Untersuchung des Dibenzofuranyl + O_2- System wie folgt vorgegangen:

- Die DFT- Methode wird verwendet. Sie ist in der Lage mit großen Spezies umgehen zu können, schnell Ergebnisse zu liefern und zufrieden stellende Energie zu liefern. Wir konnten zeigen, dass wenn die DFT- Methode mit isodesmischen Reaktionen kombiniert wird, die Genauigkeit erheblich verbessert und die Ergebnisse für thermochemische Auswertungen verwendbar werden.
- Das Dibenzofuranyl + O_2- System wird zunächst auf ein kleineres aber ähnliches System reduziert und dann wird mit der Berechnung der thermochemischen Eigenschaften fortgefahren (s. Abbildung 1,1). Basierend auf dieser Näherung kann das Dibenzofuranyl + O_2 System (a) auf das Phenyl + O_2- System (b) reduziert werden, das seinerseits wiederum zum Vinyl + O_2- System (c) vereinfacht werden kann.

Wir vergleichen und zeigen, dass das Vinylradikal eine gute Modellspezies für das Phenylradikal ist, das auch eine Modellspezies für das Dibenzofuranylradikal ist. Dieses impliziert, dass die hochgenaue Berechnungen, die für das kleinere Vinylsystem durchgeführt wurden, benutzt werden können, um die DFT- oder die ab initio- Berechnungen für das Phenyl- oder das Dibenzofuranyl- Systems zu kalibrieren. Die Ergebnisse, die von dem Phenyl- System erhalten wurden, werden auf dieselbe Weise für das Dibenzofuranyl + O_2 System verwendet.
Zu diesem Zweck wurden das Reaktionsverhalten der drei Systeme, Vinyl + O_2, Phenyl + O_2 und Dibenzofuranyl + O_2 und ihre Produkte untersucht.
Somit ist unser erstes Ziel, thermochemische Eigenschaften einer Anzahl von den Spezies, die Peroxy- und Hyperoxydgruppen (Radikale und Moleküle) mit einbeziehen und die wesentlich in dieser Arbeit benötigt werden, zu ermitteln. Wegen des erheblichen Mangels an vorhandenen Daten aus der Literatur und um ein Verständnis der Verbrennungmechanismen zu erhalten haben wir unsere Berechnungen auf eine große Auswahl von Peroxid- Spezies erweitert. Für diese Rechnungen wird die Dichtefunktionaltheorie- Methode (DFT), die mit isodesmischen Reaktionen gekoppelt wird, verwendet.

Die DFT- Methode wurde vor kurzem als eine wichtige Ergänzung zur hochgenauen ab initio Methode erhalten. Die DFT könnte eine der wenigen anwendbaren Berechnungsmethoden für große Molekülsysteme sein. Generell ist die mittels DFT erzielte Genauigkeit niedriger als die von hochgenauen ab initio- Methoden. Die Genauigkeit der Dichtefunktionaltheorie kann jedoch mit Hilfe isodesmischer Reaktionen verbessert werden. Die meisten Berechnungen basieren auf isodesmischen Reaktionen, in denen die berechnete Werte mit experimentellen oder errechneten Bildungsenthalpien von geeigneten Referenzsystemen kombiniert werden. Hier entsteht jedoch ein anderes Problem, nämlich dass die experimentellen Bildungsenthalpien, die in den isodesmischen Reaktionen benötigt werden, nicht immer bekannt sind oder durch große Fehler beeinträchtigt sind.

Zeitraubende quantenmechanische Berechnungen sind nicht immer erschwingliche Methoden um thermochemische Daten zu erhalten. Häufig wird lediglich ein schneller Schätzwert benötigt, selbst wenn die Ergebnisse weniger genau sind. Eine schnelle Methode für die Schätzung der thermochemischen Eigenschaften ist die Gruppenbeitragmethode. Besonders die von Benson entwickelte Gruppenadditivitätsmethode (GA) ist sehr populär. Diese Methode ist wegen ihrer verhältnismäßig guten Genauigkeit für die organischen und oxydierten organischen Spezies für die Schätzung der thermochemischen Daten weit verbreitet. Leider fehlen viele Gruppen, die auf Peroxyd- Spezies und Sauerstoff enthaltenden Kohlenwasserstoffe bezogen sind. Die Berechnung der thermodynamischen Daten dieser Gruppen stellt eine wichtige Zielsetzung der vorliegenden Arbeit dar.
Die Untersuchung des Phenyl- Radikals mit dem O_2- Reaktionssystem war der zweite Schritt in unserer systematischen Annäherung. Dieser Teil ist von Bedeutung, weil die elektronische Struktur des Phenyl- Radikals zu der des Dibenzofuranyl- Radikals analog ist. Das Phenyl-System kann folglich als Modell benutzt werden, wobei die Oxidation anderer heteroatomarer aromatischer Radikale untersucht werden kann. Thermodynamische Eigenschaften der Zwischenprodukte, der Übergangszustände und der Produkte, die wichtig zur Bildung und Zerstörung des aromatischen Ringes sind, wurden berechnet. Das Vinylsystem wurde benutzt, um ab initio- und DFT- Berechnungen für das Phenyl- System zu kalibrieren. Teilweise basierend auf den Ergebnissen des Vinyl- Systems werden mit der kanonischen Übergangszustands- Theorie kinetische Parameter im Hochdrucklimit erhalten. Ein elementarer Mechanismus wurde konstruiert, um die experimentellen Daten, die in einer Brennkammer bei 1 atm und in Hochdruckturbinesystemen (5-20 ATM) sowie in überkritischem Wasser erhalten wurden, zu modellieren.
Im letzten Teil der vorliegenden Arbeit wurden Bildungsenthalpien einer Reihe stabiler Spezies und Radikalen, die auf das Dibenzofuranyl + Reaktion O_2- System bezogen wurden, ausgewertet. Berechnungen wurden für den Abbau der Ringe mit DFT- und Gruppenadditivitäts-Methoden durchgeführt. Nachdem nachgewiesen wurde, dass das Phenyl- System als ein gutes Modell für das Dibenzofuranyl- System betrachtet werden kann, wurden die Übergangzustands-Spezies für das Dibenzofuranyl- Reaktion- System aus dem Phenyl- System abgeleitet.

Die vorliegende Arbeit ist folgendermaßen aufgebaut:

In Kapitel 2 werden die Berechnungsmethoden, die für die Ermittlung der thermodynamischen Daten der Spezies verwendet werden und die kinetischen Berechnungen, die in dieser Arbeit durchgeführt werden, dargestellt. Wir geben einen kurzen Bericht der Berechnungsmethoden: ab initio, Dichtefunktionaltheorie, Methoden der statistischen Mechanik, Gruppenadditivität-Methode und QRRK (Multiquantuum Rice- Ramsperger- Kassel).

In Kapitel 3 werden Enthalpien ΔH^0_{f298}, Entropien S^0_{298}, und Wärmekapazitäten $C_p(T)$ für eine Reihe ungesättigter Vinyl- und Ethynyl- Hydroperoxide sowie Phenyl- und Allyl-Hydroperoxide berechnet. Diese Hydroperoxide sowie das entsprechende Vinyl, das Ethynyl, das Allyl und die Peroxy- Radikalen werden mit DFT- Methoden mit dem B3LYP/6-311G(d, p) Basissatz berechnet.

In Kapitel 4 werden thermodynamische Eigenschaften ($\Delta H^0_{f298}, S^0_{298}$ und $Cp_{298}(T)$) einer Serie oxydierter und nicht- oxydierter Spezies berechnet. Diese Moleküle werden wegen den enthaltenden spezifischen Gruppen ausgewählt. Diese Gruppen werden berechnet um später in größeren, ungesättigten oxydierten Spezies Verwendung zu finden. Die oxydierten und nicht-oxydierten Spezies werden mit der Methode der Dichtefunktionaltheorie (DFT), die mit isodesmischen Reaktionen verbunden wurden, ermittelt.

In Kapitel 5 konnte die Zuverlässigkeit der Dichtefunktionaltheorie überprüft werden, indem für ungesättigte Oxykohlenwasserstoffe die mit genannter Methode erhaltenen Bildungsenthalpien mit denjenigen, die mit G3MP2B3 Methode ermittelt wurden, verglichen. Dadurch werden die Oxykohlenwasserstoffgruppen, die vorher durch B3LYP- Methoden entwickelt wurden, verbessert und validiert.

Kapitel 6 stellt die Berechnungen der thermochemischen Eigenschaften der Zwischenprodukte, der Übergangszustände und der Produkte, die für die Zerstörung des aromatischen Rings im Phenyl + O_2- Reaktions- System wichtig sind, dar. DFT- und hochgenaue ab initio- Methoden werden eingesetzt, um die Substituenteffekte auf eine Anzahl von chemische Reaktionen und Prozessen, die Alkyl- und peroxyl- Radikalen mit einbeziehen, zu analysieren. Kinetische Parameter für den Hochdruckbereich werden mit der kanonischen Übergangs-zustandtheorie erhalten, teilweise basierend auf den Ergebnissen, die aus dem Vinyl-System erreicht wurden. Ein elementarer Reaktionsmechanismus zur Modellierung von experimentellen Daten, die in einer Brennkammer bei 1 atm und in den Hochdruckturbinensystemen (5 - 20 atm) sowie in überkritischem Wasser erhalten werden, wird erstellt.

In Kapitel 7 werden Bildungsenthalpien für Spezies berechnet, die in die Aufspaltung des ersten aromatischen Ringes im Dibenzofuran (DBF) mit einbezogen werden. DFT- Berechnungen, Daten von dem Phenyl + O_2- und dem Vinyl + O_2- System werden verwendend. Die Kinetik jedes Reaktionweges wird als Funktion der Temperatur und des Drucks anhand der bimolekularen chemischen Aktivierungsanalyse bestimmt. Die berechneten Daten der kinetischen Parameter im Bereich des Hochdrucklimits werden wieder mit dem kanonischen Übergangszustandtheorie erhalten. Die QRRK- Analyse wird verwendet, um k(E) zu berechnen und das fall-off- Verhalten wird anhand der Mastergleichungs- Analyse untersucht, um diese komplizierte, bimolekulare, chemische Aktivierungsreaktion auszuwerten. Mit der Fertigstellung der vorliegenden Arbeit steht ein elementarer Submechanismus zur Verfügung, der zukünftig verwendet werden kann, um die Aufspaltung von Dibenzofuran im Abgas zu modellieren. Eine direkte Anwendung wäre es den Submechanismus in größere Mechanismen zu implementieren und dessen Vorhersagen mit experimentellen Ergebnissen zu vergleichen.

Zusammenfassend zeigt diese Arbeit, dass ab intio- und DFT- Methoden ausgezeichnete Werkzeuge für die Ermittlung von unbekannten thermochemischen Eigenschaften sind. Die nachgewiesene Genauigkeit der quantenmechanischen Berechnungen lassen sie als geeignete Methoden erscheinen, sie als kosteneffektive Alternativen zu zeitraubenden oder schwierigen Experimenten anzusehen.

Table of Content

1. Introduction

1.1 Background and Motivation

In the recent years, the demand of production for energy, chemicals and transportation has seen a tremendous expansion related to a better standard living. This increase has been possible through scientific expand like technical chemistry or chemical process engineering.

However, despite all the progress being made, the uncontrolled production unavoidably leads to environmental problems such as climate change or emission of toxic products. For example, dioxins and furans are unintentionally formed and released from various sources like open burning of waste, thermal processes in the metallurgical industry, residential combustion sources, motor vehicles, particularly those burning leaded gasoline, fossil fuel-fired utility and industrial boilers, waste oil refineries etc…

If released uncontrolled to the environment, dioxins and furans find their way in very low concentrations into many food sources. They are known to be toxic species, depending however, on the level of exposure. Above the "Tolerable Daily Intake" (TDI) [1] exposure to dioxins or furans may result in skin lesions and altered liver function. Long-term exposure is linked to impairment of the immune system, the developing nervous system, the endocrine system and reproductive functions. Chronic exposure of animals to dioxins has resulted in several types of cancer.

The problem with both dioxins and furans is that they are very persistent chemicals. Their presence in the environment is ubiquitous and both groups of compounds degrade very slowly, that's why it is important to reduce the emissions. This can be done in two ways:

1. to prevent the formation of these dioxins and furans
2. to destruct them before they are released to the environment

Certainly the best strategy would be to avoid the formation of dioxins and furans, however, this often is economically unfeasible. For this reason, this study focuses on the destruction of dibenzofurans, as well as its formation, because both occur under combustion conditions. We restrict ourselves to dibenzofurans which can be seen as a suitable model substance for dioxins as well.

Destruction of the aromatic compounds often starts with the loss of a phenyl-hydrogen through abstraction by an active radical pool, e.g species such as H, O, OH, or Cl. In the case of dibenzofuran, its reaction leads to a dibenzofuranyl radical. Abstraction occurs even at moderate temperatures in downstream zones of an incinerator. Dibenzofuranyl will rapidly react with molecular oxygen in the combustion environment to form an energized adduct called peroxy radical (Fig. 1.1 (A)). This implies that peroxy chemistry plays an important role in the environmental fate of this species.

Peroxy radicals ROO• are key species in the mechanisms of oxidative and combustion systems. At the same time they have been among the most difficult radicals to study experimentally. There are only limited or no thermochemical information available for unsaturated alkylperoxy and hydroperoxide species. An explanation for this paucity of data could be the fact that these species are unstable and short-lived, and therefore difficult to study and characterise by experimental methods. The difficulty arises in part from the lability of these radicals towards reversible unimolecular dissociation into R• + O_2 and then to reversible isomerization into hydroperoxy alkyl radicals •ROOH, both of these reactions occurring at comparable rates at temperatures below 450 K [2].

When reviewing the thermo database for small peroxy radicals, we observed that the thermochemistry of saturated peroxides is somewhat established but little is known on unsaturated peroxides.

Alkyl hydroperoxide species are important in limiting soot formation and in soot burnout [3, 4, 5] in combustion and we show later in this study that unsaturated hydroperoxides are even more important in soot control. Recent studies have also indicated that alkyl and peroxyl radicals are formed during oxidation of proteins in biological systems. Peroxides and peroxy species are also important intermediates in the atmospheric photochemical oxidation of hydrocarbons [6, 7] and in polymerisation initiators. Benson [2], Griffiths [8], Koert et al. [9], and Sheng and Bozzelli [10] have modelled ignition of hydrocarbons and their negative temperature behaviour. They showed that reactions of peroxy species were dominant chain branching paths. Experimental studies on reactions of vinyl radical [11] and allyl radical [12, 13] with O_2 to form the corresponding peroxy radical are reported by Gutman et al. (reviewed and summarized by Knyazev and Slagle [14]). The studies involving the negative ion chemistry of peroxy radicals are known from Clifford et al. [15] and Blanksby et al. [16]. Phenyl-peroxy radicals were reported by Lin et al. as a major product in phenyl radical reaction with O_2 at ambient temperatures [17]. Reactions of ozone with olefins also lead to vinyl peroxide formation, which is thought to rapidly undergo unimolecular dissociation to the corresponding ketene + OH.

Because of the scarcity of experimental data, computational methods need to be employed to predict molecular thermochemical data for the species. Important part of the development of such quantum chemical methods is their critical assessment by comparison to accurate experimental data. Nowadays, different theoretical schemes, such as the G2, CBS, or G3 theory, are able to predict enthalpies of formation for small systems at accuracy within a few kcal per mole. Unfortunately, these approaches, which involve the use of high-level ab initio methods, are limited by the size of molecule systems that can be studied and are still prohibitively costly. A recent review on the use of different theoretical methods for computing enthalpies of formation was published by Curtiss et al. [18, 19]. Density Functional Theory

(DFT) methods have been recently evolved as an important complement to advanced ab initio methods. DFT may be one of the few applicable calculation methods for large molecule systems and have frequently been used. In general their overall accuracy is lower than that achieved with high-level ab initio schemes. The accuracy of the Density Functional Theory can, however, be improved by use of isodesmic reactions. Most of the calculations are based on isodesmic reactions, in which calculated values are combined with experimental or computed enthalpies of formation of suitable reference systems. The use of isodesmic reaction implies that the number of bonds of each formal type is conserved. Here, another problem arises, namely the experimental enthalpies of formation of the reference species needed in isodesmic reactions are not always known or the values are affected by a large error.

Time consuming quantum mechanic calculations are not always affordable methods to obtain thermochemical data for large molecules. Frequently fast estimation is required even if the results are less accurate. A fast method for the estimation of thermochemical properties is the group contribution method. In Particularly the Group Additivity method (GA) [2, 20] developed by Benson is very popular (see section 2.1.5). This method is widely used because of its relatively good accuracy for organic and oxygenated organic compounds in the estimation of thermochemical data. Joback [21, 22] has modified Benson groups so as to include and/or require fewer overall groups and estimate a larger molecule type set; but also exhibits some loss in accuracy. Cohen and Benson have recently tabulated and improved a number of groups estimates for enthalpy and entropy at 298 K of oxygenated hydrocarbons [23, 24]. A second expansion in use of group additivity is the use of Hydrogen Bond Increments (HBI groups) for the estimation of thermochemical parameters of radical species [25]. Unfortunately, many groups related to peroxide species and oxygen- containing hydrocarbons are missing.

1.2 Objectives of the Study

In the present work we attempt to gain a better understanding of the destruction mechanism of dibenzofuranyl + O_2 system as well as the reaction pathways important in its oxygen-free decomposition. The difficulties encountered now are related to the large size of aromatic species contained in the dibenzofuranyl + O_2 system making high level calculations very costly or not possible. The approach we have taken to circumvent the problem is to reduce the system to the smallest representative unit and then to proceed with the computation of the thermochemical properties (Figure 1.1). Based on this approach dibenzofuranyl peroxy radical (A) can be concentrated around the phenyl peroxy radical reaction system (B), which itself can be reduced to vinyl peroxy radical system (C).

Consequently, our first goal is to determine thermochemical properties of a number of species involving peroxy and peroxide groups (radicals and molecules) which are essential and required in this work. Because of the drastic lack of available data in the literature, we have extended our estimation to a wide range of peroxide species in order to gain in understanding of combustion mechanisms. For these estimations the Density Functional Theory method coupled with isodesmic reactions is used.

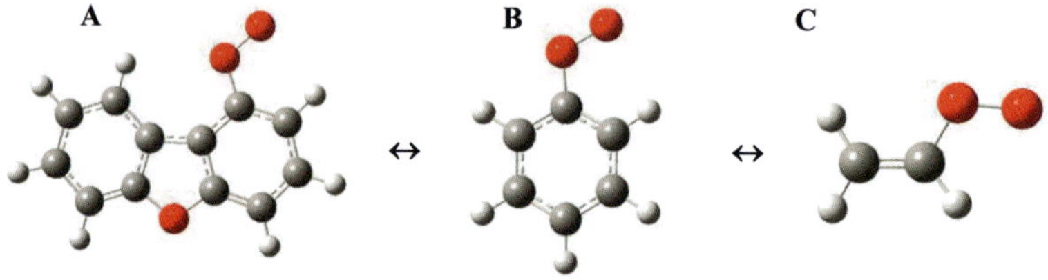

Figure 1.1: Dibenzofuranyl-peroxy / phenyl-peroxy / vinyl-peroxy systems

The decomposition reactions result in a number of unsaturated oxy-hydrocarbon intermediates and radical products, for which thermochemistry is not available. The Group Additivity method (GA) [2, 20] is a fast and reliable method to estimate or check the thermochemistry of unknown or large molecules. In this work we also develop a series of new groups (for use in group additivity (GA)) to aid our evaluation of enthalpies of formation for our system.

This thesis is structured as follows:

In chapter 2, computational methods used for the determination of thermodynamic data of the compound, and the kinetic calculations performed in this work are presented. We give a brief review of the computational methods: ab initio, and Density Functional Theory, Statistical Mechanics methods, Group Additivity method, and multifrequency Quantum Rice-Ramsperger-Kassel (QRRK).

In chapter 3, enthalpies, ΔH^0_{f298}, entropies, S^0_{298}, and heat capacities $C_p(T)$, are reported for a series of unsaturated, vinyl and ethynyl hydroperoxides plus phenyl and allyl hydroperoxide. These hydroperoxides as well as the corresponding vinyl, ethynyl, allyl and peroxy radicals are calculated using DFT calculations at B3LYP/6-311G(d,p) level of theory [26, 27, 28].

Chapter 4 is devoted to develop thermodynamic properties ($\Delta H^0_{f298}, S^0_{298}$ and $Cp_{298}(T)$) of a series of oxygenated and non-oxygenated species [29]. These molecules are selected to contain a specific group (for use in GA) which is estimated in order to use it later in larger unsaturated oxygenated intermediate. These oxygenated and non-oxygenated species are investigated with the Density Functional Theory (DFT) method coupled with isodesmic reactions.

In chapter 5, application of the Density Functional Theory is validated by comparing enthalpy ($\Delta_f H^0_{298}$) results for unsaturated oxy-hydrocarbons with those obtained with the G3MP2B3 method. The results of this study are also utilized to validate or improve the oxy-hydrocarbon groups and radical increments previously developed by B3LYP methods [30].

Chapter 6 presents estimations of thermochemical properties of intermediates, transition states and products important to destruction of the aromatic ring in the phenyl radical + O_2 reaction system. We have employed both DFT and high-level ab initio methods to analyze the substituent effects on a number of chemical reactions and processes involving alkyl and peroxyl radicals. Partially based on the results obtained in the vinyl system, high-pressure-limit kinetic parameters are obtained using canonical Transition State Theory. An elementary reaction mechanism is constructed to model experimental data obtained in a combustor at 1 atm, and in high-pressure turbine systems (5-20 atm), as well as in supercritical water [31].

Chapter 7 estimates enthalpies, $\Delta_f H^0_{298}$ for species involved in the decomposition of the first aromatic ring in dibenzofuran (DBF) using DFT calculations, by comparisons to the phenyl + O_2 data of Hadad et al. [32] and by using the vinyl + O_2 results from Mebel et al. [33]. We compare and show that the vinyl radical is a good model for phenyl, which is a model for dibenzofuranyl radical. This implies that high-level calculations on the smaller vinyl system can be used to calibrate *ab initio* and Density Functional Theory calculations on the phenyl or dibenzofuranyl + O_2 system.

Using the results obtained on the phenyl system for the dibenzofuran + O_2 system, kinetics of each path, as a function of temperature and pressure are determined using bimolecular chemical activation analysis. The high-pressure-limit kinetic parameters from the calculation results are again obtained with cannonical Transition State Theory. QRRK analysis is utilized to obtain *k(E)* and master analysis is used to evaluate the fall-off behaviour of this complex bimolecular chemical activation reaction [34].

2. Theoretical Background, Methods and Software

This chapter is devoted to the presentation of the computational methods used for the determination of the thermodynamic and the kinetic data of the compound considered in this work. We give in a first part a theoretical background of the methods used in the present work: Electronic Structure Theory (ab initio, and Density Functional Theory), Statistical Mechanics theory, Group Additivity method, and multifrequency Quantum Rice-Ramsperger-Kassel theory (QRRK).

The different software required to perform these computations are presented in a second part: Gaussian, for the ab initio and Density Functional Theory calculations, SMCPS to perform the statistical mechanistics calculations, ThermKin and Chemaster for the kinetics calculations and THERM to apply the Group Additivity method.

2.1 Theoretical Background

2.1.1 Overview of Computational Chemistry: the electronic structure theory

Based on the fundamental laws of physics [35], computational chemistry calculates properties, chemical structures and reactions numerically. Some methods can be used to model not only stable molecules, but also short-live, unstable intermediates and even transition states. In this way they can provide information about molecules and reactions which may be impossible to obtain through observation. The *electronic structure theory* (quantum mechanics) is considered as the most appropriate for calculating molecule properties, such as:

- Energy of a particular molecular structure (spatial arrangement of atom or nuclei and electrons)
- Energies and structure of transition states
- Geometries
- Vibrational frequencies
- Rotational frequencies
- Moment of inertia
- And derived properties such as bond energies, enthalpies of formation, entropies, heat capacities…

Quantum mechanics states that the energy and other related properties of a molecule may be obtained by solving the Schrödinger equation:

$$\boldsymbol{H}\Psi = E\Psi \qquad 2.1$$

Where the Hamilton operator ***H*** expresses kinetics (*T*) and potential (*V*) energy for all particles as H = T + V, Ψ is the wave function and its square is the probability of finding the particle at a given position, and *E* is the energy of the particle. Since nuclei are much heavier than electrons, their velocities are much smaller; the Schrödinger equation can therefore to a good approximation be separated into one part which describes a fixed nuclear geometry, and another part which describes the nuclear wave function, where the energy from the electronic wave function plays the role of a potential energy. This separation is called Born-Oppenheimer approximation. The electronic wave function depends parametrically on the nuclear coordinates, it depends only on the position of the nuclei, not their momenta. The picture is that the nuclei move on *Potential Energy Surfaces* (PES), which are solution to the Schrödinger equation.

Except for very simple systems, exact solutions to the Schrödinger equation are, however, only numerically feasible. Various mathematical approximations to its solution characterise the electronic structure methods.

There are two major classes of electronic structure methods [36]: *Semi-empirical methods* and *Ab initio methods*

Semi-empirical methods, use parameters derived from experimental data to simplify the computation. They solve an approximate form of the Schrödinger equation that depends on having appropriate parameters available for the type of chemical system under investigation. Semi-empirical calculations are fast and provide reasonable qualitative description of molecular system and fairly accurate quantitative predictions of energies and structures for systems where good parameter sets exist.

Ab initio methods, (latin: "from the beginning") have been developed for calculating energies and potential energy surfaces. Unlike semi-empirical methods, ab initio methods use no experimental parameters in their computations. Instead, their computations are based solely on the laws of quantum mechanics (the first principles referred to in the name ab initio) and on the values of a small number of physical constants. The calculations use modern quantum calculation procedures to determine optima structures and lowest energies for varied electron and proton arrangements.The accuracy of the calculated results is determined by the number of basis functions and the treatment of electron correlation. For many reactions the calculated structures for potential energy minima are as accurate as those found experimentally [37]. In contrast to semi-empirical, ab initio computation provide high quality quantitative predictions for a broad range of systems. The well regarded G3 method developed by Curtis et al. [38] and its modified version referred as G3MP2B3 [39] method, are used in this work to determine energy properties of molecules, intermediate radicals and transition states (chapter 5 and 6).

A third class of electronic structure methods have recently received widespread use in computational chemistry:

Density Functional Theory (DFT) [40, 41]. DFT calculations yield results comparable to those obtained from ab initio methods but at lower computational costs. DFT methods compute electron correlation via general *functionals* of the electron density (a functional is defined in mathematics as a function of a function).

All DFT computations have been performed at B3LYP/6-311G(d,p) level of theory, where B3LYP refers to the hybrid functionals of Becke's three-parameters formulation [42] and the gradiant correlation function of Lee, Yang and Parr [43]. 6-311G(d,p) is the basis sets which is the mathematical description of the atomic orbitals within a molecule or a system. The basis set can be interpreted as restricting each electron to a particular region of space. For more in depth discussion of the basis sets, see Appendix I and chapter 3.

2.1.2 Statistical Mechanics Method

In a physical system, there are many molecules, which have some probability of occupying any one of the allowable discrete energy levels; this implies that there is a distribution of molecules among the different allowable energy levels. The method of determining this probability distribution is known as statistical mechanics. Once the probability distribution is obtained, the associated energies can be calculated and derivation of thermodynamic properties achieved. The principle of statistical mechanics or chemical statistics is application of statistics to determine thermodynamic properties, e.g. enthalpy, entropy, heat capacities, free energies ... etc. The equilibrium distribution of the system is defined as the most probable distribution of the system at a given condition. The basic foundation in statistical mechanics is the partition function; once the partition function is determined, all thermodynamic properties can be determined. The partition function describes the distribution of the energy of the system over all the individual quantum levels. The canonical ensemble of the partition function is expressed as follows,

$$Q = \sum_i g_i e^{E_i / RT} \qquad 2.2$$

where Q is the partition function, g_i is the degeneracy, E_i is the energy of the ith quantum level.

The total energy of the system is the sum of all its parts that contribute to the energy, *i.e.* translational, E_t, vibrational, E_v, external rotational, $E_{ext.rot}$, internal rotational, $E_{int.rot}$ and electronic energies, E_e;

$$E_{tot} = E_t + E_v + E_{ext.\,rot} + E_{int.\,rot} + E_e \qquad 2.3$$

By substitution of the different constituents for energy, E_{tot} and through some derivation, the partition function expression can be defined as follows:

$$Q_{tot} = Q_t\, Q_v\, Q_{ext.rot}\, Q_{int.rot}\, Q_e \qquad 2.4$$

This allows one to separately calculate contributions of the individual modes to the thermodynamic properties. Once the partition functions are known, the thermodynamic properties from different modes can be determined and the thermodynamic values of interest can be obtained. The goal is then to solve the individual partition function components, and then using the above product formulation of the individual partition function to obtain the total thermodynamic value desired. The translational partition function can be determined by solving the Schrödinger Equation for a system like "particle in a box." External rotation partition functions are difficult to derive, and as a result simplifications are made, *e.g.* rigid rotator or linear rotator. Vibration partition functions are determined from assumptions that the vibrations behave as non-coupled harmonic oscillators. Excited Electronic level contributions, in general, are negligible compared to the ground state electronic state and other modes at room temperature. However, one can readily include the electronic contribution, providing energies in the low-lying electronic states are known. Electronic contribution may be significant for radical species with strong spin-orbit coupling or having several low-lying electronic states. Ideal gas behaviour is assumed in these cases. Final terms for thermodynamic properties, after derivation from partition functions, for entropy are

$$S_{tot} = S_{trans} + S_{rot} + S_{vib} + S_e \qquad 2.5$$

where

$$S_{trans} = R\left[\frac{3}{2}\ln M_w + \frac{5}{2}\ln T - \ln P + \frac{5}{2} + \frac{3}{2}\ln\left(\frac{2\pi k}{h^2}\right) - \frac{5}{2}\ln N_A + \ln R\right] \qquad 2.6$$

$$S_{rot}^{linear} = R\left[1 + \ln\left(\frac{8\pi^2 kT}{h^2\sigma} I\right)\right] \qquad 2.7$$

$$S_{rot}^{non-linear} = \frac{3}{2}R + \frac{1}{2}R\ln\left[\frac{\pi}{\sigma^2}\left(\frac{8\pi^2 kT}{h^2}\right)^3 I_x I_y I_z\right] \qquad 2.8$$

$$S_{vib} = R\sum_i\left[\frac{x_i e^{-x_i}}{1-e^{-x_i}} - \ln(1-e^{-x_i})\right] \text{ where } x_i = \frac{h\nu_i}{kT} \qquad 2.9$$

$$S_e = R\ln\omega_0 \qquad 2.10$$

Where k is the Boltzman's constant, h is the Planck's constant, N_A is the Avogadro's number. I_x, I_y , I_z are the moments of inertia about the principle axis and υ_i is the ith vibrational frequency both data are obtained from the electronic structure theory calculations. σ is the symmetry, and ω_0 is the spin degeneracy.

The molar heat capacity at constant volume and at constant pressure with assumption of ideal gas behaviour, are expressed as follows:

$$C_{V,tot} = C_{V,trans} + C_{V,rot} + C_{V,vib} \tag{2.11}$$

or

$$C_{P,tot} = C_{P,trans} + C_{P,rot} + C_{P,vib} \tag{2.12}$$

$$C_{V,trans} = \frac{3}{2}R \text{ and with assumption of ideal gas behaviour, } C_{P,trans} = \frac{5}{2}R \tag{2.13}$$

$$C_{V,rot}^{linear} = C_{P,rot}^{linear} = \frac{2}{2}R \text{ and } C_{P,rot}^{non-linear} = C_{P,rot}^{non-linear} = \frac{3}{2}R \tag{2.14}$$

$$C_{V,vib} = C_{P,vib} = R\sum_i \left[\frac{x_i^2 e^{-x_i}}{(1-e^{-x_i})^2}\right] \text{ where } x_i = \frac{h\nu_i}{kT} \tag{2.15}$$

An important energy component in these computation chemistry calculations is the zero point energy (ZPE). ZPE is the energy of the specie at absolute zero temperature. Quantum chemical calculations determine the minimum absolute energy in the potential curve from the arrangement of the electrons and nuclei. However, these do not include the vibrational energy of the molecule. Each non-linear molecule has 3n-6 (3n-5 for linear molecules) vibrations in their active ground state at zero K. A molecular species at 0 K has its vibration levels populated in their lowest level. The treatment of ZPE is not from statistical mechanics, but from quantum mechanics. ZPE can be determined by the following simple expression,

$$ZPE = \frac{1}{2}\sum_i h\nu_i \tag{2.16}$$

There are two modes that are not applied to the above statistical treatment of thermodynamic properties that may be a major factor in some molecular species. The first is contribution to entropy, S, and heat capacity, $Cp(T)$ from internal rotors, which for some species can be significant. Molecular species that do not have internal rotors are represented by the above statistical analysis representation. However, for molecular species that have hindered internal rotors, the contributions to S and $Cp(T)$ need to be separately calculated and incorporated into the thermodynamic properties. One method to estimate the hinder rotor contributions is by using the vibration frequency for the torsion.

Optical isomers are molecules that cannot be superimposed onto each other, but are a "mirror" of each other and these also require an added consideration for the entropy calculation. Hydroperoxy species have one pair of optical isomers. The thermodynamic property that is affected by optical isomers is entropy. The correction for a molecule with two optical isomers is:

$$S_{opt.iso} = R\ ln\ \sigma_{opt.iso} \qquad 2.17$$

Where $\sigma_{opt.iso}$ is the optical isomer. This correction entropy value is added to the total entropy S_{tot}

$$S_{corrected} = S_{tot} + S_{opt.iso} \qquad 2.18$$

2.1.3 Contribution of Hindered Rotations

Many authors [44] have approached the hindered rotor problem by attempting to calculate the potential energy surface of the hindered rotor (under varying assumptions about the geometry of both the rotor and the rest of the molecule) [45, 46]. This surface (potential) is then fitted to a Fourier series and the Schrödinger equation solved numerically with this description of the potential surface. A better description of this method is given by Lay et al. [47]. Explicit calculation of the energy levels gives a more accurate evaluation of the hindered rotor partition function than any of the forms previously considered. These current calculations do not require the potential surface to be symmetrical as the early pioneering work of Pitzer and Gwinn. The potential surface of the internal rotation is also calculated in the quantum calculations where the molecule geometry is permitted to be optimized at each position of the rotor. This geometry optimization (relaxation) permits a more accurate potential surface to be obtained. Many researchers fix the entire geometry except the rotation angle when evaluating the potential surface). Boltzmann-averaged partition functions can be used to account for some of the entropy and heat capacity contributions from the internal rotors in place of estimated harmonic torsion frequencies. This is an improvement on the Pitzer-Gwinn assumption of a symmetrical potential [46].

In this work, a truncated Fourier series is used to represent the torsional potential calculated at the discrete torsional angles.

The potential surface is typically described by

$$V(\Phi) = a_0 + \sum a_i \cos(i\Phi) + \sum b_i \sin(i\Phi) \qquad 2.19$$

where i runs from 1 to 7

The values of the coefficients (a_0, a_i and b_i) are calculated to provide the minimum and maximum of the torsional potentials with allowance of shift of the theoretical extreme angular positions.

2.1.4 Isodesmic Reactions

It is possible to set up reactions (hypothetical) where not only the number of electron pairs is constant (conserved), but also the number of bonds is the same on both sides of the reaction. That implies that the reaction benefit from cancellation of errors, and therefore achieves higher accuracy in terms of theoretical level. Consider for example calculating the stability of propene as shown in Scheme 2.1. In this reaction the number of C=C, C—C and C—H bond is the same on both sides, and the "reaction" energy is therefore easy to calculate since the electron correlation to a large extend is the same on both sides. In other words, comparing very similar systems enables us to take maximum advantage of the cancellation of errors [48, 49]. Such reactions which conserve both the number and type of bonds are called *isodesmic reactions*.

$$CH_2{=}CH{-}CH_3 + CH_4 \rightarrow CH_2{=}CH_2 + CH_3{-}CH_3$$

Scheme 2.1: Example of isodesmic reaction

Combining the calculated energy difference for the left- and right-hand sides with experimental values for CH_4, $CH_2{=}CH_2$ and CH_3CH_3, the (absolute) stability of propene can be obtained reasonably accurately at a quite low levels of theory. This does, however, require that the experimental values for the chosen reference compounds are available (section 3.3).

In addition to bond balance, we try to use working reactions with group balance for maximum cancellation of error. A number of work reactions in this study do not conserve group balance; but they do conserve a majority of the groups and thus have better cancellation of errors than a conventional isodesmic reaction. Density Functional Theory calculations which included the corresponding ZPE and thermal corrections are performed for all species in reactions set, and the enthalpy change $\Delta H^0_{rxn,298}$ of each reaction is calculated.

The enthalpy of formation, ΔH^0_{f298}, of the target species is determined from the calculated $\Delta H^0_{rxn,298}$ and Hess's Law with the known enthalpies of the other reactants, (reference species) according to the following equation:

$$\Delta H^0_{rxn,298} = \Sigma(\text{total energies at 298 K of products}) - \Sigma(\text{ total energies at 298 K of reactants})$$

$$\Delta H^0_{rxn,298} = \Sigma(\text{experimental } \Delta H^0_{f298} \text{ of products}) - \Sigma\ (\text{experimental } \Delta H^0_{f298} \text{ of reactants})$$

In a reaction set, the arithmetic mean is the final value obtained for ΔH^0_{f298} of a species.

2.1.5 Group Additivity Method

A straight forward and a reasonably accurate method for estimating the thermochemical properties of hydrocarbons and oxygenated hydrocarbons are found in "additivity rules" [20]. It is particularly useful for larger molecules where high-level ab initio or Density Functional Theory calculations are not practical. Benson [20] mentioned that most molecular properties of larger molecules can be considered, roughly, as being made up of additive contribution from the individual atoms or bonds in the molecule [20]. He explained that the physical basis of additivity rules appears to reside in the fact that the forces between atoms in the same or different molecules are very "short range"; that is, they are appreciable only over distances of the order of 1-3 Å. Benson and Buss [50] showed that it was possible to make a hierarchical system of such additivity laws in which the simplest- or "zeroth-"order law would be the law of additivity of atom properties. It should be noted that with the exception of a few properties such as the molecular weight, zero-order atomic contributions are of little value. The next higher or "first-order" approximation is the additivity of bond properties, e.g. $C_p(CHCl_3) = C_p(C—H) + 3C_p(C—Cl)$. This methods work well for simple molecules such as normal hydrocarbons but less accurate for more complex molecules.

The next approximation to additivity behaviour is to treat a molecule property as being composed of contributions due to groups. A group is defined as a polyvalent atom (ligancy ≥ 2), in a molecule together with all of its ligands. In an atom-additivity scheme, one assigns partial values for the property in question to each atom in the molecule. The molecular property, thereby, is the sum of all atom contributions, the other is the additivity of bond properties. These bond properties contribute to the estimation of $\Delta H^0_{f\,298}$ S^0_{298} and $Cp_{298}(T)$ of ideal gas at 25°C. The THERM (see section 2.2.4) computer code was developed to aid estimating this data. Benson's group additivity is referred to as a second-order estimation technique since it incorporates non next nearest neighbour corrections and steric effects. Second-order group contribution techniques incorporate important corrections for cyclization, gauche interactions, steric effects, repulsive and attractive effects for aromatic substituents, etc. In principle, there is no limit to the number of interaction groups which can be included, or to the accuracy which can be obtained when these effects are taken into consideration. There are two limitations to this approach, however. First, there are only limited thermodynamic data available to determine the interaction contributions. Secondly, one must recognize the interactions of importance a priori or resulting estimates will be less accurate than anticipated. Nevertheless, this method of group contributions has been embraced as the best all-around method for estimating ideal gas thermodynamic properties [29].

To get a better understanding of the meaning of a group, let's take a simple molecule like $CH_3—CH=CH_2$. Propene consists of three groups, CH_3, CH and CH_2. The first group CH_3 is described as C/CD/H_3, which means that the first carbon (in sp^3) (C) is connected to another carbon (CD, which is itself connected to double bond carbon), and three hydrogens (H_3). The second group CH is CD/C/H, meaning that the carbon has a double bond with one carbon, one single bond with another carbon and is connected to one hydrogen. The third group, CH_2, called CD/H2, means that the carbon has a double bond with one carbon and two single bond with two hydrogens. The sum of the enthalpy of the groups is the enthalpy of the molecule or species.

An example of group additivity calculation of ΔH^0_{f298} for $CH_3CH{=}CH_2$ is as follow:

$$\Delta_f H^0_{298}(CH_3CH{=}CH_2) = (\Delta_f H^0_{298}\,C/CD/H3) + \Delta_f H^0_{298}(CD/C/H) + \Delta_f H^0_{298}(CD/H2) = 4.65\ \text{kcal mol}^{-1} \qquad 2.20$$

which is in agreement with the experimental value of Furuyama et al. at 4.87 kcal mol^{-1} [51]. The heat of formation of the reference groups are taken from the literature.

2.1.6 Transition State Theory Method

The transition State Theory (TST) applies the principle of statistical mechanics and thermodynamic to a system in which activated complexes are effectively in equilibrium with the reactant molecules. For general reaction:

$$A + B \rightarrow C \qquad 2.22$$

Then the reaction scheme is:

$$A + B \leftrightarrow AB^{\ddagger} \rightarrow C \qquad 2.23$$

Where $AB^{\ddagger}$ is the transition state, A and B the reactants, and C is the product.

The transition state rate constant is evaluated from statistical-mechanical equations and can be formulated in thermodynamic terms (equation 2.24).

$$k = \frac{-d[A]/dt}{[A][B]} = \frac{k_B T}{h}\frac{Q^{\#}}{Q_A Q_B}e^{-E_0/k_B T} \qquad 2.24$$

where Q_A and Q_B are the complete partition functions for the reactantsA and B, and $Q^{\#}$ is the partition function for all the degrees of freedom of the activated complex (Transition State) excepting the reaction coordinate.

Equation 2.24 can be written as:

$$k = \frac{k_B T}{h} K_C^{\#} \qquad 2.25$$

where

$$K_c^{\#} = \frac{Q^{\#}}{Q_A Q_B} e^{-E_0/k_B T} \qquad 2.26$$

which is the equilibrium constant for formation of the transition state. If the equilibrium constant is expressed in terms of the molar Gibbs standard free energy using van t'Hoff relation:

$$\Delta G_0^{\#} = -RT \ln K_C^{\#} \qquad 2.27$$

Then equation 2.24 can be written

$$k = \frac{k_B T}{h} e^{-\Delta G_0^{\#}/RT} \qquad 2.28$$

From the electronic structure theory calculations and with the statistical mechanic, $\Delta G_0^{\#}$ may be expressed in terms of standard enthalpy and entropy changes by

$$\Delta G_0^{\#} = \Delta H_0^{\#} - T\Delta S_0^{\#} \qquad 2.29$$

Allowing equation 2.28 to be separated into two terms:

$$k = \frac{k_B T}{h} e^{\Delta S_0^{\#}/R} e^{-\Delta H_0^{\#}/RT} \qquad 2.30$$

Equation 2.30 is similar to the Arrhenius equation

$$k = Ae^{-E_a/RT} \qquad 2.31$$

2.1.7 RRK and RRKM

2.1.7.1 Lindemann Theory

A general theory for thermal unimolecular reactions that forms the basis for the current theory of thermal unimolecular rates was proposed by Lindemann [52] in 1922. He proposed that molecules becomes energized by bimolecular collisions, with a time lag between the moment of collisional energy transfer and the time the molecule decomposes. Energized molecules could then undergo deactivating collisions before decomposition occurred. Steinfeld et al. [53] indicated that "A major achievement of Lidmann's theory is its ability to explain the experimental finding that the reaction rate changes from first to second order in going from high- to low-pressure limit."

Steinfeld et al. [53] and Robinson et al. [54] explained concepts of the Lindemann theory as follows:

- A certain fraction of the molecules become energized by collision, i.e. gain energy in excess of a critical quantity E_0. M represents an added "inert" gas molecule (collision), or a second molecule of reactant. k_1 is taken to be energy-independent and is calculated from the simple collision theory equation.

$$A + M \xrightarrow{k_1} A^* + M \qquad 2.32$$

- Energized molecules are de-energized by collision, which is a reverse reaction of process (2.32). The rate constant k_{-1} is taken to be energy-independent, and is equated with collision number Z_1 by assuming that every collision A* leads to a de-energized state. This is known as "strong collision assumption" for the de-energizing collisions.

$$A^* + M \xrightarrow{k_{-1}} A + M \qquad 2.33$$

- There is a time-lag between the energization and unimolecular dissociation or isomerization of the energized molecule. This unimolecular dissociation process also occurs with a rate constant k_2 independent of the energy content of A*.

$$A^* \xrightarrow{k_2} products \qquad 2.34$$

If the steady-state hypothesis is applied to the concentration of A*, the overall rate of reaction becomes

$$Rate = k_{uni}[A] = k_2[A^*] = \frac{k_1 k_2 [A][M]}{k_{-1}[M] + k_2} \qquad 2.35$$

The overall concept can be expressed by the equations below, where M can represent a generic bath gas molecule, an added "inert" gas molecule; it may also represent a second molecule of reactant or product. In this Lindemann theory k_1, along with k_{-1} and k_2 are taken to be energy-independent and are calculated from the simple collision theory equation.

Application of the steady-state hypothesis to the concentration of A*, allows the unimolecular rate constant and the high-pressure and low-pressure limit rate and rate constant to be determined as follows:

$$k_{uni} = \frac{k_1 k_2 [M]}{k_{-1}[M] + k_2} \qquad 2.36$$

First order at high-pressure limit rate, $[M] \rightarrow \infty, k_{uni} = k_\infty = k_1 k_2 / k_{-1}$

Second order at low-pressure limit rate, $[M] \rightarrow 0, k_{uni} = k_0 = k_1[M]$

The unimolecular rate constant is then written as

$$k_\infty = \frac{k_\infty}{1 + k_\infty / k_1[M]} \qquad 2.37$$

One can expect the Lindemann theory to predict a linear change in the initial rate of a unimolecular reaction with respect to concentration of M at low pressure. The transition from high-pressure rate constant to low-pressure is called "fall off region".

The Lindemann theory, unfortunately, predicts the falloff in k_{uni} to occur at much higher pressures than what is observed experimentally.

2.1.7.2 Hinshelwood Refinement

Based on the Lindemann's suggestion that k_1 could be increased by assuming that the required energy (energized molecules) could be drawn in part from the internal degrees of freedom (mainly vibration) of the reactant molecule, Hinshelwood [55] increase k_1 by using a much higher probability of a molecule possessing total energy $\geq E_0$ in s classical degrees of freedom.

Since k_1 increases with s classical degrees of freedom in the Lindemann-Hinshelwood theory, then $k_2 = k_\infty k_{-1}/k_1$ should decrease with s. Thus the lifetime of the energized molecule t $\approx 1/k_2$ increases when the molecule can store energy among a greater number of degrees of freedom. Then k_2 is expected to depend on the energy of A*. Making k_2 energy-dependent, expressed as $k(E)$, the energy interval from E to $E + dE$ is considered:

$$A + M \xrightarrow{dk_1} A^*(E, E + dE) + M \qquad 2.38$$

$$A^*(E, E + dE) + M \xrightarrow{k-1} A + M \qquad 2.39$$

$$A^*(E, E + dE) \xrightarrow{k(E)} products \qquad 2.40$$

Then

$$dk_{uni}\,(E,\ E+dE) = k(E)(dk_1/k_{-1})\ /\ (1 + k(E)/k_{-1}\ [M]) \qquad 2.41$$

It is assumed that for all pressure dk_1/k_{-1} represents the equilibrium probability and that the A* has the energy between E and $E + dE$. This probability may be denoted $P(E)dE$. Also, k_{-1} [M] is the collision frequency ω between A* and M, then

$$k_{uni} = \omega \int_{E_0}^{\infty} k(E)P(E)dE/(k(E) + \omega) \qquad 2.42$$

In order to make accurate quantitative predictions of the fall-off behaviour of a unimolecular reaction it is essential to take into account the energy dependence of the rate constant $k(E)$ for the conversion of energized molecules into activated complexes where products results from decomposition or reaction of the energized complex.

2.1.7.3 RRK Theory

Steinfeld et al. [53] noted that two quite different approaches may be taken to determine $k(E)$. One is to consider the explicit nature of the intramolecular motion of highly energized molecules, such as Slater theory [56]. The other approach is based on statistical assumptions, such as RRK theory and its extention, RRKM (Marcus) theory.

The RRK theory was developed independently by Rice and Ramsperger [57] and Kassel [58, 59, 60]. Both Rice and Ramsperger theory and Kassel theory consider that for reaction to occur a critical energy E_0 must become concentrated in one part (specific vibration) of the molecule. They used the basic Lindemann-Hinshelwood mechanism of collision energy transfer and de-energization, but assumed more realistically that the rate constant for conversion of an energized molecule into products is proportional to a specific probability. There is a finite statistical probability that energy, E_0, is found in the relevant part of the energized molecule which contains total energy, E, is greater than E_0 since E of the molecule under consideration is assumed to be rapidly redistributed around the molecule. This probability will increase with E and make k_2 a function of its energy content. This probability is given by the number of ways to attain this particular distribution divided by the total number of ways to distribute E between the s oscillators. The quantum RRK rate constant is simply this probability multiplied by the vibrational frequency (ν) for the critical oscillator. In the high-pressure limit, the quantum RRK expression for k_{uni} becomes the Arrhenius equation:

$$k_\infty = A\ exp\ (-E_0 / k_B T) \qquad 2.43$$

The RRK theory assumes that the Arrhenius high-pressure thermal A-factor is given by the frequency for the critical oscillator, which is in the range of 10^{13} to 10^{14} sec^{-1}. However, for many reactions Arrhenius high-pressure thermal A-factors are in fact larger than 10^{14} sec^{-1}. This inability is overcome by the RRKM (Marcus) theory, or by use (in this study) of a pre-exponential factor determined by quantum calculation of the thermochemical properties of the transition state structure.

2.1.7.4 RRKM Theory

The RRKM (Marcus [61]) theory was developed using the RRK model and extending it to consider explicitly vibration and rotation energies and to include zero point energies. Several minor modifications of the theory have been made, primarily as a result of improved treatments of external degrees of freedom.

RRKM theory is a microcanonical transition state theory and as such, it gives the connection between statistical unimolecular rate theory and the transition state theory of thermal chemical reaction rates. Isomerization or dissociation of an energized molecule A* is assumed in RRKM theory to occur via the mechanism

$$A^* \xrightarrow{k(E)} A^{\neq} \rightarrow \text{Products} \qquad 2.44$$

where $A^{\neq}$ is the transition state. The energized molecule A^* contains both vibrational and external rotational energies denoted by E_v and E_r, respectively.

Different experimental techniques, including static pyrolysis, carrier (flow) techniques, shock tube methods, and very-low-pressure-pyrolysis, have been used to measure k_{uni} as a function of temperature and pressure. One of the most significant achievements of RRKM theory is its ability to match measurements of k_{uni} with pressure.

2.1.8 QRRK

2.1.8.1 Master Equation

Consider a system in which molecules are distributed among a series of energy levels i with a concentration of the molecules ion the ith energy level. The processes by which molecules enter or leave this energy level are shown in Scheme 2.2.

Collisional transfer into i → ith energy level → Collisional transfer out of i

→ Unimolecular reaction

Scheme 2.2: Transport of molecules into and out of a given energy level

Scheme 2.2 means that there is a collisional transfer of molecules into and out of each level and an unimolecular reaction occurs with a rate-constant k_i for molecules with energy $\geq E_0$.

Consider now the Hinshelwood model to describe a pressure-dependent unimolecular reaction:

$A\,(E, J) + M \rightarrow A\,(E', J') + M$

$A\,(E', J') + M \rightarrow A\,(E, J) + M$

$A\,(E', J') \rightarrow$ Products

Where, J is the rotational quantum number for a given energy level and E is the energy.

[A (E, J)] is the concentration of the reactants

[M] is the concentration of the collision partner

Starting with this model, the Master Equation has to be solved to express the rate constant. The master equation describes the differential equation of the distribution of the concentration as a function of all rotational quantum number. The time rate of change of the reactants is given by the following expression:

$$\frac{\partial A(E,J)}{\partial t} = +[M]Z \int_{E'=0}^{\infty} \sum_{J'} P(E',J',E,J)[A(E',J')]dE'$$
$$-[M]Z \int_{E'=0}^{\infty} \sum_{J'} P(E,J,E',J')[A(E,J)]dE'$$
$$-k(E,J)[A(E,J)] \qquad 2.45$$

where,

P (E', J', E, J): probability of collision induced transition for a transition state of E,J to E′,J′

k (E,J): rate constant

dA(E,J)/dt describes the concentration change with time of the reactants A in a (*E, J*) state. The first term of equation 2.1 describes the population-win through the collision energy transfer, the second and third terms describe the population-lost through the collision energy transfer and the reaction, respectively.

2.1.8.2 QRRK with Master Equation

Multifrequency Quantum Rice-Ramsperger-Kassel (QRRK) is a method used to predict temperature and pressure-dependent rate coefficients for complex bimolecular chemical activation and unimolecular dissociation reactions. Both the forward and reverse paths are included for adducts, but product formation is not reversible in the analysis. A three-frequency version of QRRK theory is developed coupled with a Master Equation model to account for collisional deactivation (fall-off). The QRRK/Master Equation analysis is described thoroughly by Chang et al. [62, 63].

Molecular density-of-state functions are constructed through direct convolution of single frequency density functions. The functions corresponding to each reduced frequency are explicitly convoluted into a relative density of states, which is normalized by the partition function (*ρ(E)/Q*). The inclusion of one external rotation, corresponding to the symmetric top, is incorporated into the calculations by convoluting the vibration density function with the proper rotational density function.

Rate constant results from QRRK/Master Equation analysis are shown to accurately reproduce (model) experimental data on several complex systems. They also provide a reasonable method to estimate rate constants for numerical integration codes by which the effects of temperature and pressure can be evaluated in complex reaction systems.

2.2 Software used in this Work

2.2.1 Gaussian 98

All ab initio and DFT calculations are performed using the Gaussian 98 program suite [64, 65, 36] which runs on Windows systems as well as the Linux workstations or supercomputers. Gaussian computes the energy of a particular molecular structure (spatial arrangement of atoms or nuclei and electrons). Properties related to the energy may also be predicted. The geometry optimizations, which locate the lowest energy molecular structure in close proximity to the specified starting structure are performed, however optimizations do not always find minimum energy structures. Geometry optimizations depend primarily on the gradient of the energy, the first derivative of the energy with respect to atomic positions.

The determination of vibrational frequencies by ab initio computational methods is important in many areas of chemistry. One such area is the identification of experimentally observed reactive intermediates for which the theoretically predicted frequencies can serve as fingerprints. Another important area is the derivation of thermochemical and kinetic information through statistical thermodynamics. The vibrational frequencies of molecules resulting from interatomic motion within the molecules are computed. Frequencies depend on the second derivative of the energy with respect to atomic structure, and frequency calculations may also predict other properties which depend on the second derivative.

Ab initio harmonic vibrational frequencies are typically larger than the fundamentals observed experimentally [37]. The overestimation of ab initio harmonic vibrational frequencies is, however, found to be relatively uniform, and as a result generic frequency scaling factors are often applied. The determination of appropriate scale factors for estimating experimental fundamental frequencies from theoretical harmonic frequencies has received considerable attention in the literature [66, 67, 68, 69].

Scheme 2.3: Input file calculation for formaldehyde (CH_2=O)

```
%chk=ch2o
%NoSave
%mem=350mb
# B3LYP/6-311G(d,p) opt freq

CH2=O

0 1
C
O   1  r21
H   1  r31        2  a312
H   1  r41        2  a412        3  d4123       0

r21      1.202132
r31      1.091201
r41      1.091201
a312     121.793480
a412     121.793593
d4123    -180.002554
```

Gaussian offers the entire range of electronic structure methods: *Semi-empirical* methods, *Ab initio* methods and *Density Functional Theory* (DFT).

Scheme 2.3 shows a simple example of Gaussian input, where the optimized (opt) geometry of formaldehyde, and the frequencies (freq) are calculated with a DFT method (B3LYP/6-311G(d,p)). The geometry of formaldehyde is here given in a so called Z-matrix format.

2.2.2 SMCPS

SMCPS (**S**tatistical **M**echanics for Heat Capacity and Entropy $\mathbf{C_p}$ and **S**) is a Fortran program written by Sheng [74]. This program is useful for users of computational chemistry, such as Gaussian 98 [64], to calculate thermodynamic properties for molecules with hindered rotations. Required input parameters to SMCPS can be extracted from Gaussian98 calculations to determine the desired thermodynamic properties. The thermodynamic properties can then be directly applied to kinetic models, or other systems requiring thermodynamic properties. The program assumes the system of interest is at one atmospheric pressure, where most of literature thermodynamic data are found.

The statistical mechanics treatment to determine thermodynamic properties for *Cp*, *S* and *ΔH(T-0°K)*, as described above and in Sheng's thesis [74], has been fully implemented.

The results from SMCPS compare very well with both the NIST [75] and TRC [76] reported values for both the entropy *S* and heat capacity *Cp*.

The maximum number of different temperatures allowed is 40 and the maximum number of frequencies allowed is 500, which is sufficient for 166 atoms, *i.e.* for most applications (this maximum can be readily expanded).

Accurate treatment of hindered internal rotators is not included in this program; for hindered rotor barriers of 4 kcal mol^{-1} or less a more accurate treatment is recommended. More accurate methods are available and include those of Pitzer and Gwinn [77], McClurg , et al. [78, 79], Knyazev [80] or Rotator [47];

Scheme 2.4 shows an input file for CH_2=$CHOOCH_3$. The key word "ROTOR" indicates the number of rotors contained in CH_2=$CHOOCH_3$ and means that the first three first frequencies (hindered internal rotators) are not considered in the computation of *S* and *Cp*.

An output file format can be easily converted to the NASA polynomial data format, which is used in other computational packages, using THERM (see section 2.2.4). Scaling factors are optional for vibrational frequencies.

Scheme 2.4: SMCPS input file for the estimation of *S* and *Cp* for $CH_2CHOOCH_3$

```
NAME
CH2=CHOOCH3

COMMENTS:
Done with Gaussian results-B3LYP.

TEMPERATURE
31
0.1 10 50 100 150 200 298.1 300 400 500 600 700 800 900 1000 1100  1200  1300 1400 1500
1600 1700 1800 1900 2000  2500  3000  3500  4000  4500  5000

ROTOR
3

MOLECULAR WT
74

OPTICAL ISOMER
1

MULTIPLICITY
1      multiplicity of molecular specie of interest

HF298
-10.04

STOICHIOMETRY (in form of "atom x" "number of atom x")
C 3  H 6 O 2
rem (do not put any comments on same line as stoichiometry info)
rem (The stoichiometry is NOT sorted.  Will write to *.lst file as is).

rem RSCALING FACTOR  (Uses Scott & Radom's scaling factors integer inpt)
0.975

USCALING FACTOR  (User define scaling factors: ZPE, Hvib, Svib, Cpvib)
0.97 0.98 1.0 1.0   (include decimal input)

MOMENT (1)=10 e-40 g*cm^2  (2)=GHz  (3)=amu-Bohr^2 (4)=amu-Angstrom^2
2   choice of moment of inertia units
12.97302    3.32420    2.90595

SYMMETRY
3

NON-LINEAR

FREQ (The format for the frequencies is not important.  Units are cm-1)
27

 30.1213      204.8584     219.9317     318.2377
436.5902      604.3580     710.1240     846.4357
853.5620      969.0757     971.7280    1029.6845
1153.9169    1174.5748    1215.1864    1322.0966
1420.4431    1451.9299    1460.4156    1511.0245
1699.9275    3014.1281    3084.4701    3115.1385
3167.0819    3182.0940    3270.3654
```

2.2.3 Rotator

As mentioned in section 2.2.2, treatment of hindered internal rotators is not included in SMCPS; instead a more exact contribution from hindered rotations is calculated by use of the ROTATOR program. The calculated entropy and heat capacities are then added to those calculated with SMCPS. The example input file shown in Scheme 2.5 evaluates the rotation of the first oxygen about the next carbon (C=C—OOC). It is important to emphasis that each rotor is computed separately. Required data are all extracted from Gaussian computation, geometry of the optimized structure and the rotational energy from which the parameters of the Fourier equation are derived.

Scheme 2.5: Rotator input file for the calculation of the contribution of the hindered rotors for *S* and *Cp* for CH_2=CH—$OOCH_3$

```
C=C--OOC
11
                x          y          z
 1   6    -1.815699  -0.741194   0.073174
 2   6    -1.292850   0.477553   0.130026
 3   8    -0.011668   0.877052  -0.124844
 4   8     0.803296  -0.254540  -0.533273
 5   6     1.931762  -0.247447   0.328735
 6   1    -1.232085  -1.603674  -0.210609
 7   1    -2.862680  -0.870217   0.309432
 8   1    -1.856363   1.367013   0.395197
 9   1     1.643718  -0.428094   1.368691
10   1     2.485157   0.692429   0.245966
11   1     2.549944  -1.071027  -0.035348

3 2 (rotation of bond  between atoms 2 and 3)
3 5  (5 atoms attached to atom number 3)
4 5 9 10 11  (IDs of atoms attached to 3)
2 4  (4atoms attached to atom number 2)
1 8 6 7 (IDs of atoms attached to 2)

V(x)=a0 +B*COS(nx)+C*SIN(nx)  b3lyp/6-311g(d.p)
0 0 1
1
200 (size of the matrix)
7   (a0, a1-7, a1-7)
2.5966
0.4989          -0.0764
-2.2682          0.0391
-1.1177          0.0219
0.2178           4.1829e-3
0.0472          -0.0453
0.0187           5.7525e-3
-2.8699e-3       0.0429

31 (temperature)
0.1 10. 50. 100. 150. 200. 298.15 300. 400. 500. 600. 700. 800. 900. 1000. 1100. 1200.
1300. 1400. 1500. 1600. 1700. 1800. 1900. 2000. 2500. 3000. 3500. 4000. 4500. 5000.
```

2.2.4 THERM

The THERM code [81, 82], acronym for (**Th**ermodynamic property **E**stimation for **R**adicals and **M**olecules) is a versatile computer code designed to automate the estimation of ideal *gas* phase thermodynamic properties for radicals and molecules, which are important for combustion and reaction modelling studies. The calculated thermodynamic properties include heat of formation and entropies at 298 K and heat capacities from 300 to 1500 K (scheme 2.6). Heat capacity estimates are then extrapolated to above 5000 K, and NASA format polynomial thermodynamic property representations valid from 298 to 5000 K are generated. THERM uses group additivity principles of Benson and current best values for bond strengths, changes in entropy, and loss of vibrational degrees of freedom to estimate properties for radical species from parent molecules. In addition, heats of reaction, entropy changes, Gibbs free-energy changes, and equilibrium constants can be calculated as functions of temperature from a NASA format polynomial database.

Scheme 2.6: Thermodynamic properties estimated by THERM

```
SPECIES      Hf      S       Cp 300  400    500    600    800    1000   1500   DATE           ELEMENTS
Y(C3O2)     -25.52   71.38   18.08   24.11  30.02  34.80  41.40  46.79  .00    8/19/ 5        C  3 H  6 O  2   0 G 0
Y(C2O2)      11.30   66.51   12.95   16.57  20.14  23.25  29.02  32.45  .00    8/19/ 5        C  2 H  4 O  2   0 G 0
```

THERM includes groups for hydrocarbons, oxygen, nitrogen, halogen, and sulfur-containing species. Additional groups exist for ring strain correction, gauche interactions or optical isomers. The bond dissociation group data file is included for the estimation of radical species important to high-temperature reaction modelling and combustion studies. Groups include bond dissociation energy, BDE, which reflects the change of entropy and heat capacities from the loss of an H atom from a parent molecule.

2.2.5 ThermKin

ThermKin (**Th**ermodynamic **E**stimation of **R**adical and **M**olecular **Kin**etics) evolved (see Sheng's thesis [74]) from a previously developed computer code, *i.e.* THERMRXN (included in THERM) [82] which calculates equilibrium thermodynamic properties for any given reaction. Additionally, ThermKin determines the forward rate constants, k(T), based on the canonical transition state theory (CTST).

The modified Arrhenius parameters are determined from regression analysis with application of the principle of least squares. CTST describes the forward rate constant from reactant to the transition state (TS) s a function of the equilibrium between reactant and TS. ThermKin requires thermodynamic properties in the NASA polynomial format, needs to know whether the reaction is uni- or bimolecular, and either a two-parameter fit or a three-parameter fit is desired. Finally, the reaction to be calculated has to be given in the form:

[reactant] = [transition state] for an unimolecular reaction

[reactant 1] + [reactant 2] = [transition state] for a bimolecular reaction

The output file generated by THERMKIN is divided into three sections. In the example output file shown in scheme 2.7, the first section of the output file identifies the reaction, the second section is the thermochemistry of the reaction as a function of temperature and the last section describes the kinetics.

Scheme 2.7: Output file for the reaction $C_6H_5OO\bullet$ = TS[$C_6H_5OO\bullet$]

```
THERMODYNAMIC ANALYSIS for REACTION

Rx        PHOOJ  =  TS1*8
Hf {Kcal/mol}  31.300    60.700
S {cal/mol K}   86.200    79.700

dHr {kcal/mol}  (298K)  =      29.40   dHr avg (298., 1500. K) =      30.45
dU (dE) {kcal/mol} (")  =      29.40   dUr avg (298., 1500. K) =      30.45
dSr {cal/mol K}  ( " )  =      -6.50   dSr avg (298., 1500. K) =      -5.35
dGr  {kcal/mol}  ( " )  =      31.34   dGr avg (298., 1500. K) =      35.26
         Af/Ar  ( " ) =  3.796E-02   Af/Ar avg (298., 1500. K ) = 6.781E-02
Fit Af/Ar     : A = 5.183E-03 n =  .30 alpha =-5.916E-04 avg error   7.08 %
Fit Af/Ar w/ddU: A = 3.149E-04 n =  .78 alpha =-5.778E-04 avg error  13.27 %

  T (K)  dH(Kcal/mol) dU(Kcal/mol) dS(cal/mol K)  (Af/Ar)   dG(Kcal/mol)
  300.00  2.940E+01  2.940E+01  -6.502E+00  3.793E-02  3.135E+01
  400.00  2.941E+01  2.941E+01  -6.483E+00  3.829E-02  3.200E+01
  500.00  2.948E+01  2.948E+01  -6.331E+00  4.132E-02  3.264E+01
  600.00  2.960E+01  2.960E+01  -6.110E+00  4.618E-02  3.326E+01
  800.00  2.996E+01  2.996E+01  -5.590E+00  6.000E-02  3.444E+01
 1000.00  3.042E+01  3.042E+01  -5.082E+00  7.751E-02  3.550E+01
 1200.00  3.089E+01  3.089E+01  -4.653E+00  9.616E-02  3.647E+01
 1500.00  3.150E+01  3.150E+01  -4.194E+00  1.211E-01  3.780E+01
 2000.00  3.229E+01  3.229E+01  -3.740E+00  1.523E-01  3.977E+01

 The model fitted is for uni-molecular reaction.

 The 3 parameters for the model equation of A(T) = Aprime * T^n * exp(-Ea/RT)
   Aprime = 1.7523E+06       n =  1.86486     Ea =   2.8683E+04

 Temp(K)       AF(T)         T_K^n        k_calc(T)        k_fit
 300.00      2.371E+11     4.164E+04     9.028E-11      9.250E-11
 400.00      3.191E+11     7.120E+04     2.723E-05      2.651E-05
 500.00      4.305E+11     1.079E+05     5.607E-02      5.478E-02
 600.00      5.774E+11     1.517E+05     9.535E+00      9.462E+00
 800.00      1.000E+12     2.593E+05     6.513E+03      6.624E+03
 1000.00     1.615E+12     3.932E+05     3.625E+05      3.708E+05
 1200.00     2.404E+12     5.524E+05     5.685E+06      5.776E+06
 1500.00     3.786E+12     8.375E+05     9.720E+07      9.709E+07
 2000.00     6.345E+12     1.432E+06     1.879E+09      1.841E+09
```

2.2.6 Chemaster

Chemaster code is based on the quantum Rice-Ramsperger-Kassel (QRRK) analysis for *k(E)* and Master Equation analysis for fall off. Chemaster will be employed to determine kinetic parameters in complex reaction systems, such as C_6H_5 + O_2. The source code for the QRRK analysis was originally developed by Dean (1985) and later modified for multi frequency by Ritter and Bozzelli (1990) and converted to PC DOS and Windows. Chang, Bozzelli, and Dean made further improvements to calculate systems with multiple wells. Recently, Sheng, et al. [74] updated the QRRK with master equation to express rate coefficients in Chebyshev polynomial form to account for pressure dependence. The current version of the QRRK computer code utilizes a reduced set of vibration frequencies, which accurately reproduce the heat capacity data of the molecule (adduct). Molecular density of state functions are constructed through direct convolution of single frequency density functions on a 10 cm^{-1} grid. The functions corresponding to each reduced frequency are explicitly convoluted into a relative density of states ($\rho(E)$), which is normalized by the partition function (Q). The inclusion of one external rotation is incorporated into the calculations by convoluting the vibration density function with the proper rotational density function. A detailed description of this and comparisons of ratios of these $\rho(E)/Q$ with direct count $\rho(E)/Q$ are shown to be in good agreement [83]. Nonlinear Arrhenius effects resulting from changes in the thermodynamic properties of the respective transition states with temperature are incorporated using a two parameter Arrhenius pre-exponential A-factor (A, n) in AT^n. Fall-off is incorporated using master equation. An input file example of Chemaster for the Phenyl + O_2 system is given in Appendix F.

3. The Vinyl System

In this chapter, thermodynamic data of a series of unsaturated hydroperoxides and methylhydroperoxide are calculated for the first time. Vinyl, allyl, and ethynyl are here considered.
We first justify the use of the density functional calculation method (B3LYP/6-311G(d,p) coupled with isodesmic reactions through a number of comparisons with higher level calculations and recent experimental data.

Enthalpies of formation, entropies, and heat capacities are then calculated for these hydro- and methyl-peroxides as well as for the corresponding peroxy radicals. These calculations require computation of a number of unsaturated ethers and alcohols enthalpies, because these values are needed as reference species.

Finally the calculated hydroperoxide enthalpies are utilized to determine internal rotations and bond energies such as R—OOH, RO—OH, ROO—H. these are compared with the corresponding methylperoxides, R—OOCH$_3$, RO—OCH$_3$, ROO—CH$_3$.

3.1 Introduction

Peroxides and peroxy radicals are important intermediates in the photochemical oxidation of hydrocarbons in the atmosphere [6, 7], in polymer initiations, and in low temperature calculations. Reactions of peroxides control the chemistry of intermediate temperature (below 750 K) hydrocarbon self-ignition. Reactions of ozone with olefins lead to formation of vinyl peroxide, which is thought to undergo rapid unimolecular dissociation to ketenyl radical plus OH [84]. The reaction scheme in Figure 3.1 [84] shows a simplified mechanism for the 2-butene + O_3 reaction, including the formation of vinylic peroxy species as intermediates. Olzman et al. [84] have shown that the reactions of ozone with substituted olefins form vinyl hydroperoxides, in which the vinyl hydroperoxide undergoes rapid cleavage of the weak peroxide bond to form a OH radical plus a ketenyl radical. There is a limited number of recent experimental studies on peroxides, primarily those using flow reactors with photoionization those using mass spectrometric detection (Gutman et al. [11] reviewed and summarized by Knyazev and Slagle [14]), and two studies involving negative ion chemistry from Clifford et al. [15] and Blanksby et al. [16].

There is a substantial number of computational chemistry or evaluation studies on peroxy and hydroperoxy alkyl radicals [2, 8, 9, 85, 10]. Bozzelli and co-workers [86, 87, 88, 89] have reported enthalpies based on the use of isodesmic reactions. Entropies and heat capacity values where obtained from statistical mechanics and the detailed analysis of internal rotors. They have also determined group additivity (GA) contributions.

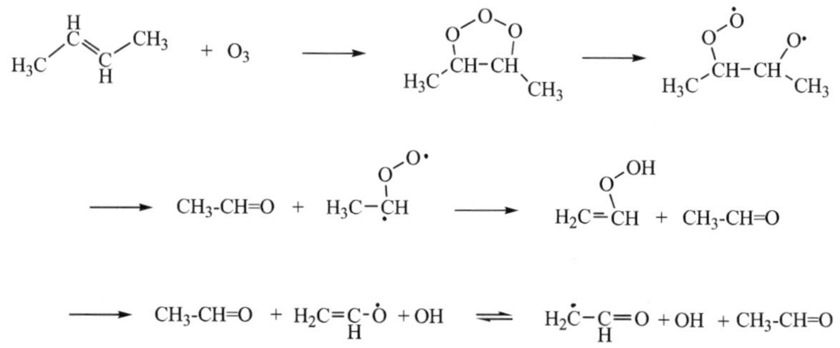

Figure 3.1: Schematic representation of the mechanism for the ozonolysis of 2-butene

Jonsson and co-workers [90, 91] have also reported enthalpy values from semiempirical and a number of higher level ab initio calculations. Sheng et al. [10, 92] report data on the ethyl peroxy and 2-hydroperoxide ethyl radicals and on the stable ethylhydroperoxide. Schaefer and co-workers [93, 94] have reported results on the ethyl peroxy and 2-hydroperoxide ethyl radicals. Sumathi and Green [95] have reported results from ab initio molecular orbital calculations at the Hartree-Fock level for *S* and *Cp*(*T*) and complete basis set (CBS-Q atomization reaction) level for enthalpies on methyl, ethyl, isopropyl, and tertbutyl hydroperoxides and their corresponding methylperoxides.

There is little or no thermochemical property data available for unsaturated alkylperoxy and peroxide species. Peroxides are often impure and/or instable, and therefore difficult to isolate and characterize by experimental methods. There is no experimental data on vinyl, phenyl, ethynyl, or allyl peroxides that we are aware of. Experimental studies on the reaction of vinyl radical and allyl radical with O_2 to form the corresponding peroxy radical have been reported by Gutman et al. [11, 12]. The phenyl-peroxy radical was reported by Lin's group as a major product in the phenyl radical reaction with O_2 at ambient temperatures [17].

This study reports bond energies, enthalpy, entropy, heat capacity, internal rotation potential, and structure data for a series of unsaturated peroxides. Thermochemical property groups are developed as well for future use in group additivity estimation methods.

Overall the vinyl, phenyl, and ethynyl-alkyl hydroperoxides and methylperoxides do not exist as stable species under atmospheric conditions and its life time scales of seconds. Determination of the enthalpies of formation provides analysis of the relative bond strengths in these species and subsequent information concerning the reactions of peroxide radicals, which are important in both combustion and atmospheric chemistry of hydrocarbons. Reactions of peroxide species and the corresponding radicals are important to chain-branching in combustion and oxidation that controls ignition in new generation, high-compression nonspark ignition engines. The unsaturated vinyl, phenyl, and ethynyl hydroperoxides and methylperoxides are products from reactions of peroxide alkyl radicals.

3.2 Results and Dicussion

All of the calculations are performed using the Gaussian 98 program suite [64]. As mentioned in chapter 2, optimized geometries of species are first needed to perform determine thermodynamic data of these species. These geometries were calculated with Density Functional Theory at the B3LYP/6–311G(d,p) level. The geometries for vinyl, allyl, ethynyl hydroperoxides and methylperoxides are listed in Appendix B. Also given are harmonic vibrational frequencies [96, 97, 98].

3.2.1 Enthalpy of Formation

3.2.1.1 Selection of the Calculation Method

The Density Functional method or commonly called the DFT method is, as mentioned in chapter 2, one with the lowest computational costs though its similarity in many ways to ab initio methods. DFT has the big advantage to be usable for large systems where high level computations are costly or not possible. In this section, we give the evidence of the suitability of the DFT method through comparisons with high-level calculation methods.

The B3LYP/6-31G(d,p), which differs from the B3LYP/6-311G(d,p) by only split valence basis (see Appendix F) is taken as example because it is widely reported in the literature to yield accurate geometries and reasonable energies when used with isodesmic or homodesmic reactions [99, 100]. This implies that the basis set used in this work (6-311G(d,p)), can provide the same or even better accuracy. Byrd et al. [100] and Curtis et al. [101] report that B3LYP/6-31G(d,p) provides accurate structures for compounds with elements up to atomic number 10. Durant [99, 102] has compared Density Functional calculations B3LYP and hybrid (BH and H) with MP2 and Hartree-Fock methods for geometry and vibration frequencies. He reports that these density functional methods provide excellent geometries and vibration frequencies, relative to MP2 at a reduced computational expense. Peterson et al. [103] currently recommends the use of B3LYP for geometry and frequencies in several of his CBS calculation methods. The Density Functional calculated geometry is often selected and serves as an accurate structure for higher-level composite method calculations.

Table 3.1: Comparison of Enthalpy of Formation of some peroxides, determined by different methods.

Species	ΔH^0_{f298} (kcal mol^{-1})	Method
CH_3OOH	-31.8 ± 0.94[86,104]	MP2//MP2 (full)/6-31G(d)
	-30.67[92]	CBS-APNO
	-30.9 ± 0.7[16]	CBS-APNO
	-33[105]	MM and ab initio calculation
	-31.3[106]	Heat of equilibrium measurement
	-30.75[107]	CBS-Q
	-31.36 ± 0.81[a]	***B3LYP/6-311G(d,p)***
CH_3CH_2OOH	-39.52 ± 0.53[86]	CBS-q//MP2 (full)/6-31G(d)
	-40.05[86]	MP2//MP2 (full)/6-31G(d)
	-39.9[104]	MP4SDTQ/6-31G*//MP2/6-31G*
	-39.9 ± 1.5[92]	CBS-Q //B3LYP/6-31G(d,p)
	-40.1 ± 1.8[92]	G2
	-39.5 ± 0.7[16]	Derived from experiment
	-40[105]	MM and ab initio calculation
	-38.78[107]	CBS-Q
	-39.28 ± 0.01[a]	***B3LYP/6-311G(d,p)***
$CH_3CH_2CH_2OOH$	-45.19[86]	CBS-q//MP2 (full)/6-31G(d)
	-44.35[86]	MP2//MP2 (full)/6-31G(d)
	-44.05 ± 0.14[a]	***B3LYP/6-311G(d,p)***
$CH_2{=}C(CH_3)OOH$	-24.38[87]	CBS-4//MP2 (full)/6-31G(d)
	-24.51[87]	CBS-q//MP2 (full)/6-31G(d)
	-24.52[87]	MP2//MP2 (full)/6-31G(d)
	-19.6[87]	CCSD(T)
	-21.8 ± 0.06[a]	***B3LYP/6-311G(d,p)***
$CH_2{=}C(CH_3)OOCH_3$	-19.94[87]	CBS-4//MP2 (full)/6-31G(d)
	-20.14[87]	CBS-q//MP2 (full)/6-31G(d)
	-19.61[87]	MP2//MP2 (full)/6-31G(d)
	-20.79 ± 0.42[a]	***B3LYP/6-311G(d,p)***
$CH_3OO\bullet$	1.2[92]	CBS-APNO
	2.24[87]	G2
	2.7 ± 0.8[108]	Experiment $CH_3 + O_2 \leftrightarrow CH_3OO$
	2.9 ± 1.5[109]	Experiment $CH_3 + O_2 \leftrightarrow CH_3OO$
	2.15 ± 1.22[110]	Experiment $CH_3 + O_2 \leftrightarrow CH_3OO$
	2.07 ± 0.7[16]	CBS/APNO
	2.02 ± 0.1[a]	***B3LYP/6-311G(d,p)***
$CH_3CH_2OO\bullet$	-6.7[92]	CBS-Q //B3LYP/6-31G(d,p)
	-6.8[92]	G2
	-6.5 ±2.4[110]	Experiment $CH_3CH_2 + O_2 \leftrightarrow CH_3CH_2OO$
	-6.8 ±0.7[16]	CBS/APNO
	-6.8 ±2.3[16]	Negative-ion/acidity/CBS
	-5.75 ± 0.1[a]	***B3LYP/6-311G(d,p)***
$CH_2\bullet CH_2OOH$	11.12[86]	CBS-q//MP2 (full)/6-31G(d)
	10.85[86]	MP2//MP2 (full)/6-31G(d)
	11.2[92]	CBS-Q //B3LYP/6-31G(d,p)
	10.5[92]	G2
	11.34 ± 1[a]	***B3LYP/6-311G(d,p)***
$CH_2{=}CHOO\bullet$	24.45[33]	G2M(RCC,MP2)
	24.34 ± 0.42[a]	***B3LYP/6-311G(d,p)***
$CH_2{=}C(CH_3)COO\bullet$	11.34[87]	CBS-4//MP2 (full)/6-31G(d)
	11.06[87]	CBS-q//MP2 (full)/6-31G(d)
	12.25[87]	MP2//MP2 (full)/6-31G(d)
	10.91 ± 0.34[a]	***B3LYP/6-311G(d,p)***
$HC\bullet{=}C{=}O$	41.98[111]	CBS-QCI/APNO
	40.40[a]	***B3LYP/6-311G(d,p)***

[a]This work

The CBS-Q/B3 composite calculation method uses this level for structure calculation, based on studies showing it results in improved accuracy [112].

Chen and Bozzelli [86] have used the density functional calculations for oxygenated hydroperoxides and peroxy radical species; they show that two different levels of B3LYP calculations, combined with use of isodesmic reaction analysis, result in good agreement with

MP2, MP4 ab initio and CBSlq composite methods. B3LYP/6-311G(d.p) was also shown by Mebel et al. [33] to give similar enthalpies to values calculated by the G2M composite method, on a number of unsaturated peroxide peroxy species.

B3LYP/6-311G(d,p) is chosen in the present work because it has a somewhat larger basis set than the commonly used B3LYP/6-31G(d,p) and it is still accurate for structure optimisation and force constant calculation on the molecules considered in this study.

We further justify the use of B3LYP/6-311G(d,p) as a suitable calculation method for these unsaturated oxygenated molecules and radicals through a number of comparisons with higher level calculations and recent experimental data presented in Table 3.1. Calculation of the enthalpies of methyl, ethyl, and propyl hydroperoxides, CH_3OOH, CH_3CH_2OOH, $CH_3CH_2CH_2OOH$ as well as three radicals, $CH_3OO•$, $CH_3CH_2OO•$ and $CH_2•CH_2OOH$ with B3LYP/6-311G(d,p) are included in Table 3.1 to illustrate the accuracy of our methods and provide comparison of results with literature values. This is achieved by combination of the B3LYP calculation data with isodesmic work reactions chosen for similarity in bond environment on both sides.

The data in Table 3.1 shows good agreement through a range of peroxy and peroxide species with a number of high-level calculation methods and values reported in recent experimental studies. From 43 literature values listed in Table 3.1, 34 of them agree within less than 1 kcal mol^{-1}, 5 of them agree within 1.5 kcal mol^{-1} and there are 4 values with deviations near 2.5 kcal mol^{-1}.

Ethyl-hydroperoxide, CH_3CH_2OOH (Table 3.1) is an example for which DFT calculations result in -39.28 kcal mol^{-1} [26], which is good agreement with the recent experimental studies of Blanksby et al. [16] –39.5 kcal mol^{-1}. The use of groups developed by Lay and Bozzelli [104] yields a value of –39.9 kcal mol^{-1}. All other high-level methods used to calculate the enthalpy of ethyl-hydroperoxide agree within less than 0.8 kcal mol^{-1} with the DFT calculations. GA and CBSQ atomization reaction values from ref. 107 are –38.9 kcal mol^{-1} and –38.78 kcal mol^{-1} respectively.

Sebbar *et al.* [26] report density function calculations for the enthalpy of 1-methyl vinyl hydroperoxide ($CH_2{=}C(CH_3)OOH$). The result, -21.8 kcal mol^{-1}, is in good agreement with Olzman's [84] value (–19.6 kcal mol^{-1}). Schlegel et al. [113] report the values of the O—O bond energy in $CH_2{=}C(CH_3)OOH$ to be 22.7 kcal mol^{-1} by using G2MP2 ab initio method, while Sebbar et al. [26] reported 21.2 kcal mol^{-1}. Lee and Bozzelli [111] have determined the value of the ketenyl radical HC•=C=O to be 42.0 kcal mol^{-1} using the high-level and accurate

CBS-QCI/APNO method. The result by Sebbar et al. [26] of 40.4 kcal mol^{-1} with DFT differs slightly.

Ethers are reported by Chen and Bozzelli [114] and by Lee et al. [115 , 116]. They are reported in Tables 3.2 and 3.3. The data illustrate good agreement through a range of ether species with a number of high-level calculation methods.

Table 3.2: Comparison of the Enthalpies of Formation ($\Delta_f H_{298}$ (kcal mol^{-1}), 298 K) for Ethers and the corresponding Radicals revealed from the literature[a]

	//B3LYP/6-31G(d,p)			**//MP2/631G(d,p)**		
	G3MP2	**CBSQ**	**B3LYP/ 6-311+G(3df,2p)**	**G3MP2**	**CBSQ**	**MP2/ 6-311+G(2df,2p)**
$CH_3CH_2OCH_3$	-52.28	-52.32	**-52.04**	-52.31	-52.35	-52.32
$(CH_3)_2CHOCH_3$	-60.04	-60.23	**-59.54**	-60.37	-60.26	-60.44
$(CH_3)_3COCH_3$	-67.93	-68.04	**-66.51**	-68.02	-68.10	-68.06
$C{\bullet}H_2CH_2OCH_3$	-1.76	-1.77	**-1.70**	-1.71	-1.71	-1.64
$C{\bullet}H_2CH(CH_3)OCH_3$	-10.03	-10.27	**-9.70**	-10.30	-10.27	-10.28
$C{\bullet}H_2(CH_3)_2COCH_3$	-17.91	-18.05	**-16.83**	-17.94	-18.07	-17.83
$CH_3CH_2OC{\bullet}H_2$	-8.02	-8.06	**-7.89**	-8.02	-8.06	-8.01
$(CH_3)_2CHOC{\bullet}H_2$	-16.71	-16.86	**-16.37**	-17.02	-16.89	-17.11
$(CH_3)_3COC{\bullet}H_2$	-24.60	-24.87	**-23.46**	-24.68	-24.95	-24.70
$CH_3C{\bullet}HOCH_3$	-9.17	-9.49	**-9.86**	-9.12	-9.42	-8.89
$(CH_3)_2C{\bullet}OCH_3$	-16.87	-17.46	**-18.09**	-17.20	-17.51	-16.83

[a]Ref.114

Table 3.3: Comparison of the Enthalpies of Formation (ΔH_{f298} (kcal mol^{-1}), 298 K) for Ethers and the corresponding Radicals revealed in the literature[a]

Species	**CBSQ**	**B3LYP/6-31G(d)**
$C(OO{\bullet})H_2CHO$	-21.01± 0.24	-21.26 ± 0.47
$C(OOH)H_2C{\bullet}O$	-19.64 ± 0.04	-20.33 ± 0.09
$C(OOH)H_2CHO$	-56.19	-56.02
$CH_3C(=O)OOH$	-84.80	-83.21
$C{\bullet}H_2C(=O)OOH$	-32.95	-32.67
$CH_3C(=O)OO{\bullet}$	-38.57	-37.52

[a]Ref. 115 and 116

3.2.1.2 Selection of Isodesmic Reactions

Isodesmic reactions may be used to predict the heat of formation for compounds of interest by first calculating the enthalpy of the reaction and then computing the desired heat of formation by subtracting the known heats of formation for the other compounds from this quantity. This implies that the accuracy of this method is controlled by several factors:

- The choice of the working chemical reactions used to cancel systematic calculation errors.
- The accuracy of enthalpy values $\Delta_f H^0_{298}$ of the standard (reference) species.
- The level of sophistication (method + basis set) applied to calculate the electronic energy.

A number of work reactions have been selected in this study to calculate unknown enthalpies of formation of selected species. For maximum cancellation of error, we have attempted to construct reactions which conserve not only bonds but also groups. In this work, not all reactions conserve group balance; but they do conserve a majority of the groups and, thus, most likely provide better cancellation of errors than conventional isodesmic reactions. Density Functional Theory calculations including ZPVE (zero-point vibrational energies) and thermal corrections are performed for all species in the reactions set, and the enthalpy change $\Delta H^0_{rxn,298}$ of each reaction is calculated. The chose of the reference values is also crucial, since it has a direct impact on the final value of the unknown enthalpy. For this reason, we have carefully selected our reference values, by choosing them either from well established databank, or from accurate experimental values.

Additionally, several working reactions, whenever possible, were used for each enthalpy calculation of a target species. Finally, it is important to use different reference species to avoid that one appears in many reactions and consequently influences strongly the enthalpy result of the target species.

Nevertheless, despite the above described cautions, deviations in the calculated enthalpies are still observable. We can explain these deviations by the fact that:

- It is not easy to find working reactions which always conserve group balance. If no or only partial group conservation is achieved, the reaction doesn't benefit as much from cancellation of errors as one would hope which implies a possible deviation in the enthalpy value of the desired compound.
- Despite its accuracy, the method used (B3LYP/6-311G(d,p)) is responsible for some deviation as well.

3.2.1.3 Reference Species

Examination of the isodesmic reactions used below for the calculation of the enthalpies, (see Table 3.6) reveals that accurate enthalpies for a number of vinyl and ethynyl ethers and alcohols are needed so they can be used as reference species. Enthalpies of the species, which were not found in the literature, were calculated as well with DFT combined with isodesmic reactions. To validate the accuracy of the data for reference species, they were also estimated with the group additivity method (GA) with help of "THERM" program. Enthalpies for these reference species which can be found in the literature are listed in Appendix A.

The calculated enthalpy values for the alcohols and ethers are listed in Table 3.4. The calculated DFT values of the enthalpies of the ethers and the alcohols show good agreement with the group additivity method. The group additivity data result from several studies including those of Turecek et al. [117], who calculated standard enthalpy of formation of a series of alcohols. They also investigated substitution effects by methyl groups on the heat of formation using a series of methyl-substituted enols [118].

Table 3.4: Calculated ΔH^0_{f298} for Vinyl and Ethynyl Alcohols and Ethers needed as Reference

Reactions Series	$\Delta_f H^0_{298}$	Error limit[a]	Therm[b]
trans-$CH_3CH{=}CHOCH_3 + CH_3OH \rightarrow$ *trans*-$CH_3CH{=}CHOH + CH_3OCH_3$	-34.33	± 0.95	-33
cis-$CH_3CH{=}CHOCH_3 + CH_3OH \rightarrow$ *cis*-$CH_3CH{=}CHOH + CH_3OCH_3$	-36.24	± 0.95	-33
$CH_2{=}CHCH_2OCH_3 + CH_3OH \rightarrow CH_2{=}CHCH_2OH + CH_3OCH_3$	-25.68	± 1.30	-25
$CH_2{=}C(CH_3)OCH_3 + CH_3OH \rightarrow CH_2{=}C(CH_3)OH + CH_3OCH_3$	-32.55	± 0.95	-35
$CH_3(CH_3)C{=}CHOCH_3 + CH_4 \rightarrow CH_3(CH_3)C{=}CH_2 + CH_3OCH_3$	-43.72	± 1.23	-41
$CH_3(CH_3)C{=}COH + CH_4 \rightarrow CH_3(CH_3)C{=}CH_2 + CH_3OH$	-49.31	± 1.16	-46
$CH{\equiv}COCH_3 + CH_3OH \rightarrow CH{\equiv}COH + CH_3OCH_3$	26.08	± 0.95	24[c]
$CH{\equiv}COH + CH_2{=}CH_2 \rightarrow CH{\equiv}CH + CH_2{=}CH\text{-}OH$	20.22	± 1.62	20[c]
$CH_3\text{-}C{\equiv}COH + CH_3CH_3 \rightarrow CH_3CH{\equiv}CH + CH_3CH_2OH$	9.84	± 1.31	10[c]
$CH_3\text{-}C{\equiv}COCH_3 + CH_3CH_3 \rightarrow CH_3CH{\equiv}CH + CH_3CH_2OCH_3$	15.93	± 1.27	15[c]

[a]reported errors for each of the standard species (see Appendix A), and estimate error due to the method (Appendix C) [b]values are from group additivity unless noted otherwise
[c]Groups developed in this work for ethynes, not previously available

The isodesmic reactions formulated to determine $\Delta_f H^0_{298}$ of the target methylperoxides (Table 3.9), reveals that knowledge of the enthalpy on $CH_3CH_2OOCH_3$ is needed to be used as reference species. The isodesmic reactions used to determine enthalpy of $CH_3CH_2OOCH_3$ are listed in Table 3.5. The value estimated with group additivity (-39.3 kcal mol^{-1}) is in good agreement with the calculated one. A value of –37.0 kcal mol^{-1} is reported by Sumathi and Green [107]. This deviation may be explained by the different calculation method used by Sumathi et al., and also by the fact that the authors didn't use isodesmic reactions in their evaluation. Our recommended value is –39 kcal mol^{-1}.

The enthalpy value of CH_3OOCH_3 which is frequently used in the working reactions has been determined by different groups, making its choice difficult. To help in the selection of the most reliable CH_3OOCH_3 enthalpy value, we have recalculated (see Table 3.5). The enthalpy determined for CH_3OOCH_3 using DFT (B3LYP) coupled with isodesmic reactions is -30.77 kcal mol^{-1}. Lay and Bozzelli [104] and Carballeira et al. [119] reported -31.0 kcal mol^{-1}. The value obtained with group additivity using the O/C/O group developed by Lay and Bozzelli is -31 kcal mol^{-1} as well. In a recent study, Sumathi and Green [107] have developed groups for alcohols, O/C/H, and for hydro and alkyl peroxides, O/H/O and O/C/O using enthalpy values from CBS-Q calculations with heats of atomization and multivariate linear regression. The use of the two groups from Sumathi and Green for the calculations of enthalpy values on species in this study consistently results in differences (higher values) in enthalpy of about 1 kcal mol^{-1} relative to data in this work. The group additivity (GA) values of Sumathi and Green result in enthalpy for CH_3OOCH_3 of -29.1 kcal mol^{-1}. As mentioned above, this difference may be caused by the calculation methods.

Table 3.5: Calculated $\Delta_f H^0_{298}$ Alkyl Peroxides used in Reference Reactions; Comparison with Literature

Reactions Series	$\Delta_f H^0_{298}$	Error limit[a]	Therm[b]
$CH_3CH_2OOCH_3 + CH_4 \rightarrow CH_3CH_3 + CH_3OOCH_3$	-39.29	± 1.26	
$CH_3CH_2OOCH_3 + CH_3OOH \rightarrow CH_3CH_2OOH + CH_3OOCH_3$	-39.09	± 3.34	
$CH_3CH_2OOCH_3 + CH_3OO\bullet \rightarrow CH_3CH_2OO\bullet + CH_3OOCH_3$	-39.28	± 4.63	
$CH_3CH_2OOCH_3 + CH_3O\bullet \rightarrow CH_3CH_2O\bullet + CH_3OOCH_3$	-38.91	± 2.46	
$CH_3CH_2OOCH_3 + HOOH \rightarrow CH_3CH_2OOH + CH_3OOH$	-38.71	± 3.46	
Average	**-39.0 ± 0.24**		-39.3 / -37.02[c]
$CH_3OOCH_3 + CH_3CH_2OH \rightarrow CH_3CH_2OCH_3 + CH_3OOH$	-30.18	± 2.79	
$CH_3OOCH_3 + CH_3OH \rightarrow CH_3OOH + CH_3OCH_3$	-30.17	± 2.60	
$CH_3OOCH_3 + HOOH \rightarrow CH_3OOH + CH_3OOH$	-30.62	± 3.37	
$CH_3OOCH_3 + CH_3CH_2O\bullet \rightarrow CH_3OO\bullet + CH_3CH_2OCH_3$	-30.98	± 3.24	
$CH_3OOCH_3 + CH_3O\bullet \rightarrow CH_3OO\bullet + CH_3OCH_3$	-31.90	± 3.81	
$CH_3OOCH_3 + CH_3CH_2OH \rightarrow CH_3CH_2OOH + CH_3OCH_3$	-30.79	± 3.16	
Average	**-30.77 ± 0.64**		-31 /-31[d,e]/-29.17[c]

[a]reported errors for each of the standard species (Appendix A), and estimate error due to the method (App. C)
[b]values are from group additivity unless noted otherwise; [c]value from Sumathi and Green Ref. 107; [d]value from Lay and Bozzelli Ref. 104; [e]value from Carballeira et al. Ref. 119

Peroxy methyl and ethyl peroxy radicals are needed as reference species as well and are calculated with the isodesmic reactions listed in Table 3.6. The accuracy of the results has already been discussed in section 3.2.1.2.

Table 3.6: Calculated ΔH^0_{f298} for Peroxy Radicals

Radicals	Average	Error limit[a]	Therm[b]
$CH_3OO\bullet + CH_3CH_3 \rightarrow CH_3CH_2OO\bullet + CH_4$	1.95	± 3.33	
$CH_3OO\bullet + CH_3CH_2OCH_3 \rightarrow CH_3CH_2OO\bullet + CH_3OCH_3$	2.09	± 3.42	
Average	**2.02** ± 0.1		2.40
$CH_3CH_2OO\bullet + CH_3OH \rightarrow CH_3CH_2OH + CH_3OO\bullet$	-5.87	± 2.23	
$CH_3CH_2OO\bullet + CH_3OOH \rightarrow CH_3CH_2OOH + CH_3OO\bullet$	-6.3	± 3.45	
$CH_3CH_2OO\bullet + CH_3OCH_3 \rightarrow CH_3CH_2OCH_3 + CH_3OO\bullet$	-5.69	± 2.26	
Average	**-5.95** ± 0.31		-5.90

[a]Reported errors for each of the standard species (Appendix A) + estimated error due to the method (Appendix C)
[b]values are from Group Additivity.

3.2.1.4 Hydroperoxides

Enthalpies of formation, of different vinyl, allyl, and ethynyl hyroperoxides (target species) are determined from energies calculated at the B3LYP/6-311G(d,p) level and a series of working reactions. The enthalpies of reactions ($\Delta H^0_{rxn,298}$) are estimated as explained previously using total energies obtained by DFT calculations, with zero-point energies (ZPVE) scaled by 0.97 as recommended by Scott and Radom [120] and thermal corrections to 298.15 K [37]. The target hydroperoxides are:

Vinyl hydroperoxides: *CH₂=CHOOH, trans-CH₃CH=CHOOH, cis-CH₃CH=CHOOH, CH₃(CH₃)C=CHOOH, CH₂=C(CH₃)OOH* and *CH₃CH=C(CH₃)OOH.*

Allyl hydroperoxide: $CH_2{=}CHCH_2OOH$.

Ethynyl hydroperoxide: $CH{\equiv}COOH$ and $CH_3C{\equiv}COOH$.

Before moving to unsaturated species we use the value of methyl hydroperoxide to test our calculation method on ethylhydroperoxide (CH_3CH_2OOH) and provide some comparison of our results with available data. The calculated ethylhydroperoxide value is included in Table 3.7 and is found to be within 0.3 kcal mol^{-1} of the selected literature value of 39.52 kcal mol^{-1}, given in Appendix A (Table A.3).

$\Delta_f H^0_{298}$ of each hydroperoxide obtained from isodesmic reactions is given in Table 3.7 and the arithmetic mean is reported in units of kcal mol^{-1}. The $\Delta_f H^0_{298}$ values for most species show good precision within the different working reactions.

The good reproducibility can be partially attributed to our choice of working reactions and the corresponding error cancellation. For example, if we consider the enthalpy calculation of vinyl-hydroperoxide $CH_2{=}CHOOH$ (scheme 3.1), we observe, for each isodesmic reaction used, a perfect conservation of bonds and groups between the right and left side of the reactions.

$\boldsymbol{CH_2{=}CHOOH} + CH_3OH \rightarrow CH_3OOH + CH_2{=}CHOH$ -9.58 kcal mol^{-1}

$\boldsymbol{CH_2{=}CHOOH} + CH_3CH_2OH \rightarrow CH_3CH_2OOH + CH_2{=}CHOH$ -9.73 kcal mol^{-1}

$\boldsymbol{CH_2{=}CHOOH} + CH_3OCH_3 \rightarrow CH_3OOH + CH_2{=}CHOCH_3$ -9.59 kcal mol^{-1}

Scheme 3.1: Example of enthalpy of formation calculation with isodesmic reactions having bonds and groups conservation.

Table 3.7: Calculated $\Delta_f H^0_{298}$ for Viny, Allyl, Ethynyl and phenyl Hydroperoxides vs Reaction

Reactions Series	$\Delta_f H^0_{298}$ (kcal mol^{-1})	Error Limit[a]	Therm[b]
$CH_3OOH + CH_3CH_3 \rightarrow CH_3CH_2OOH + CH_4$	-31.54	± 1.50	
$CH_3OOH + CH_3CH_2OH \rightarrow CH_3CH_2OOH + CH_3OH$	-32.23	± 1.56	
$CH_3OOH + CH_3CH_2OCH_3 \rightarrow CH_3CH_2OOH + CH_3OCH_3$	-32.4	± 1.59	
Average	**-31.72** ± **1.02**		-31.8
$CH_3CH_2OOH + CH_3OH \rightarrow CH_3OOH + CH_3CH_2OH$	-39.27	± 1.98	
$CH_3CH_2OOH + CH_3OCH_3 \rightarrow CH_3OOH + CH_3CH_2OCH_3$	-39.29	± 2.01	
Average	**-39.28** ± 0.01		-39.52[c]
$CH_2=CHOOH + CH_3OH \rightarrow CH_3OOH + CH_2=CHOH$	-9.58	± 2.43	
$CH_2=CHOOH + CH_3CH_2OH \rightarrow CH_3CH_2OOH + CH_2=CHOH$	-9.73	± 2.17	
$CH_2=CHOOH + CH_3OCH_3 \rightarrow CH_3OOH + CH_2=CHOCH_3$	-9.59	± 1.95	
Average	**-9.63** ± **0.08**		-10
$\text{Trans-}CH_3\text{-}CH=CHOOH + CH_3OH \rightarrow CH_3OOH + CH_3CH=CH\text{-}OH$	-20.39	± 1.88	
$\text{Trans-}CH_3\text{-}CH=CHOOH + CH_3CH_2OH \rightarrow CH_3CH_2OOH + CH_3CH=CHOH$	-20.54	± 1.62	
$\text{Trans-}CH_3\text{-}CH=CHOOH + CH_3OCH_3 \rightarrow CH_3OOH + CH_3CH=CHOCH_3$	-20.41	± 1.61	
Average	**-20.44** ± **0.08**		-19
$\text{Cis-}CH_3\text{-}CH=CHOOH + CH_3OH \rightarrow CH_3OOH + CH_3\text{-}CH=CHOH$	-21.63	± 1.87	
$\text{Cis-}CH_3\text{-}CH=CHOOH + CH_3CH_2OH \rightarrow CH_3CH_2OOH + CH_3CH=CHOH$	-21.75	± 1.61	
$\text{Cis-}CH_3\text{-}CH=CHOOH + CH_3OCH_3 \rightarrow CH_3OOH + CH_3CH=CHOCH_3$	-21.62	± 1.94	
Average	**-21.66** ± **0.07**		-19
$CH_3(CH_3)C=CHOOH + CH_3OH \rightarrow CH_3OOH + CH_3(CH_3)C=CHOH$	-30.73	± 1.86	
$CH_3(CH_3)C=CHOOH + CH_3CH_2OH \rightarrow CH_3CH_2OOH + CH_3(CH_3)C=COH$	-30.86	± 1.60	
$CH_3(CH_3)C=CHOOH + CH_3OCH_3 \rightarrow CH_3OOH + CH_3(CH_3)C=CHOCH_3$	-30.80	± 1.93	
Average	**-30.79** ± **0.06**		-26
$CH_2=C(CH_3)OOH + CH_3OH \rightarrow CH_3OOH + CH_2=C(CH_3)OH$	-21.76	± 1.76	
$CH_2=C(CH_3)OOH + CH_3CH_2OH \rightarrow CH_3CH_2OOH + CH_2=C(CH_3)OH$	-21.88	± 1.60	
$CH_2=C(CH_3)OOH + CH_3OCH_3 \rightarrow CH_3OOH + CH_2=C(CH_3)OCH3$	-21.76	± 1.93	
Average	**-21.80** ± **0.06**		-21
$CH_3\text{-}CH=C(CH_3)OOH + CH_3OH \rightarrow CH_3OOH + CH_3CH=C(CH_3)OH$	-29.87	± 3.20	
$CH_3\text{-}CH=C(CH_3)OOH + CH_3CH_2OH \rightarrow CH_3CH_2OOH + CH_3CH=C(CH_3)OH$	-30.35	± 2.94	
$CH_3\text{-}CH=C(CH_3)OOH + CH_2=CHOH \rightarrow CH_2=CHOOH + CH_3CH=C(CH_3)OH$	-29.93	± 1.64	
Average	**-30.03** ± **0.23**		-28
$CH_2=CHCH_2OOH + CH_3OH \rightarrow CH_3OOH + CH_2=CHCH_2OH$	-13.5	± 2.26	
$CH_2=CHCH_2OOH + CH_3CH_2OH \rightarrow CH_3CH_2OOH + CH_2=CHCH_2OH$	-13.76	± 2.0	
$CH_2=CHCH_2OOH + CH_3OCH_3 \rightarrow CH_3OOH + CH_2=CHCH_2OCH_3$	-13.52	± 3.27	
Average	**-13.59** ± 0.14		-13
$CH\equiv C\text{-}OOH + CH_3OH \rightarrow CH_3OOH + CH\equiv COH$	42.54	± 1.87	
$CH\equiv C\text{-}OOH + CH_3CH_2OH \rightarrow CH_3CH_2OOH + CH\equiv CHOH$	42.36	± 1.61	
$CH\equiv C\text{-}OOH + CH_3OCH_3 \rightarrow CH_3OOH + CH\equiv COCH_3$	42.53	± 1.94	
Average	**42.25** ± 0.13		52[d]
$CH_3\text{-}C\equiv COOH + CH_3OH \rightarrow CH_3OOH + CH_3C\equiv COH$	30.34	± 1.94	
$CH_3\text{-}C\equiv COOH + CH_3CH_2OH \rightarrow CH_3CH_2OOH + CH_3C\equiv COH$	30.16	± 1.72	
$CH_3\text{-}C\equiv COOH + CH_3OCH_3 \rightarrow CH_3OOH + CH_3\text{-}C\equiv COCH_3$	30.51	± 2.01	
Average	**30.26** ± 0.13		41[d]

[a]Reported errors for each of the standard species (see App. A), and estimate error due to the method (see App. C);
[b]THERM = Group Additivity; [c]ref. 86
[d]Groups for general alkyl-peroxy radical not for ethynyl – we do not expect agreement (see text);

The magnitude of small fluctuations shown in the working reaction sets of Table 3.7, demonstrates internal consistency of the enthalpy values of the standard specie. Heat of formation of each hydroperoxide was also estimated with group additivity. The values determined with GA (Table 3.7) are in good agreement with the DFT values.

3.2.1.5 Unsaturated Radicals ROO•, RO• and R•

The enthalpy of reaction, ΔH^0_{rxn}, is an important parameter in the kinetics related to hydrocarbon oxidation for the hydrocarbon radical reaction with O_2 to form a peroxy radical. This ΔH^0_{rxn} determines the amount of extra energy the formed peroxy radical gains from the formation of the new R—OO• bond. The association reaction generates an energised peroxy radical, which can undergo reaction to new products before it is stabilised.

It is difficult to determine these thermochemical parameters from experiment, because it is hard to monitor the precursor hydrocarbon radical and the formed peroxy radical. The experiment is further complicated by the presence of reactions to new products by the energized peroxy radicals; which can prevent the monitoring of equilibrium. Experiments on ion methods using proton affinity or basicity, often with mass spectrometric analysis, are also utilized to determine enthalpies of formation of radicals. Our methods rely heavily on experimentally determined thermochemical data and we would like to point out that this data is very valuable to validate the computational methods.

An alternate method is the use of computational chemistry to estimate these ΔH^0_{rxn} values from the enthalpy of formation of the alkyl radical, and the respective peroxy radical.

As with the stable hydroperoxide molecules above, the isodesmic reactions for the unsaturated alkoxy and peroxy radicals resulting from C—OOH, CO—OH and COO—H bond cleavage reveal that enthalpies on a number of vinyl and ethynyl, alkoxy and peroxy radicals are needed for use as reference species. Limited data are available for these radical species enthalpies [47] and this data is listed Appendix A.

The peroxy radicals listed in Table 3.8 are calculated with DFT and use of the following isodesmic reactions:

R1 = ROO• + $CH_3OH \rightarrow$ ROH + CH_3OO•

R2 = ROO• + $CH_3CH_2OH \rightarrow$ ROH + CH_3CH_2OO•

R3 = ROO• + $CH_3OOH \rightarrow$ ROOH + CH_3OO•

R4 = ROO• + $CH_3CH_2OOH \rightarrow$ ROOH + CH_3CH_2OO•

R5 = ROO• + CH_2=CHOOH $\rightarrow$ ROOH + CH_2=CHOO•

Table 3.8: Calculated ΔH^0_{f298} for Peroxy Radicals

	ΔH^0_{f298} (kcal mol^{-1})							
Radicals	R1	R2	R3	R4	R5	Average	Error limit[a]	Therm[b]
CH_2=CHOO•	24.72	23.9	24.72	24.07	-	**24.34** ± 0.42	± 4.42	24.45[c]
trans-CH_3CH=CHOO•	14.70	13.83	13.06	12.47	12.84	**13.38** ± 0.88	± 4.11	16
cis-CH_3CH=CHOO•	14.73	13.91	11.94	11.35	11.72	**12.73** ± 1.49	± 4.11	16
$CH_3(CH_3)$C=CHOO•	2.48	1.67	2.43	1.84	2.21	**2.12** ± 0.35	± 4.09	8
CH_2=C(CH_3)OO•	11.25	10.48	11.21	10.62	10.99	**10.91** ± 0.34	± 4.07	13
CH_3-CH=C(CH_3)OO•	6.02	5.21	3.59	2.99	2.93	**4.15** ± 1.39	± 5.62	6
CH_2=CHCH_2OO•	22.14	21.33	20.11	19,53	19.98	**20.61** ± 1.08	± 6.18	2
CH≡COO•	84.47	83.65	84.39	83.79	84.37	**84.11** ± 0.36	± 4.16	85[d]
CH_3C≡COO•	71.07	70.26	71.00	70.41	70.78	**70.70** ± 0.35	± 4.20	75[d]

[a]Reported errors for each of the standard species (App. A) + estimated error due to the method (App. C)
[b]values are from Group Additivity unless noted otherwise
[c]value from Mebel et al. Ref. 33
[d]Groups developed in this work, not previously available

The enthalpies of the peroxy radicals are conveniently estimated by a work reaction which use enthalpies of the parent hydroperoxides that we have calculated above. Reactions R3, R4, and R5 from Table 3.8 have similar bonding environments on both sides of the work reaction and results in better cancellation of errors, than in reactions that conserve bonds only. Comparisons to Group Additivity estimates are given as well. The values recommended are noted in bold.

The alkoxy and hydrocarbon radicals listed in Tables 3.9 are estimated by two work reactions R6 - R7 and R8 - R9, respectively which involve the parent molecules as well. We note very good agreement with the group additivity values.

R6 = RO• + CH_3OH → ROH + CH_3O•

R7 = RO• + CH_3CH_2OH → ROH + CH_3CH_2O•

R8 = R• + CH_2=CH_2 → RH + CH_2=CH•

Except for *CH_2=CH•* where CH_2=CH_2 is replaced by CH_3O• (R• + CH_3OH → RH + CH_3O•)

R9 = R• + CH_3CH_3 → RH + CH_3CH_2•

The methoxy CH_3O• and ethoxy CH_3CH_2O• radicals given in Table 3.9 are calculated only for bond energy validation to support our validation and discussion (see section 3.2.3.2).

Table 3.9: Calculated ΔH^0_{f298} for Unsaturated Alkoxy and Alkyl Radicals

Radical	ΔH^0_{f298} (kcal mol^{-1})				
	R6	R7	Average	Error limit[a]	Therm[b]
trans-$CH_3CH{=}CHO\bullet$	-7.95	-8.39	-8.17 ± 0.31	± 2.14	-7[c]
cis-$CH_3CH{=}CHO\bullet$	-8.68	-9.13	-8.90 ± 0.31	± 2.14	-8[d]
$CH_3(CH_3)C{=}CHO\bullet$	-22.90	-23.35	-23.12 ± 0.31	± 2.14	-19[d]
$CH_2{=}C(CH_3)O\bullet$	-9.39	-9.84	-9.61 ± 0.31	± 2.14	-9[c]
$CH_3CH{=}C(CH_3)O\bullet$	-18.70	-19.15	-18.92 ± 0.31	± 2.14	-18[d]
$CH_2{=}CHO\bullet$	4.66	4.22	4.44 ± 0.31	± 1.79	4.0[e]
$CH_2{=}CHCH_2O\bullet$	23.71	23.26	23.48 ± 0.31	± 2.49	22
$CH{\equiv}CO\bullet$	40.63	40.18	40.40 ± 0.31	± 2.14	41[f]
$CH_3C{\equiv}CO\bullet$	29.36	28.90	29.13 ± 0.32	± 2.14	33[g]

$CH_3O\bullet + CH_3CH_2OCH_3 \rightarrow CH_3OCH_3 + CH_3CH_2O\bullet$ **3.66**
$CH_3CH_2O\bullet + CH_3OOCH_3 \rightarrow CH_2CH_3OCH_3 + CH_3OO\bullet$ **-2.75**

Radical	R8	R9	Average	Error limit[a]	Therm[b]
Trans-$CH_3CH{=}CH\bullet$	64.39	64.21	64.30 ± 0.12	± 1.68	64
cis-$CH_3CH{=}CH\bullet$	63.75	63.57	63.66 ± 0.12	± 1.68	64
$CH_3(CH_3)C{=}CH\bullet$	54.02	53.83	53.92 ± 0.13	± 1.79	55
$CH_2{=}C(CH_3)\bullet$	58.98	58.8	58.89 ± 0.12	± 1.68	61
$CH_3\text{-}CH{=}C(CH_3)\bullet$	53.74	53.55	53.64 ± 0.13	± 1.77	53
$CH_2{=}CH\bullet$	72.02	71.43	71.72 ± 0.4	± 1.90	71.62[h]
$CH_2{=}CHCH_2\bullet$	39.23	39.04	39.13 ± 0.13	± 1.69	41 (allyl)
$CH{\equiv}C\bullet$	140.56	140.94	140.75 ± 0.26	± 1.67	134
$CH_3C{\equiv}C\bullet$	128.69	128.50	128.59 ± 0.13	± 1.74	125

[a]Reported errors for each of the standard species (App. A) + estimated error due to the method (App. C);
[b]values are from Group Additivity unless noted otherwise;
[c]ketenyl radical;
[d]Groups developed in this work, not previously available;
[e]Lee et al. Ref. 96;
[f]From parent alcohol in Table 2, and vinoxy (phenoxy) bond energy of 85 kcal mol^{-1};
[g]Estimate does not account for radical on secondary or tertiary alkyl site;
[h]Mebel et al. Ref. 33

3.2.1.6 Methylperoxides

Enthalpy of formation of target methylperoxides are calculated as described in section 3.2.1.3. For each species four to five isodesmic reaction are used to determine the enthalpy of formation value, $\Delta_f H^0_{298}$. The following methylperoxides have been investigated

Vinyl peroxides: $CH_2{=}CHOOCH_3$, *trans*-$CH_3CH{=}CHOOCH_3$, *cis*-$CH_3CH{=}CHOOCH_3$, $CH_3(CH_3)C{=}CHOOCH_3$, $CH_2{=}C(CH_3)OOCH_3$ and $CH_3CH{=}C(CH_3)OOCH_3$;

Allyl-peroxide: $CH_2{=}CHCH_2OOCH_3$;

Ethynyl-peroxide: $CH{\equiv}COOCH_3$ and $CH_3C{\equiv}COOCH_3$.

The $\Delta_f H^0_{298}$ value of each peroxide is obtained from the isodesmic reactions illustrated in Table 3.10 and the arithmetic mean is reported in kcal mol^{-1}. The $\Delta_f H^0_{298}$ values for many given species show good precision through the different work reactions.

Table 3.10: Calculated $\Delta_f H^0_{298}$ for Vinyl, allyl and ethynyl methylperoxides vs Reaction

Reactions Series	$\Delta_f H^0_{298}$ [a]	Error limits[b]	Therm
$CH_2{=}CHOOCH_3 + CH_3OH \rightarrow CH_3OOCH_3 + CH_2{=}CHOH$	-9.56	± 2.04	
$CH_2{=}CHOOCH_3 + CH_3CH_2OH \rightarrow CH_3CH_2OOCH_3 + CH_2{=}CHOH$	-10.08	± 2.48	
$CH_2{=}CHOOCH_3 + CH_3OOH \rightarrow CH_3OOCH_3 + CH_2{=}CHOOH$	-9.60	± 2.46	
$CH_2{=}CHOOCH_3 + CH_3CH_3 \rightarrow CH_3CH_2OOCH_3 + CH_2{=}CH_2$	-11.17	± 1.95	
$CH_2{=}CHOOCH_3 + CH_3OO\bullet \rightarrow CH_3OOCH_3 + CH_2{=}CHOO\bullet$	-9.83	± 3.79	
Average	**-10.04 ± 0.66**		-11.91
$trans\text{-}CH_3CH{=}CHOOCH_3 + CH_3OH \rightarrow CH_3OOCH_3 + tr\text{-}CH_3CH{=}CHOH$	-19.87	± 1.50	
$trans\text{-}CH_3CH{=}CHOOCH_3 + CH_3CH_2OH \rightarrow CH_3CH_2OOCH_3 + tr\text{-}CH_3CH{=}CHOH$	-20.39	± 1.94	
$trans\text{-}CH_3CH{=}CHOOCH_3 + CH_3OCH_3 \rightarrow CH_3OOCH_3 + tr\text{-}CH_3CH{=}CHOCH_3$	-20.67	± 1.57	
$trans\text{-}CH_3CH{=}CHOOCH_3 + CH_3CH_3 \rightarrow CH_3CH_2OOCH_3 + CH_3CH{=}CH_2$	-18.91	± 2.02	
$trans\text{-}CH_3CH{=}CHOOCH_3 + CH_3OOH \rightarrow CH_3OOCH_3 + tr\text{-}CH_3CH{=}CHOOH$	-19.91	± 2.47	
Average	**-19.95 ± 0.67**		-19.78
$cis\text{-}CH_3CH{=}CHOOCH_3 + CH_3OH \rightarrow CH_3OOCH_3 + cis\text{-}CH_3CH{=}CHOH$	-21.14	± 1.30	
$cis\text{-}CH_3CH{=}CHOOCH_3 + CH_3CH_2OH \rightarrow CH_3CH_2OOCH_3 + cis\text{-}CH_3CH{=}CHOH$	-21.66	± 1.64	
$cis\text{-}CH_3CH{=}CHOOCH_3 + CH_3OCH_3 \rightarrow CH_3OOCH_3 + cis\text{-}CH_3CH{=}CHOCH_3$	-22.13	± 1.38	
$cis\text{-}CH_3CH{=}CHOOCH_3 + CH_3OOH \rightarrow CH_3OOCH_3 + cis\text{-}CH_3CH{=}CHOOH$	-21.16	± 2.26	
Average	**-21.27 ± 0.47**		-21[c]
$CH_3(CH_3)C{=}CHOOCH_3 + CH_3OH \rightarrow CH_3OOCH_3 + CH_3(CH_3)C{=}CHOH$	-30.22	± 1.73	
$CH_3(CH_3)C{=}CHOOCH_3 + CH_3OCH_3 \rightarrow CH_3OOCH_3 + CH_3(CH_3)C{=}CHOCH_3$	-31.94	± 1.80	
$CH_3(CH_3)C{=}CHOOCH_3 + CH_3CH_3 \rightarrow CH_3CH_2OOCH_3 + CH_3(CH_3)C{=}CH_2$	-29.97	± 2.35	
$CH_3(CH_3)C{=}CHOOCH_3 + CH_3OOH \rightarrow CH_3OOCH_3 + CH_3(CH_3)C{=}CHOOH$	-30.28	± 2.68	
Average	**-30.60 ± 0.90**		-28.23
$CH_3CH{=}C(CH_3)OOCH_3 + CH_3OH \rightarrow CH_3OOCH_3 + CH_3CH{=}C(CH_3)OH$	-29.49	± 1.25	
$CH_3CH{=}C(CH_3)OOCH_3 + CH_3CH_2OH \rightarrow CH_3CH_2OOCH_3 + CH_3CH{=}C(CH_3)OH$	-30.02	± 1.59	
$CH_3CH{=}C(CH_3)OOCH_3 + CH_3OOH \rightarrow CH_3OOCH_3 + CH_3CH{=}C(CH_3)OOH$	-29.25	± 2.27	
$CH_3CH{=}C(CH_3)OOCH_3 + CH_3CH_2OOH \rightarrow CH_3CH_2OOCH_3 + CH_3CH{=}C(CH_3)OOH$	-29.19	± 2.15	
Average	**-29.58 ± 0.32**		-30.38
$CH_2{=}C(CH_3)OOCH_3 + CH_3OH \rightarrow CH_3OOCH_3 + CH_2{=}C(CH_3)OH$	-20.99	± 1.25	
$CH_2{=}C(CH_3)OOCH_3 + CH_3CH_2OH \rightarrow CH_3CH_2OOCH_3 + CH_2{=}C(CH_3)OH$	-21.52	± 1.69	
$CH_2{=}C(CH_3)OOCH_3 + CH_3OCH_3 \rightarrow CH_3OOCH_3 + CH_2{=}C(CH_3)OCH3$	-20.37	± 1.32	
$CH_2{=}C(CH_3)OOCH_3 + CH_3CH_3 \rightarrow CH_3CH_2OOCH_3 + CH_2{=}CHCH_3$	-20.69	± 1.77	
$CH_2{=}C(CH_3)OOCH_3 + CH_3OOH \rightarrow CH_3OOCH_3 + CH_2{=}C(CH_3)OOH$	-21.04	± 2.20	
Average	**-20.79 ± 0.42**		-22.51
$CH_2{=}CHCH_2OOCH_3 + CH_3OH \rightarrow CH_3OOCH_3 + CH_2{=}CHCH_2OH$	-10.70	± 2.23	
$CH_2{=}CHCH_2OOCH_3 + CH_3CH_2OH \rightarrow CH_3CH_2OOCH_3 + CH_2{=}CHCH_2OH$	-11.23	± 2.67	
$CH_2{=}CHCH_2OOCH_3 + CH_3OCH_3 \rightarrow CH_3OOCH_3 + CH_2{=}CHCH_2OCH3$	-12.67	± 1.95	
$CH_2{=}CHCH_2OOCH_3 + CH_3CH_3 \rightarrow CH_3CH_2OOCH3 + CH_2{=}CHCH_3$	-13.11	± 2.40	
$CH_2{=}CHCH_2OOCH_3 + CH_3OOH \rightarrow CH_3OOCH_3 + CH_2{=}CHCH_2OOH$	-12.73	± 2.91	
Average	**-12.08 ± 1.05**		-12.91
$CH{\equiv}COOCH_3 + CH_3OH \rightarrow CH_3OOCH_3 + CH{\equiv}C\text{-}OH$	41.89	± 1.53	
$CH{\equiv}COOCH_3 + CH_3CH_2OH \rightarrow CH_3CH_2OOCH_3 + CH{\equiv}CHOH$	41.37	± 1.97	
$CH{\equiv}COOCH_3 + CH_3CH_3 \rightarrow CH_3CH_2OOCH_3 + HC{\equiv}CH$	40.06	± 2.08	
$CH{\equiv}COOCH_3 + CH_3OOH \rightarrow CH_3OOCH_3 + CH{\equiv}COOH$	41.6	± 2.55	
$CH{\equiv}COOCH_3 + CH_3CH_2OOH \rightarrow CH_3CH_2OOCH_3 + CH{\equiv}CHOOH$	41.66	± 3.23	
Average	**41.31 ± 0.72**		41.45
$CH_3C{\equiv}COOCH_3 + CH_3OH \rightarrow CH_3OOCH_3 + CH_3\text{-}C{\equiv}COH$	29.78	± 1.10	
$CH_3C{\equiv}COOCH_3 + CH_3CH_2OH \rightarrow CH_3CH_2OOCH_3 + CH_3\text{-}C{\equiv}COH$	29.26	± 1.54	
$CH_3C{\equiv}COOCH_3 + CH_3CH_3 \rightarrow CH_3CH_2OOCH_3 + CH_3\text{-}C{\equiv}CH$	29.31	± 1.67	
$CH_3C{\equiv}COOCH_3 + CH_3OOH \rightarrow CH_3OOH + CH_3\text{-}C{\equiv}COOH$	29.72	± 2.12	
Average	**29.51 ± 0.27**		31.87

[a]units: kcal mol^{-1};
[b]reported errors for each of the standard species (see App. A) + estimated error due to the method (see App. C).

As for the hydroperoxides, this is partially attributed to our choice of working reactions and the corresponding error cancellation. $\Delta_f H^0_{298}$ of each peroxide was also estimated by use of the group additivity. The values determined with GA (Table 3.10) are in good agreement with the DFT values.

3.2.2 Entropy and Heat Capacity

3.2.2.1 Hydroperoxides

Thermodynamic properties S^0_{298} and $Cp_{298}(T)$ referred to a standard state of an ideal gas at 1 atm are calculated for vinyl, and allyl hydroperoxides, using B3LYP/6-311G(d, p) determined geometries and frequencies. Results are summarized in Table 3.11. TVR represent the sum of the contributions from translations, vibrations and external rotations for S^0_{298} and $Cp_{298}(T)$. The torsion frequencies calculated for the internal rotors are not included in TVR. Instead, a more exact contribution from hindered rotations is calculated. I.R, represents the contributions from the internal rotation about C—O, O—O and C—C bonds to S^0_{298} and $Cp(T)$.

Table 3.11: Ideal Gas-Phase Thermodynamic Properties of vinyl-hydroperoxides

Species			ΔH^0_{f298} [a]	S^0_{f298}	Cp (T) cal/mole K						
					300 K	**400 K**	**500 K**	**600 K**	**800 K**	**1000 K**	**1500 K**
CH_2=CHOOH	TVR[d]			64.41	14.33	17.83	20.79	23.19	26.81	29.43	33.57
(2)[b]	IR[e]	C=C—OOH		4.79	3.03	2.57	2.33	2.19	1.97	1.77	1.30
(1)[c]		C=CO—OH		3.74	1.34	1.35	1.36	1.37	1.36	1.33	1.25
	Total		**-9.63**	**72.94**	**18.70**	**21.75**	**24.48**	**26.75**	**30.14**	**32.53**	**36.12**
trans-CH_3CH=CHOOH	TVR[d]			69.13	17.90	22.69	26.98	30.59	36.16	40.25	46.59
(2)[b]	IR[e]	C—C=COOH		4.64	1.97	1.81	1.64	1.50	1.32	1.22	1.10
(3)[c]		CC=C—OOH		4.66	2.06	1.98	1.90	1.82	1.65	1.51	1.29
		CC=CO—OH		3.45	1.52	1.52	1.53	1.54	1.53	1.48	1.34
	Total		**-20.44**	**81.88**	**23.45**	**28.00**	**32.05**	**35.45**	**40.66**	**44.46**	**50.32**
cis-CH_3CH=CHOOH	TVR[d]			68.59	17.61	22.53	26.88	30.52	36.12	40.20	46.56
(2)[b]	IR[e]	C—C=COOH		5.17	1.77	1.54	1.38	1.28	1.16	1.11	1.04
(3)[c]		CC=C—OOH		5.66	2.22	2.39	2.38	2.24	1.82	1.43	0.81
		CC=CO—OH		3.40	1.57	1.55	1.55	1.56	1.53	1.48	1.33
	Total		**-21.66**	**82.82**	**23.17**	**28.01**	**32.19**	**35.60**	**40.63**	**44.22**	**49.74**
$(CH_3)_2$C=CHOOH	TVR[d]			71.10	21.63	27.86	33.47	38.24	45.67	51.14	59.61
(2)[b]	IR[e]	(C)C—C=COOH		4.88	1.95	1.74	1.56	1.42	1.26	1.17	1.08
(9)[c]		(C—)CC=COOH		4.88	1.95	1.74	1.56	1.42	1.26	1.17	1.08
		(C)2C=C—OOH		3.68	2.19	2.24	2.23	2.16	1.96	1.76	1.43
		(C)2C=CO—OH		4.76	1.78	1.68	1.63	1.60	1.53	1.45	1.29
	Total		**-30.79**	**89.30**	**29.5**	**35.26**	**40.45**	**44.84**	**51.68**	**56.69**	**64.49**
CH_3CH=C(CH_3)OOH	TVR[d]			71.04	21.65	27.93	33.55	38.31	45.71	51.16	59.60
(2)[b]	IR[e]	C—C=C(C)OOH		5.88	1.83	1.58	1.42	1.31	1.18	1.12	1.02
(9)[c]		CC=C(—C)OOH		4.49	1.87	1.72	1.57	1.45	1.29	1.20	1.09
		CC=C(C)—OOH		3.63	2.43	2.56	2.47	2.30	1.94	1.67	1.33
		CC=C(C)O—OH		3.42	1.52	1.49	1.50	1.51	1.51	1.48	1.37
	Total		**-30.03**	**88.46**	**29.30**	**35.28**	**40.51**	**44.88**	**51.63**	**56.63**	**64.41**

Table 3.11 (con't): Ideal Gas-Phase Thermodynamic Properties of vinyl and allyl hydroperoxides

Species			ΔH^0_{f298} [a]	S^0_{f298}	Cp (T) cal/mole K						
					300 K	**400 K**	**500 K**	**600 K**	**800 K**	**1000 K**	**1500 K**
CH_2=C(CH_3)OOH	TVR[d]			68.20	18.07	23.08	27.39	30.95	36.39	40.37	46.61
(2)[b]	IR[e]	C=C—C(OOH)		4.99	1.92	1.72	1.55	1.43	1.27	1.18	1.08
(3)[c]		C=C(C)—OOH		3.83	3.70	3.97	3.72	3.31	2.54	1.98	1.16
		C=C(C)O—OH		3.41	1.35	1.39	1.43	1.46	1.51	1.51	1.43
	Total		**-21.80**	**80.43**	**25.04**	**30.16**	**34.09**	**37.15**	**41.71**	**45.04**	**50.28**
CH_2=CHCH_2OOH	TVR[d]			70.09	17.18	22.19	26.64	30.35	36.03	40.16	46.54
	IR[e]	C=C—COOH		1.68	2.12	2.61	2.89	2.99	2.97	2.87	2.51
		C=CC—OOH		3.99	3.12	2.72	2.41	2.18	1.86	1.66	1.38
		C=CCO—OH		3.48	1.40	1.41	1.44	1.46	1.48	1.47	1.37
	Total		**-13.59**	**77.54**	**23.82**	**28.94**	**33.38**	**36.98**	**42.34**	**46.16**	**51.80**

[b] ΔH^0_{f298} in kcal mol^{-1}; S^0_{298} in cal mol^{-1} K^{-1}; [b]optical isomers number; [c]symmetry number; [f]the sum of contributions from translations, external rotations and vibrations; [e]Contribution from internal rotation about the C—O, O—O and C—C bonds

3.2.2.2 Methylperoxides

Entropies, and heat capacities, of vinyl and allyl methylperoxides are summarized in Table 3.12. As described above, TVR represent the sum of the contributions from translations, vibrations and external rotations for S^0_{298} and $Cp(T)$. I.R represents the contributions from the internal rotation about C—O, O—O, O—C and C—C bonds for S^0_{298} and $Cp(T)$.

Table 3.12: Ideal Gas-Phase Thermodynamic Properties of vinyl and allyl Methylperoxides[a]

Species			$\Delta_f H^0_{298}$ [b]	S^0_{298}	*Cp (T)* cal mol^{-1} K^{-1}						
					300 K	**400 K**	**500 K**	**600 K**	**800 K**	**1000 K**	**1500 K**
CH_2=CHOOCH_3	TVR[f]			65.73	16.70	21.82	26.35	30.13	35.94	40.17	46.65
(2)[d]	IR[g]	C=C—OOC		5.38	2.71	2.35	2.18	2.08	1.90	1.75	1.46
(3)[e]		C=CO—OC		6.73	1.32	1.30	1.31	1.33	1.38	1.42	1.45
		C=COO—C		4.42	2.15	2.14	2.01	1.86	1.60	1.43	1.21
	Total		**-10.04**	**82.26**	**22.88**	**27.61**	**31.85**	**35.4**	**40.82**	**44.77**	**50.77**
trans-CH_3CH=CHOOCH_3	TVR[f]			68.39	20.45	26.78	32.60	37.56	45.33	51.01	59.70
(2)[d]	IR[g]	C—C=COOC		4.88	2.00	1.83	1.65	1.51	1.33	1.22	1.10
(9)[e]		CC=C—OOC		6.76	1.97	1.92	1.85	1.76	1.58	1.45	1.24
		CC=CO—OC		6.66	1.44	1.40	1.41	1.44	1.50	1.55	1.55
		CC=COO—C		4.54	2.13	2.08	1.93	1.78	1.53	1.38	1.18
	Total		**-19.95**	**91.23**	**27.99**	**34.01**	**39.44**	**44.05**	**51.27**	**56.61**	**64.77**
cis-CH_3CH=CHOOCH_3	TVR[f]			68.05	20.17	26.64	32.51	37.51	45.29	50.97	59.67
(2)[d]	IR[g]	C—C=COOC		5.23	1.77	1.54	1.39	1.29	1.17	1.11	1.05
(9)[e]		CC=C—OOC		7.13	2.28	2.18	2.00	1.83	1.56	1.39	1.19
		CC=CO—OC		6.55	1.47	1.42	1.42	1.44	1.51	1.56	1.55
		CC=COO—C		4.53	2.13	2.09	1.94	1.79	1.54	1.38	1.19
	Total		**-21.27**	**91.49**	**27.82**	**33.87**	**39.26**	**43.86**	**51.07**	**56.41**	**64.65**

[a]Thermodynamic properties are referred to a standard state of an ideal gas at 1 atm.;

[b] $\Delta_f H^0_{298}$ in kcal mol^{-1}; [c] S^0_{298} in cal mol^{-1} K^{-1};

[d]optical isomers number; [e]symmetry number;

[f]The sum of contributions from translations, external rotations and vibrations; [g]Contribution from internal rotation about the corresponding C—O, O—O and C—C bonds.

Table 3.12 (con't): Ideal Gas-Phase Thermodynamic Properties of vinyl and allyl Methylperoxides[a]

Species			$\Delta_f H^0_{298}$ [b]	S^0_{298}	Cp (T) cal mol^{-1} K^{-1} 300 K	400 K	500 K	600 K	800 K	1000 K	1500 K
$(CH_3)_2C$=$CHOOCH_3$	TVR[f]			70.25	24.20	31.97	39.09	45.21	54.83	61.91	72.72
(2)[d]	IR[g]	(C)C—C=COOC		5.01	1.97	1.75	1.56	1.43	1.26	1.18	1.08
(27)[e]		(C—)CC=COOC		5.01	1.97	1.75	1.56	1.43	1.26	1.18	1.08
		(C)2C=C—OOC		6.72	2.22	2.19	2.16	2.09	1.91	1.72	1.41
		(C)2C=CO—OC		6.69	2.28	2.17	1.99	1.82	1.55	1.39	1.19
		(C)2C=COO—C		4.56	2.13	2.07	1.92	1.77	1.52	1.37	1.18
	Total		**-30.60**	**98.24**	**34.77**	**41.9**	**48.28**	**53.75**	**62.33**	**68.75**	**78.66**
CH_3CH=$C(CH_3)OOCH_3$	TVR[f]			70.43	24.23	32.03	39.17	45.28	54.87	61.92	72.71
(2)[d]	IR[g]	C—C=C(C)OOC		5.21	1.77	1.55	1.40	1.30	1.18	1.11	1.05
(27)[e]		CC=C(—C)OOC		4.96	1.93	1.76	1.60	1.47	1.30	1.20	1.09
		CC=C(C)—OOC		6.84	2.66	2.49	2.27	2.05	1.71	1.50	1.24
		CC=C(C)O—OC		6.13	1.59	1.70	1.78	1.82	1.84	1.82	1.69
		CC=C(C)OO—C		4.57	2.13	2.07	1.92	1.76	1.52	1.36	1.17
	Total		**-29.58**	**98.14**	**34.31**	**41.6**	**48.14**	**53.68**	**62.42**	**68.91**	**78.95**
CH_2=$C(CH_3)OOCH_3$	TVR[f]			67.76	20.64	27.18	33.00	37.91	45.53	51.12	59.71
(2)[d]	IR[g]	C=C—C(OOC)		4.98	1.93	1.75	1.59	1.46	1.29	1.20	1.09
(9)[e]		C=C(C)—OOC		5.12	3.80	3.64	3.22	2.82	2.24	1.88	1.43
		C=C(C)O—OC		5.29	3.01	2.83	2.51	2.24	1.91	1.75	1.61
		C=C(C)OO—C		4.55	2.13	2.07	1.93	1.77	1.53	1.37	1.18
	Total		**-20.79**	**87.7**	**31.51**	**37.47**	**42.25**	**46.2**	**52.5**	**57.32**	**65.02**
CH_2=$CHCH_2OOCH_3$	TVR[f]			69.64	19.68	26.23	32.21	37.28	45.16	50.90	59.64
(2)[d]	IR[g]	C=C—COOC		4.88	2.00	1.83	1.65	1.51	1.33	1.22	1.10
(3)[e]		C=CC—OOC		6.76	1.97	1.92	1.85	1.76	1.58	1.45	1.24
		C=CCO—OC		6.66	1.44	1.40	1.41	1.44	1.50	1.55	1.55
		C=CCOO—C		4.54	2.13	2.08	1.93	1.78	1.53	1.38	1.18
	Total		**-12.08**	**92.48**	**27.22**	**33.46**	**39.05**	**43.77**	**51.1**	**56.5**	**64.71**

[a]Thermodynamic properties are referred to a standard state of an ideal gas at 1 atm.;
[b] $\Delta_f H^0_{298}$ in kcal mol^{-1}; [c] S^0_{298} in cal mol^{-1} K^{-1}; [d]optical isomers number; [e]symmetry number;
[f]The sum of contributions from translations, external rotations and vibrations; [g]Contribution from internal rotation about the corresponding C—O, O—O and C—C bonds.

In this study, we have determined entropies and heat capacities of eight vinyl and allyl hydroperoxides listed above in Table 3.7. We have reported these thermodynamic properties for the corresponding methylperoxides as well (Table 3.10). To evaluate relative differences in the contributions to S^0_{298} and $Cp_{298}(T)$ at 298 K between the C—OOC and the C—OOH rotor and between the O—OC and O—OH rotor values, we list the respective internal rotor contributions of these rotors in Tables 3.11 and 3.12. The contributions to the entropies for the R—OOC are higher by 0.6 to 2.7 cal mol^{-1} K^{-1} than those for the analogue entropies. The RO—OC rotors are higher by 1.9 to 3.2 cal mol^{-1} K^{-1} than the RO—OH rotors.

The contribution to the heat capacities for the R—OOC and RO—OH differs by ca. 1 cal mol^{-1} K^{-1} for the temperature range 300 K to 1500 K.

3.2.2.3 Internal Rotational Barriers

Potential barriers versus torsion angle for internal rotation are calculated for the following hydroperoxides and their corresponding methylperoxides:

CH_2=CHOOH / CH_2=CHOOCH_3

trans-CH_3CH=CHOOH / *trans*-CH_3CH=CHOOCH_3

cis-CH_3CH=COOCH / *cis*-CH_3CH=COOCH_3

CH_3(CH_3)C=CHOOH / CH_3(CH_3)C=CHOOCH_3

CH_3CH=C(CH_3)OOH / CH_3CH=C(CH_3)OOCH_3

CH_2=C(CH_3)OOH / CH_2=C(CH_3)OOCH_3

CH_2=CHCH_2OOH / CH_2=CHCH_2OOCH_3.

Potential barriers for internal rotations about the C_d—O ($C_d \equiv C_{double\ bond}$), O—OC, O—OH, OO—$CH_3$, and C_d—C internal rotations for vinyl, allyl, and ethynyl hydroperoxides and methylperoxides were computed at the B3LYP/6-311G(d,p) calculation level, by scanning individual torsion angles from 0° to 360° at 15° intervals of the corresponding dihedral angle, and allowing the remaining molecular structural parameters to be optimized.

Potential energy curves for the, C_dO—OH, C_d—OOH, and C_d—C rotors for the hydroperoxides and C_dOO—C, C_dO—OC, C_d—OOC and C_d—C rotors for the peroxides can be found Appendix D. Points are calculated values and lines are the results from the Fourier expansion fit to the data (Figures 3.1, 3.2, and 3.3). The coefficients, a_i and b_i in equation 2.19 (chapter 2), are listed in appendix B. Values for the coefficients of the Fourier expansion, a_i and b_i in equation I, are obtained from the fitting modus in SigmaPlot version 2.0. They are used in "ROTATOR" to calculate the contribution of internal rotors to S^0_{298} and $Cp^0(T)$ ($0.1 \leq T/K \leq 5000$). The Rotator program takes into account the entropy of mixing and the optical isomer corrections.

Data on the rotational barriers in the C_d—OOC, C_dO—OC, C_dOO—C and C_d—C bonds for the peroxides are summarized in Table 3.12, along with data on the corresponding barriers in hydroperoxides for comparison purposes.

C_dOO—C Internal Rotor Potential

A potential energy diagram for internal rotation versus C_dOO—C torsion angle is shown in Figure 3.1 for CH_2=CHOOCH_3. The C_dOO—CH_3 rotor potentials for the other peroxides are similar and typical for alkyl methyl rotations with 3-fold symmetry and barriers ranging between 2.7 and 3.0 kcal mol^{-1}.

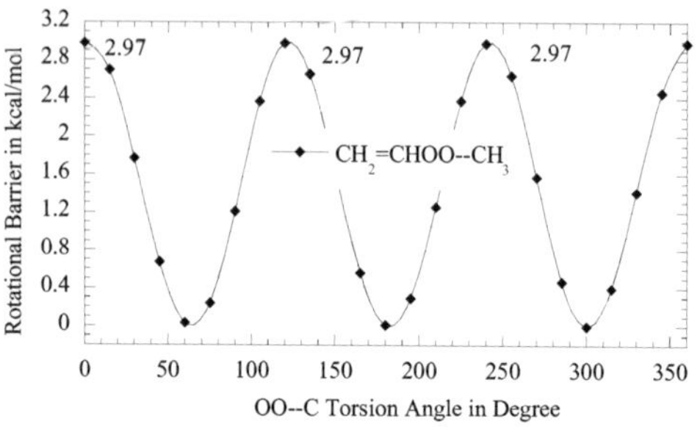

Figure 3.1: Potential for internal rotation about CH_2=CHOO—CH_3 bond in vinyl peroxide.

C_d—C Internal Rotor Potential

Rotation barriers for C_d—CH_3 bonds for all peroxides (hydro and methyl-peroxides) have the same 3-fold symmetry and have all relatively low barriers, 1.3 to 2.1 kcal mol^{-1}.

The C_d—C rotation barrier in allyl hydroperoxide (CH_2=CH—CH_2OOH) is higher (6.4 kcal mol^{-1}) which is 3 times the corresponding rotational barrier for allyl peroxide (CH_2=CH—CH_2OOCH_3). Both have one-fold symmetry.

C_d—OOR Internal Rotor Potentials

The rotation barriers about C_d—OOC are similar to or lower than those in C_d—OOH; the barriers vary from 3.4 to 5.5 kcal mol^{-1} (see figure 3.2 as example).

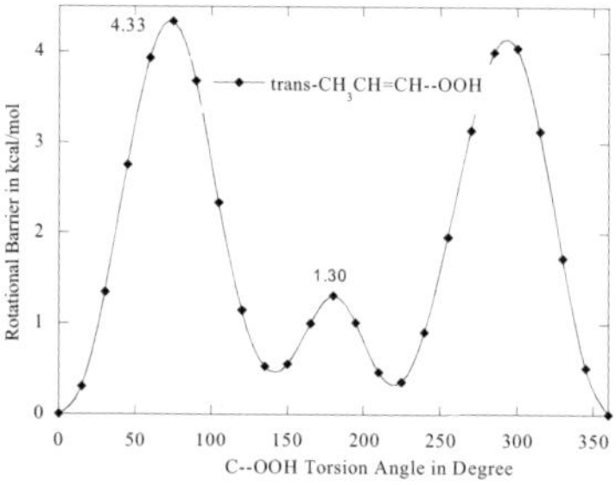

Figure 3.2: Potential barrier for the internal rotation about the transCH_3CH=CH—OOH bond in vinyl hydroperoxide. The minimum barrier is at a dihedral angle of OOC_dC_d = 0 degrees. The maximum barrier is at a dihedral angle of OOC_dC_d = 75 degrees.

C_dO—OR Internal Rotor Potentials

Figure 3.3 shows examples of some CO—O internal rotor potentials. Rotation about C_dO—OC peroxide bonds shows a significant increase relative to the C_dO—OH hydroperoxides. The barrier in CH_2=CHO—OCH_3 is 9.75 kcal mol^{-1}, some 4 kcal mol^{-1} higher than that in the corresponding vinyl hydroperoxide, CH_2=CHO—OH. The CH_2=C(CH_3)$OOCH_3$ barrier is 13 kcal mol^{-1}, some 6.1 kcal mol^{-1} higher than the barrier in the corresponding hydroperoxide $CH_2=C(CH_3)O—OH$. The $C_dO—OH$ rotation barriers in the remaining peroxides are in the range of 8.3 - 9.7 kcal mol^{-1} and all are about 3 kcal mol^{-1} higher than the corresponding hydroperoxide (see Table 3.8). The rotation barrier for the C=O—OC bond in $CH_3CH=C(CH_3)O—OCH_3$ is 10.1 kcal mol^{-1}, some 3.8 kcal mol^{-1} higher than that in the corresponding hydroperoxide.

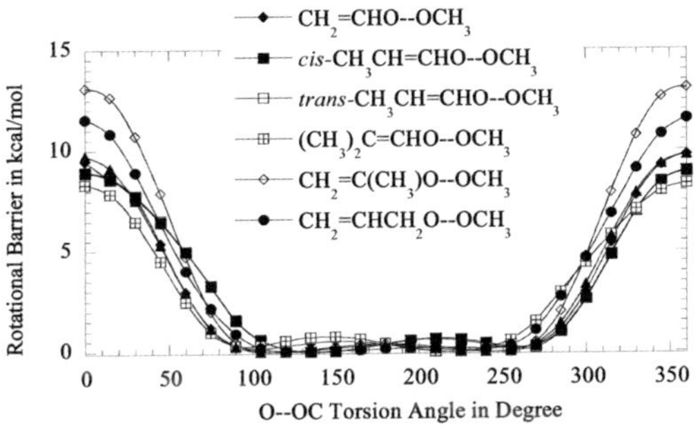

Figure 3.3: Potential barriers for internal rotations about O—OC bond in studied methylperoxides.

Each of these C=O—OC internal rotations have one relatively high barrier, ranging from 8.5 to13 kcal mol^{-1}; the PE curve also shows one very low barrier at a fraction of 1 kcal mol^{-1}. Examination of the vibrational frequency motions suggests that the frequency determined in the B3LYP calculation only represents the low barrier torsion. We use the *S* and *Cp(T)* contributions from the internal rotor analysis, and not from the torsion frequency. The RO—OC barrier in allyl methyl peroxide $CH_2=CHCH_2O—OCH_3$ is also higher, 11.55 kcal mol^{-1}, than that in the hydroperoxide, 6.3 kcal mol^{-1}.

Table 3.13: Comparison of maxima rotational barrier: ROOH versus ROOC (298 K) Units: kcal mol^{-1}

Species	C_d—OO	O—O	C—C_d	C(—C)C_d / C_d(—C)O	OO—C
$CH_2=CHOOCH_3$	5.57	9.75	-	-	2.97
$CH_2=CHOOH$	6.22	5.76	-	-	-
trans-$CH_3CH=CHOOCH_3$	3.58	8.90	2.09	-	2.77
trans-$CH_3CH=CHOOH$	4.33	5.59	2.04	-	-
cis-$CH_3CH=CHOOCH_3$	3.65	8.91	1.44	-	2.80
cis-$CH_3CH=CHOOH$	3.67	5.45	1.33	-	
$CH_3(CH_3)C=CHOOCH_3$	4.63	8.32	-	1.82	2.73
$CH_3(CH_3)C=CHOOH$	4.40	5.01	-	1.82	-
$CH_2=C(CH_3)OOCH_3$	5.19	13.11	-	1.97	2.75
$CH_2=C(CH_3)OOH$	5.93	6.97	-	1.89	-
$CH_3CH=C(CH_3)OOCH_3$	3.44	10.1	1.48	2.03	2.71
$CH_3CH=C(CH_3)OOH$	3.44	6.24	1.49	2.02	-
$CH_2=CHCH_2OOCH_3$	4.69	11.55	2.09	-	2.91
$CH_2=CHCH_2OOH$	6.49	6.30	6.40	-	-

3.2.3 Bond Energy

3.2.3.1 Bond Strengths in Hydroperoxides

The ROO—H, RO—OH, and R—OOH bonds energies as well as the bond energies for radicals RO—O• and R—OO• are estimated using ΔH^0_{f298} from Tables 3.6, 3.7, and 3.8. Literature values for the radicals are used whenever possible. Otherwise radical enthalpies are calculated. These bond energy results are listed in Table 3.14. The similarity in the calculated bond energies through a given class of radical and the close agreement to literature values are discussed below and provide support for accuracy in the absolute values.

Table 3.14: R—OOH, RO—OH, ROO—H, R—OO•, and RO—O• Bond Energy (298 K) in kcal mol^{-1}

	ROO—H	RO—OH	R—OOH		R—OO•	RO—O•
CH_3OOH	85.84	44.34	-	CH_3OO•	-	61.19
CH_3CH_2OOH	85.43	45.49	-	CH_3CH_2OO•	-	62.75
CH_2=CHOOH	86.07	23.03	84.6	CH_2=CHOO•	47.38	39.65
trans-CH_3CH=CHOOH	85.92	21.23	87.99	*trans*-CH_3CH=CHOO•	50.92	38.0
cis-CH_3CH=CHOOH	86.49	21.72	88.57	*cis*-CH_3CH=CHOO•	50.93	37.92
$CH_3(CH_3)C$=CHOOH	85.01	16.63	87.96	$CH_3(CH_3)C$=CHOO•	51.8	34.31
CH_2=$C(CH_3)OOH$	84.81	21.15	83.94	CH_2=$C(CH_3)OO$•	48.01	39.03
CH_3CH=$C(CH_3)OOH$	85.52	19.31	86.16	CH_3CH=$C(CH_3)OO$•	49.49	36.48
CH_2=$CHCH_2OOH$	86.3	46.03	55.97	CH_2=$CHCH_2OO$•	18.52	62.42
CH≡COOH	93.96	7.11	94.0	CH≡COO•	48.89	15.84
CH_3C≡COOH	92.54	7.83	101.58	CH_3C≡COO•	57.81	17.90

3.2.3.2 Validation of Calculated Bond Strengths in CH_3OOH and CH_3CH_2OOH

Methyl and ethyl-hydroperoxides enthalpy values are among the rare peroxides for which data are available in the literature. To test and validate the accuracy of our calculated bond energies for unsaturated hydroperoxides, we have compared our estimated bond strengths for CH_3OOH and CH_3CH_2OOH with the bond energies given in the literature.

The derived ROO—H bond energy for the calculated CH_3OO—H and CH_3CH_2OO—H bonds, given in Table 3.6, are 85.8 and 85.4 kcal mol^{-1} respectively. These are one to two kcal mol^{-1} lower than the data by Jonsson et al. data [90, 91]. They report values for CH_3OO—H and CH_3CH_2OO—H bonds obtained with a high-level method known as G2M, to be 87.2 and 86.9 kcal mol^{-1} respectively, using atomisation reactions. The value we estimated for CH_3CH_2OO—H is also in good agreement with experimental values of Clifford et al. [15] which is 85 ± 2 kcal mol^{-1}. Blanksby et al. [16] report experiment values of 87.8 ± 1.0 kcal mol^{-1}, for CH_3OO• and 84.8 ± 2.2 kcal mol^{-1} for CH_3CH_2OO•.

Stocker and Pilling [121] derived a ROO—H value of 87.56 kcal mol^{-1} from evaluation of the literature; and Slagle and Knayzev use the value of Stocker and Pilling in their evaluation. Sun and Bozzelli [122] also report values for the methyl hydroperoxy bond energy CH_3OO—H to be 86.0 kcal mol^{-1}. Lay and Bozzelli [104] report values of 86.6 to 86.1 kcal mol^{-1}.

Differences in the ROO—H bond energy value for methyl and ethyl -hydroperoxides are only about 2 kcal mol^{-1}.

The RO—OH bonds in saturated alkyl hydroperoxides are weak, typically 44 to 46 kcal mol^{-1}, as recently reported by Mulder et al. [123]. Values in this study around 45 kcal mol^{-1} are in good agreement with the literature.

We note that dissociation of alkyl hydroperoxides ROOH $\rightarrow$ RO + OH is often attributed to be the major component of chain branching in low to moderate temperature hydrocarbon combustion. We do not see any evidence for a change in this RO—OH bond energy for the alkyl hydroperoxides in this study. Curran et al. [124, 125] suggest lower values for the dissociation reactions in order to fit experimental data for negative temperature. Modelling of oxidation of hydrocarbon and oxyhydrocarbon lubricants in engine oils as being the principle chain branching reaction and its known thermochemistry is also unsuccessful with these alkyl hydroperoxides dissociation. The lower bond energy needed by Curran et al. and the lack of sufficient model for oxidation of lubricants suggests that a lower energy branching path may exist. Indeed, we find in this study that the vinyl and ethynyl hydroperoxides are observed to have significantly lower RO—OH dissociation energies, as discussed below.

The calculated O—O bond energy in CH_3O—O• is 61.2 kcal mol^{-1} and increases slightly in CH_3CH_2O—O• (62.7 kcal mol^{-1}). There are unfortunately no published results available, that we are aware of, to validate these results.

3.2.3.3 Bond Strengths in Vinyl, Allyl, and Ethynyl Hydroperoxides

Vinyl Hydroperoxides, Effect of Methyl Substitution on Bond Energies:

The bond energy for the C=COO—H is 86.0 kcal mol^{-1}. The four other vinyl hydroperoxides (*trans*-CC=COOH, *cis*-CC=COOH and C(C)C=COOH and CC=C(C)OOH) have roughly the same ROO—H bond energies: 85.9, 86.5, 85.0 and 85.5 kcal mol^{-1}, respectively. C=C(C)OOH is lower, 84.8 kcal mol^{-1}. The ROO—H bond energies for all vinylic hydroperoxides in this study vary within a range: 84.8 to 86.5 kcal mol^{-1}, which is within the uncertainty range.

The calculated CO—OH bond energy, is 23.0 kcal mol^{-1} in C=COOH and 21.2 kcal mol^{-1} and 21.7 kcal mol^{-1} in the *trans* and *cis* CC=CO—OH respectively. It is lower, yet, in CC=C(C)OOH (19.3 kcal mol^{-1}). These values are more than 20 kcal mol^{-1} lower than the RO—OH bonds in alkyl hydroperoxides. The weak bond (low stability) in these molecules leads to lifetimes, on the order of seconds at room temperature and much shorter lifetimes under low temperature combustion conditions. We note that the range in bond energies is larger for this case than others in this study. We also note an unexpected deviation from group additivity of the enthalpies of the substituted vinylic ethers and alcohols used in the working reactions, Table 3.3.

There is one previously reported value of a vinyl hydroperoxide enthalpy, that we are aware of, that of Schlegel et al. [113]. Their data suggest a value for the O—O bond energy of a secondary methyl substituted vinyl hydroperoxide, C=C(C)OOH of 23 kcal mol^{-1}, which is in reasonable agreement with our calculated value for that species, 21.1 kcal mol^{-1}.

The C=C—OOH bond energies are 85 kcal mol^{-1}, about 15 kcal mol^{-1} stronger than the corresponding bond strength observed in the alkyl hydroperoxides. This is likely due to interaction (overlap) of the Π system in the vinyl group and the oxygen atom bonded on the carbon, along with an energy preference to form a carbonyl, relative to an olefinic bond.

The R—OOH bond energies in the vinyl hydroperoxides (C=COOH, *trans*-CC=COOH, *cis*-CC=COOH and C(C)C=COOH) are observed to increase slightly from 84.6 kcal mol^{-1} to 88.6 kcal mol^{-1}, with increasing methyl substitution on the non peroxide sp^2 carbon. This increase is observed in R—OOH bond energies for vinyl hydroperoxides with methyl substitution on the non-peroxide carbon. The *cis*-CC=C—OOH bond energy is slightly lower than the *trans*-CC=C—OOH form, probably because of the steric position of OOH to methyl radical.

The RC=C—OOH bond energies for the two secondary vinylic species C=C(C)—OOH and CC=C(C)—OOH are 83.9 and 86.2 kcal mol^{-1}, respectively.

Propenyl (Allyl) Hydroperoxide: C=CCOOH

The bond energy value found for C=CCOO—H, 86.3 kcal mol^{-1}, is in agreement with the value of G2M calculations of Jonsson [90] 86.2 kcal mol^{-1}.

The bond energy for C=CCO—OH, is 46.0 kcal mol^{-1}. This is similar to that of alkyl hydroperoxides RO—OH discussed above in section 3.6.1.1.

The bond energy for C=CC—OOH, is 55.9 kcal mol^{-1}. This is lower that in the alkyl hydroperoxides due to loss of resonance in the allylic moiety.

Ethynyl Hydroperoxides: C≡COOH and CC≡COOH:

The ethynyl ROO—H bond strength is significantly increased to a value of 93.9 kcal mol^{-1} which is nearly 10 kcal mol^{-1} higher than in the alkyl, allyl, vinyl hydroperoxides, for which bond energies between 85 – 87 kcal mol^{-1} are found. This is probably a result of the overlap with the expanded Π system of the ethynes with the oxygen π orbitals.

There is a slight decrease in bond energies with an additional carbon on the non hydroperoxide *sp* carbon, from 93.9 kcal mol^{-1} for C≡COO—H to 92.5 kcal mol^{-1} for CC≡COO—H. More data and higher-level calculations are needed to further verify this. We note that the group additivity (GA) estimation values in Table 3.7 do not include this extra bond strength in the ethyne-hydroperoxide and thus the GA estimates are some 10 kcal mol^{-1} high.

The calculated C≡CO—OH bond energy is very weak, 7.1 kcal mol^{-1} in C≡CO—OH and 7.8 kcal mol^{-1} in the methyl substituted CC≡CO—OH; the bond strengths are only on the order of a hydrogen bond. The ethynyl hydroperoxide moiety is highly unstable. This molecule will rapidly dissociate to form a ketenyl radical + OH:

HC≡CO—OH → HC•=C=O + •OH $\Delta H^0_{rxn} = 7.2$ kcal mol^{-1}.

The reasons for this low bond strength involves a strong carbonyl bond formation in the ketenyl structure and the overlap of the Π system.

The C≡C—OOH bond strength increases even further above that of the vinyl systems to ca. 100 kcal mol^{-1}. This corresponds with the increase in the C=COO—H bond compared to vinyl and alkyl systems. We observe a difference of 7 kcal mol^{-1} with an additional carbon on the non hydroperoxide sp carbon. The R—OOH bond energy increases from 94 (C≡C—OOH) to 101.58 (CC≡C—OOH) kcal mol^{-1}.

The strengthening of the C≡C—OOH and C≡COO—H bonds combined with the dramatic weakening of the C≡CO—OH bond does suggest no overlap throughout the π and oxygen system. Instead, it indicates a strong overlap of the Π system with one oxygen atom, and the O—H strength from an OH group more close to hydroxyl than a hydroperoxide.

3.2.3.4 Bond strengths in Vinyl, Allyl, and Ethynyl Peroxy Radicals

Vinyl-peroxy radicals

Bond energies for C=C—OO• and C=C(C)—OO• are 47.4 and 48.0 kcal mol^{-1}, respectively. Mebel and Lin [33] report a value of 46.5 kcal mol^{-1} for vinyl–OO•. The two other vinyl hydroperoxides radical bond energies of *trans*-CC=C—OO• and *cis*-CC=C—OO• are slightly higher, 50.92 and 50.93 kcal mol^{-1}. Similar values are obtained for C(C)C=C—OO• (51.8 kcal mol^{-1}) and for CC=C(C)—OO• (49.5 kcal mol^{-1}).

The bond energy in C(C)C=CO—O• is the lowest with 34.3 kcal mol^{-1}, while the strength of CC=C(C)O—O• is calculated at 36.5 kcal mol^{-1}. Other values range from 39.6 kcal mol^{-1} for C=CO—O•, 39.0 kcal mol^{-1} for C=C(C)O—O, 38.0 kcal mol^{-1} for *trans*-CC=CO—O to 37.9 kcal mol^{-1} for *cis*-CC=CO—O•. This is a significant spread in values and suggests that there are strong interactions resulting from the methyl group. Mebel and Lin [33] calculated a value of 33.4 kcal mol^{-1} for C=CO—O• (value converted from 0 K to 298 K).

Allyl-peroxy radicals

Jonsson [90] reports values between 13.4 and 23.4 kcal mol^{-1} for C=CC—OO•. Our value is 18.52 kcal mol^{-1} and the estimated value of Benson is 18.2 ± 0.5 kcal mol^{-1} [20]. Knyazev and Slagle [14] have found 18.42 kcal mol^{-1}.

The C=CCO—O• value is 62.4 kcal mol^{-1}, which is similar to values of the alkyl-peroxy systems.

Ethynyl-peroxy radicals

The Bond energies for C≡C—OO• and CC≡C—OO• are higher than those for vinyl and alkyl systems rely from 49 to 57 kcal mol^{-1}. Bond strengths between C≡C—OO• and CC≡C—OO• show a difference of 9 kcal mol^{-1}, from 48.9 to 57.8 kcal mol^{-1} with an additional carbon on the non-hydroperoxide *sp* carbon.

The O—O bond energies in C≡CO—O• and CC≡CO—O• are weak at about 16 kcal mol^{-1}, similar to the weak bonds (ca. 7 kcal mol^{-1}) in the ethynyl hydroperoxides. The O—O bond energies in C≡CO—O• and CC≡CO—O• show a slight increase from 15.8 to 17.9 kcal mol^{-1} with an additional carbon on the non hydroperoxide *sp* carbon.

3.2.3.5 Bond Energies in Methylperoxides

Bond energies for the C_d—OOC, C_dO—OC, and C_dOO—C bonds are reported and compared with previously determined C_d—OOH, C_dO—OH, and C_dOO—H bond energies.

These C_d—OOC, C_dO—OC, and C_dOO—C bond energy results are listed in Table 3.12 below. The data show good consistency in the calculated bond energies though a given class of radical and the agreement with literature values is also good.

Methyl Vinyl Peroxides, Effect of Methyl Substitution on Bond Energies:

Methyl vinyl peroxide, C=COOC, has the highest C_dOO—C bond energy, 69.2 kcal mol^{-1}, where C_d represents a carbon with a double bond. Methyl group substitution for a hydrogen atom on either carbon of the vinyl group, reduces the C_dOO—C bond energies by up to 2.7 kcal mol^{-1}. The C_dOO—C bonds in this category remain nearly constant with a small increase from C=C(C)OO—C (66.5 kcal mol^{-1}) to C(C)C=COO—C (67.4 kcal mol^{-1}) and CC=C(C)OO—C (68.5 kcal mol^{-1}). *Trans*-CC=COO—C and *cis*-CC=COO—C and have nearly identical but slightly higher C_dOO—C bond energies: 68.1 and 68.8 kcal mol^{-1}, respectively.

The C_dO—OC bond energies in these unsaturated peroxides, are very weak; this results in unstable molecules with lifetime of less than a second at room temperature. The C_dO—OC bond energy ranges from 11.6 to 18.6 kcal mol^{-1}, and the trends indicate that single methyl substitution on the vinyl carbon decreases the C_dO—OC bond by up to 4 kcal-mol^{-1}. The simplest methyl vinyl peroxide C=CO—OC has the highest bond strength at 18.58 kcal mol^{-1}. The bond energy decreases with increasing methyl substitution on the vinyl carbons: the *cis*-CC=CO—OC bond is 16.47 kcal mol^{-1}, slightly higher than *trans*-CC=CO—OC, 15.9 kcal mol^{-1}; C=C(C)O—OC is 15.3 and CC=C(C)O—OC is 14.7 kcal mol^{-1}.

Substitution of two methyl groups for two primary olefin hydrogen results in the weakest C_dO—OC bond determined for CC(C)=COOC to be only 11.6 kcal mol^{-1}.

The C=C—OOC bonds are similar in bond strength and range from 84 to 87 kcal mol^{-1}.

Methyl Propenyl (Allyl) Peroxide: C=CCOOCH$_3$:

The C=CCOO—C bond energy in allyl-methyl peroxide is 68.0 kcal mol^{-1}, similar to the phenyl and the vinyl peroxides above. The C=CCO—OC bond energy is 39.66 kcal mol^{-1}, about 2 times stronger than the vinyl and phenyl peroxides, but similar to the RO—OR bonds in dimethyl or ethyl-methyl (alkyl) peroxides.

Methyl Ethynyl Peroxides: C≡COOCH$_3$ and CC≡COOCH$_3$:

The C_tOO—C bond strength in ethynyl-peroxide is 77.7 kcal mol^{-1} and in CC=COO—H it is 76 kcal mol^{-1}, an increase of about 9 kcal mol^{-1} relative to the phenyl and vinyl peroxides. This is similar to differences in bond energies in the corresponding hydroperoxides.

The C≡CO—OC and CC≡CO—OC bond energies in ethynyl peroxides are 3.27 and 3.72 kcal mol^{-1} respectively but they are too small to exist physically at high temperature; the ethynyl peroxide moiety is unstable. The $HC{\equiv}COOCH_3$ and $CH_3C{\equiv}COOCH_3$ molecules will immediately dissociate to form a ketenyl radical (C•H=C=O or CC•=C=O and CH_3O) with bond energies less than 4 kcal mol^{-1}. This instability is similar in the ethynyl hydroperoxide molecules.

$$HC{\equiv}COOCH_3 \leftrightarrow HC{\equiv}CO\bullet + \bullet OCH_3 \leftrightarrow HC\bullet{=}C{=}O + \bullet OCH_3,\ \Delta H^0_{rxn,298} = 3.27 \text{ kcal mol}^{-1}$$

$$CH_3C{\equiv}COOCH_3 \leftrightarrow CH_3C{\equiv}CO\bullet + \bullet OCH_3 \leftrightarrow CH_3C\bullet{=}C{=}O + \bullet OCH_3,\ \Delta H^0_{rxn,298} = 3.72 \text{ kcal mol}^{-1}$$

This small bond strength can be explained by the facile formation of a strong carbonyl bond (ketenyl group) at the expense of cleavage of the weaker (second) π bond in the ethyne (RC≡CO—OC) and the ketenyl π system: RC≡COOC → RC•=C=O + •OC.

The ethyne peroxides, CC≡COOC and CCC≡COOC, have RC≡COOC bond energies of 94.0 and 100.0 kcal mol^{-1}, respectively, showing a significant difference resulting from the methyl substitution on the primary ethyne carbon. The increase in strength of this bond in these quasi-molecules results from the overlap of the Π electrons on the oxygen with the ethyne Π bonds.

3.2.3.6 Comparison of Bond Energies in Hydroperoxides versus Peroxides

Tables 3.15 and 3.16 list the R—OOX, RO—OX and ROO—X (X = CH_3 or H) bond energies for the different peroxides. The C_dOO—C energies for all species listed in Table 3.15, are lower than the corresponding C_dOO—H bond strengths by ca. 17 kcal mol^{-1}. This is consistent with data for alcohols where carbon – oxygen single bonds are typically ~13.5 kcal mol^{-1} lower than hydrogen–oxygen bonds. The bond energy in C_dO—OC is about 5 kcal mol^{-1} weaker than the bond in C_dO—OH, for all peroxides reported in Table 3.15.

This result is consistent with the slightly shorter bond length in the C_dO—OH series compared to the C_dO—OC bond lengths. These results are in agreement with the HO—OH/HO—OCH_3 and CH_3O—OH/CH_3O—OCH_3 systems. In both cases, the substitution of H by a methyl results in longer bond and weaker bond energy. The HO—OCH_3 bond is 5.6 kcal mol^{-1} weaker, than the HO—OH bond, and the CH_3O—OCH_3 bond is 5.66 kcal mol^{-1} weaker than CH_3O—OH (Table 3.15).

Table 3.15: Comparison of: ROO—H vs ROO—C, RO—OH vs RO—OC and R—OOH vs R—OOC Bond Energies (298 K), Units: kcal mol^{-1}

Species[a]	ROO—H	ROO—C	RO—OH	RO—OC	R—OOH	R—OOC
C=COOX	86.07	69.2	23.03	18.58	84.60	83.91
trans-CC=COOX	85.92	68.15	21.23	15.88	87.99	86.4
cis-CC=COOX	86.49	68.82	21.72	16.47	88.57	87.08
C(C)C=COOX	85.01	67.45	16.63	11.58	87.96	86.67
C=C(C)OOX	84.81	66.52	21.15	15.28	83.94	81.83
CC=C(C)OOX	85.52	68.55	19.31	14.76	86.16	85.37
C≡COOX	93.96	77.70	7.11	3.27	94	93.92
CC≡COOX	92.54	76.01	7.83	3.72	101.58	101.23
C=CCOOX	86.3	68.05	46.03	39.66	55.97	53.36

[a]X = H or CH_3

The C_d—OOC versus C_d—OOH bond energies show almost no variation in the corresponding vinyl, phenyl and ethynyl species. Difference between C_d—OOC and C_d—OOH bond energy in vinyl species values vary from 0.7 to 1.5 kcal mol^{-1}. This difference increases slightly, to 2.6 kcal mol^{-1} when we consider the allyl hydroperoxide and peroxide.

The ethynyl species do not exhibit any difference in bond energy between C_t—OOH and C_t—OOC. The C_t—OOX bond energies are found to be 94 kcal mol^{-1} for CH≡COOH and $CH≡COOCH_3$ and 101 kcal mol^{-1} for $CH_3C≡COOH$ and $CH_3C≡COOCH_3$.

The CH_3—OOH/CH_3—$OOCH_3$ system show bond strength difference of 1.9 kcal mol^{-1} (from Table 13.16).

Table 3.16: ROO—X[a] and RO—OX Bond Energies (298 K; Units: kcal mol^{-1})

	ROO—H	ROO—C	Bond difference[b]	RO—OH	RO—OC	Bond difference[c]
HOOX	87.89	69.87	18.02	50.45	44.86	5.59
COOX	86.05	67.97	18.08	44.86	39.2	5.66

[a]X= H or CH_3, [b]Bond difference between O—H and O—C, [c] Bond difference between O—OH and O—OC

3.2.3.7 Chain Branching via Cleavage of the Weak Vinyl and Ethynyl Peroxide Bonds

As discussed previously, the RCO—OH bonds are very weak in the vinyl, and ethynyl hydroperoxides. The CO—OC bonds are slightly weaker (ca. 1 - 4 kcal mol^{-1}) in the peroxides of this study. The peroxide dissociation energies range from 19.3 to 3.2 kcal mol^{-1}; and these species are unstable towards simple dissociation:

RC=CO—OC → RC=CO• + •OC → RC•C=O + •OC

Reasons for this low bond strength involve the formation of the strong carbonyl bond relative to loss of a weaker Π bond and the overlap of the Π system.

A chain branching process can result from addition reactions of an alkyl peroxy radical to olefins or acetylenes, where vinyl, or acetyl peroxides are formed. The initial radical formed by the addition of the peroxy radical would undergo an addition reaction with molecular oxygen, then undergo molecular elimination of HO_2 to form an unsaturated peroxide. The unsaturated peroxide would then undergo rapid cleavage of the weak peroxide bonds. A mechanism that implements this chain branching results from ROO• addition to olefins is illustrated by:

COO• + RC=C → RC•COOC	Addition / stabilisation
RC•OOC + O_2 → RC(OO•)COOC	(O_2 addition)
RC(OO•)COOC → HO_2 + RC=COOC	(HO_2 molecular elimination)
RC=COOC → RCO• + •OC	(Cleavage of weak vinyl peroxide bond, Chain Branching)

(cleavage of weak vinyl peroxide bond, chain branching, plus two subsequent, facile eliminations to form carbonyls)

3.3 Conclusion

For the first time, thermodynamic Properties ($\Delta_f H^0_{298}$, S^0_{298} and $Cp(T)$ ($300 \leq T \leq 1500$)), bond energies and rotational barriers potentials of vinyl hydroperoxides and their corresponding peroxides: CH_2=CHOOH / CH_2=CHOOCH_3, *trans*-CH_3CH=CHOOH / *trans*-CH_3CH=CHOOCH_3, *cis*-CH_3CH=CHOOH / *cis*-CH_3CH=CHOOCH_3, $(CH_3)_2$C=CHOOH / $(CH_3)_2$C=CHOOCH_3, CH_2=C(CH_3)OOH / CH_2=C(CH_3)OOCH_3 and CH_3CH=C(CH_3)OOH / CH_3CH=C(CH_3)OOCH_3 are reported. Values for allyl peroxides CH_2=CHCH_2OOH / CH_2=CHCH_2OOCH_3, two ethynyl peroxides, CH≡COOH / CH≡COOCH_3 and CH_3C≡COOH / CH_3C≡COOCH_3 are also determined.

Standard enthalpies of formation, ΔH^0_{f298} are calculated using the B3LYP/6-311G(d,p) calculation level of density functional theory and isodesmic reaction schemes for cancellation of errors. The recommended ΔH^0_{f298} of each value is the average value of data obtained from several isodesmic reactions. Comparison of the thermodynamic properties between the peroxides and corresponding hydroperoxides shows uniformity in property trends. A set oxygenated hydrocarbon groups are determined that result in consistent prediction of thermodynamic properties for the substituted alkyl and vinyl peroxides and hydroperoxides.

Analysis of the bond energies shows that ethynyl (HC≡C•), vinyl (C_2H_3•) hydroperoxides are unstable towards dissociation to the corresponding ketenyl, vinoxy radicals + OH, respectively. The low RO—OH bond energies of vinyl, phenyl and ethynyl hydroperoxides results in new chain branching paths that may be important in low to moderate temperature hydrocarbon oxidation. The olefinic and acetylene peroxides and hydroperoxides are unstable to cleavage of the very weak (22 to 3 kcal mol^{-1}) RO—OH or RO—OR peroxide bonds due to formation of strong carbonyl bond (ca. 20 to 40 kcal mol^{-1} stronger than the hydrocarbon Π bonds) in the vinyl, phenyl and acetylenic peroxides and hydroperoxides.

A mechanism that implements the chain branching resulting from ROO• addition to olefins is illustrated by the cleavage of the weak vinyl peroxide bond, followed by chain branching resulting in facile eliminations to form carbonyls.

4. Group Additivity Calculations

Oxidation of unsaturated and aromatic hydrocarbons in atmospheric and combustion processes results in formation of linear and cyclic unsaturated, oxygenated-hydrocarbon intermediates. The thermochemical parameters, $\Delta_f H^0_{298}$, S^0_{298} *and* $Cp_{f298}(T)$ *for these intermediates are needed in order to understand their stability and reaction paths in further oxidation. These properties are not available for a majority of these unsaturated oxy-hydrocarbons and their corresponding radicals, even via group additivity method.*

In the previous chapter, Density Functional calculations have been used to determine the thermodynamic properties of specific species. A second application is considered in this chapter which allows developing new group for group additivity application. A consistent set of oxygenated peroxy-hydrocarbon and acetylene-alcohols groups are calculated.

The thermochemistry of a series of unsaturated oxygenated and non-oxygenated species and radicals is first determined. The molecules are selected because they are small and contain the unknown groups which are our target groups in this chapter.

This thermochemistry is then used to calculate thirty-one new groups which are needed to evaluate large molecules system (larger unsaturated oxygenated or multi-oxygenated intermediates) where possible calculation techniques are less accurate.

4.1 Introduction

The initial steps in oxidative reaction of aromatic, poly-aromatic and other cyclic and linear unsaturated hydrocarbons in the atmosphere or in combustion involve radical formation. These radicals react with molecular oxygen. The subsequent reactions of these peroxy radicals, as shown e.g. in Figure 4.1, result in unsaturated linear or cyclic, oxygenated or multi-oxygenated hydrocarbon intermediates. The thermochemistry for these unsaturated – oxygenated species is needed to evaluate their stability and likely reaction paths in the environment, in combustion and in other thermal and oxidative processes.

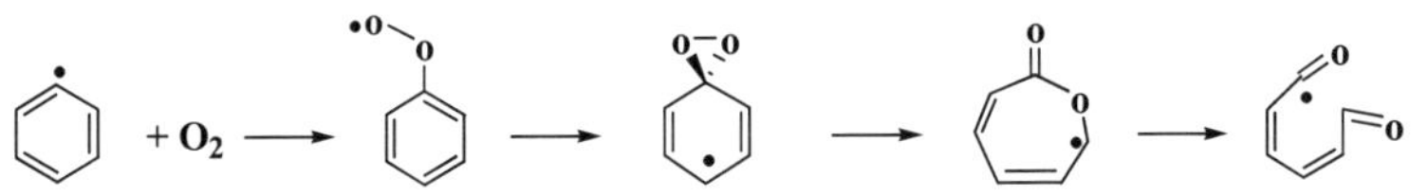

Figure 4.1: Oxidation of phenyl radical with O_2

While one may readily visualize a series of beta scission reactions (unzipping of the above ring opening product), the reactions of substituted aromatics and the reactions of stabilized, unsaturated intermediates with O_2 and the radical pool are more complex. The thermochemical parameters are not available for a many of the unsaturated, oxy-hydrocarbon intermediates in such reaction paths. One possibility to obtain these data is the use of group additivity (see chapter 2). However, inspection of the available groups shows that many groups needed for these molecules are missing. This provides the motivation of this part of this thesis to develop new groups

This study estimates thermochemical properties for a number of unsaturated oxy-hydrocarbons using DFT combined with isodesmic work reactions to determine enthalpy of formation, entropy, and heat capacity values. The molecules are selected to contain a specific group needed in unsaturated oxygenated intermediates resulting from aromatic species decomposition. We develop the thermochemical properties for the molecules containing the target groups, and then use the thermochemistry along with known groups to calculate the values for the unknown groups.

The GA method serves as a fast and accurate estimation technique for many scientists and engineers whose work involves thermodynamic characterization of elementary and overall reaction processes. One convenient way to perform Group additivity calculations is by using the “THERM” code for hydrocarbons distributed by Bozzelli [81, 82].

In this work, we calculate enthalpy, entropy and heat capacity values for thirty groups found in unsaturated and oxygenated hydrocarbons and in addition the C/CD2/C/H; the groups are listed in Table 4.1. We have focussed on groups needed to estimate the molecules formed during the oxidation and pyrolysis decomposition of polyaromatics, dioxins and furans.

Table 4.1: Groups Developed

		phenyl containing groups	
CT/O	CD/O2	O/CB/O	CO/CB/H
O/C/CT	CD/CD/CO	C/CB/C/O2	C/CB/H2/O
O/CT/O	CD/CO/CT	CD/CB/CD	
O/CD/O	C/CD2/H/O	C/CB/C2/CO	
C/CD/CO/CO/H	C/CD/CO/H2	CB/CO	
C/CD/CO/CT/H	O/CD2	CD/CB/CO	
C/CO/O2/H	CO/CD/H	O/CB/CO	
C/C/CO/O2	CD/CD/O	CD/C/CB	
C/CD2/CO/H	C/CD2/C/H	CD/CB/O	
C/CD2/O2		CD/CB/H	

We develop nineteen new groups, listed in the first two columns of Table 4.1 all based on unsaturated, oxygenated species with one exception the C/CD2/C/H group which has no oxygen. Twelve additional groups involve a bond to a phenyl ring carbon (CB for carbon benzene). The phenyl groups are listed in columns 3-4 of Table 4.1. Each group is estimated from only one reference molecule. The calculation method is explained in the next sections.

In this chapter, the calculation procedure is as follows:

- Calculations of *reference species* required in the working reactions, if these species can not be found in the literature.
- Calculations of a series of new species (stable and radicals) which contain our target groups. These species are called *target species* in the text.
- Calculations of new groups or *target groups* in the text.

4.2 Results and Discussion

4.2.1 Enthalpies of the Target Species

A series of 28 stable molecules and 12 radicals containing the groups listed in table 4.1 are calculated and reported below. For these calculations, enthalpy values for standard reference species are needed in the working reactions. In some cases the enthalpy value of a reference molecule is also determined from own calculation. Since no reliable literature data was available, the propenoxyl radical CH_2=CHC(O•)=O and formyloxyl radical CH(O•)=O used as reference species were estimated with the following isodesmic reactions:

R—C(O•)=O + CH_3CH_3 → R—C(OH)=O + CH_3CH_2• and

R—C(O•)=O + CH_2=CH_2 → R—C(OH)=O + CH_2=CH• where R ≡ H or CH_2=CH

2-propyl radical CH_2=C•CH_3 is estimated according to the reaction below:

CH_2=C•CH_3 + R' → CH_2=$CHCH_3$ + R• where R' ≡ CH_2=CH_2 and C_6H_6

The calculated and literature reference species are listed in appendix A along with literature references for reported enthalpies.

Standard enthalpies of formation for the oxygenated and non-oxygenated hydrocarbons compounds were calculated, using the isodesmic work reactions in Tables 4.2, 4.3 and 4.4. $\Delta_f H^0_{298}$ of the target species are estimated using two to five isodesmic reaction schemes for each stable molecule or radical. The Density Functional energy, E, the ZPVE and the thermal correction to 298.15 K are calculated for each molecule, and $\Delta H^0_{rxn,298}$ is calculated. Thermal corrections account for translation, external rotation, and vibrations, and include the torsion frequencies, for all molecules in the work reactions. Selection of the work reaction with similar internal rotors on both sides allows for cancellation of this error.

The isodesmic reactions used to determine $\Delta_f H^0_{298}$ values for the linear, unsaturated oxygenated hydrocarbons are illustrated in Table 4.2 and for aromatic and related unsaturated oxy-hydrocarbons in Table 4.3. The first column of Tables 4.2 and 4.3 also lists the target group to be estimated using the enthalpy of the calculated specie. Fourteen unsaturated C_2-C_6 species are listed in Table 4.2; they all contain an oxygenate functionality (alcohol, peroxide, and aldehyde). There is one non-oxygenated species, CH_2=$CHCH(CH_3)CH$=CH_2, in Table 4.2. Table 4.3 lists nine unsaturated oxygenated species; all of them contain a phenyl group. The several work reaction calculations for each species show good precision, which provides an internal data check.

Table 4.2: Calculated $\Delta_f H^0_{298}$ of the target species

Groups to be Calculated	Reactions Series	$\Delta_f H^0_{298}$ (kcal mol^{-1})	Error Limit[b]
C/CD/CO/CO/H	***CH₂=CHCH(CH=O)CH=O*** + CH_3CH_3 → CO + CH_2=O + CH_2=$CHCH_2CH_2CH_3$	-40.38	± 2.14
	CH₂=CHCH(CH=O)CH=O + CH_3CH_3 → $2CH_2$=O + CH_2=$CHCH_2CH$=CH_2	-40.08	± 2.1
	CH₂=CHCH(CH=O)CH=O + CH_4 → CO + CH_2=O + CH_2=$CHCH_2CH_3$	-38.93	± 1.97
	CH₂=CHCH(CH=O)CH=O + CH_4 → CO + CH_3CH=O + CH_2=$CHCH_3$	-38.85	± 2.27
	Average	**-39.56 ± 0.78**	
C/CD/CO/CT/H	***CH₂=CHCH(CH=O)C≡CH*** + CH_3CH_3 → CO + CH≡CH + CH_2=$CHCH_2CH_2CH_3$	45.07	± 1.77
	CH₂=CHCH(CH=O)C≡CH + CH_4 → CO + CH_3C≡CH + CH_2=$CHCH_3$	45.39	± 1.59
	Average	**45.23 ± 0.22**	
C/CO/O2/H	***CH(OH)₂CH=O*** + CH_4 → $2CH_3OH$ + CO	-119.4	± 1.54
	CH(OH)₂CH=O + CH_3CH_3 → CH_3CH_2OH + CH_3OH + CO	-118.77	± 1.66
	Average	**-119.08±0.44**	
C/C/CO/O2	***CH₃C(OH)₂CH=O*** + $2CH_4$ → CH_3CH_2CH=O + $2CH_3OH$	-131.55	± 2.73
	CH₃C(OH)₂CH=O → CH_3C(=O)OH + CH_2=O	-129.08	± 2.08
	CH₃C(OH)₂CH=O + $2CH_3CH_3$ → CH_3CH_2CH=O + $2CH_3CH_2OH$	-131.09	± 2.97
	Average	**-130.57±1.31**	
C/CD2/CO/H	***CH₂=CHCH(CH=O)CH=CH₂*** + CH_2CH_2 → CH_2=$CHCH$=O + C_5H_8	-1.74	± 3.43
	CH₂=CHCH(CH=O)CH=CH₂ + $2CH_4$ → CH_3CH=O + $2CH_2$=$CHCH_3$	-0.049	± 2.66
	CH₂=CHCH(CH=O)CH=CH₂ + CH_4 → CH_3CH=O + CH_2=$CHCH_2CH$=CH_2	-0.69	± 2.36
	Average	**-0.83 ± 0.85**	
C/CD2/O2	***CH₂=CHC(OH)₂CH=CH₂*** + CH_4 → $2CH_2$=$CHCH_2OH$	-61.01	± 1.96
	CH₂=CHC(OH)₂CH=CH₂ + $2CH_4$ → $CH_2(OH)_2$ + $2CH_2$=$CHCH_3$	-61.59	± 1.31
	Average	**-61.30 ± 0.41**	
CD/O2	***CH₂=C(OH)₂*** + CH_4 → $CH_2(OH)_2$ + CH_2=CH_2	-74.42	± 1.6
	CH₂=C(OH)₂ + CH_2CH_2 → $2CH_2$=CHOH	-73.49	± 5.53
	Average	**-73.95 ± 0.65**	

[b]reported errors for each of the standard species (see App. A) + estimated error due to the method (see App. C).

Table 4.2 (con't): Calculated $\Delta_f H^0_{298}$ of the target species

Groups to be Calculated	Reactions Series	$\Delta_f H^0_{298}$ (kcal mol^{-1})	Error Limit[b]
C/CD/CO/H2	$\mathit{CH_2{=}CHCH_2CH{=}O} + CH_4 \rightarrow CH_2{=}CHCH_3 + CH_3CH{=}O$	-19.96	± 2.1
	$\mathit{CH_2{=}CHCH_2CH{=}O} + CH_4 \rightarrow CH_2{=}CHCH_2CH_3 + CH_2{=}O$	-20.04	± 1.78
	$\mathit{CH_2{=}CHCH_2CH{=}O} + CH_2{=}CH_2 \rightarrow CH_2{=}CHCH_3 + CH_2{=}CHCH{=}O$	-21.01	± 2.73
	$\mathit{CH_2{=}CHCH_2CH{=}O} + CH_3CH_3 \rightarrow CH_2{=}CHCH_2CH_2CH_3 + CH_2{=}O$	-21.19	± 1.95
	Average	**-20.62 ± 0.74**	
C/CD2/C/H	$\mathit{CH_2{=}CHCH(CH_3)CH{=}CH_2} + CH_2CH_2 \rightarrow CH_2{=}CHCH_3 + CH_2{=}CHCH_2CH{=}CH_2$	17.75	± 1.55
	$\mathit{CH_2{=}CHCH(CH_3)CH{=}CH_2} + CH_3CH_3 \rightarrow CH_3CH_2CH_3 + CH_2{=}CHCH_2CH{=}CH_2$	18.10	± 1.55
	$\mathit{CH_2{=}CHCH(CH_3)CH{=}CH_2} + CH_4 \rightarrow CH_3CH_3 + CH_2{=}CHCH_2CH{=}CH_2$	18.09	± 1.48
	Average	**17.98 ± 0.20**	
CD/CD/CO	$\mathit{CH_2{=}C(CH{=}O)CH{=}CH_2} + CH_2CH_2 \rightarrow CH_2{=}CHCH{=}O + CH_2{=}CHCH{=}CH_2$	-2.93	± 2.67
	$\mathit{CH_2{=}C(CH{=}O)CH{=}CH_2} + CH_3CH_3 \rightarrow CH_3CH_2CH{=}CH_2 + CH_2{=}CHCH{=}O$	-2.83	± 2.59
	$\mathit{CH_2{=}C(CH{=}O)CH{=}CH_2} + CH_4 \rightarrow CH_3CH{=}O + CH_2{=}CHCH{=}CH_2$	-1.93	± 1.94
	Average	**-2.56 ± 0.55**	
CD/CO/CT	$\mathit{CH_2{=}C(CH{=}O)C{\equiv}CH} + CH_3CH_3 \rightarrow CH_3CH_2C{\equiv}CH + CH_2{=}CHCH{=}O$	38.33	± 3.15
	$\mathit{CH_2{=}C(CH{=}O)C{\equiv}CH} + CH_2CH_2 \rightarrow CH_2{=}C(CH{=}O)CH{=}CH_2 + CH{\equiv}CH$	40.04	± 2.73
	Average	**39.18 ± 1.21**	
C/CD2/H/O	$\mathit{CH_2{=}CHCH(OH)CH{=}CH_2} + CH_4 \rightarrow CH_2{=}CHCH_2OH + CH_3CH{=}CH_2$	-13.18	± 2.14
	$\mathit{CH_2{=}CHCH(OH)CH{=}CH_2} + CH_2{=}O \rightarrow CH_2{=}CHCH_2OH + CH_2{=}CHCH{=}O$	-12.25	± 2.86
	$\mathit{CH_2{=}CHCH(OH)CH{=}CH_2} + CH_3CH_3 \rightarrow CH_2{=}CHCH_2OH + CH_2{=}CHCH_2CH_3$	-11.63	± 2.22
	Average	**-12.35 ± 0.78**	
O/CD2	$\mathit{CH_2{=}CHOCH{=}CH_2} + CH_4 \rightarrow CH_3CH{=}O + CH_3CH{=}CH_2$	-6.19	± 1.64
	$\mathit{CH_2{=}CHOCH{=}CH_2} + CH_2{=}O \rightarrow CH_2{=}CHOH + CH_2{=}CHCH{=}O$	-6.45	± 4.04
	$\mathit{CH_2{=}CHOCH{=}CH_2} + CH_3CH_3 \rightarrow CH_2{=}CHOH + CH_2{=}CHCH_2CH_3$	-5.83	± 3.4
	Average	**-6.15 ± 0.31**	
CO/CD/H	$\mathit{CH_2{=}CHCH{=}O} + CH_3CH_3 \rightarrow CH_2{=}CHCH_2CH_3 + CH_2{=}O$	-18.03	± 2.04
	$\mathit{CH_2{=}CHCH{=}O} + CH_2{=}CH_2 \rightarrow CH_2{=}CHCH{=}CH_2 + CH_2{=}O$	-18.16	± 2.02
	$\mathit{CH_2{=}CHCH{=}O} + CH_4 \rightarrow CH_2{=}CHCH_3 + CH_2{=}O$	-19.75	± 1.96
	Average	**-18.65 ± 0.95**	

[b]reported errors for each of the standard species (see Appendix A) + estimated error due to the method (see Appendix C).

Table 4.3: Calculated $\Delta_f H^0_{298}$ of the target species

groups to be Calculated	Reactions Series		$\Delta_f H^0_{298}$ kcal mol^{-1}	error limit[b]
	$C_6H_5C(OH)_2CH_3$ + CH_4	→ $C_6H_5CH_2CH_3$ + $CH_2(OH)_2$	-80.10	± 1.86
C/C/CB/O2	$C_6H_5C(OH)_2CH_3$ + 2CH_4	→ $C_6H_5CH_3$ + $CH_2(OH)_2$ + CH_3CH_3	-78.95	± 2.0
		Average	**-79.52 ± 0.81**	
	$C_6H_5C(CH_2)CH{=}CH_2$ + CH_2CH_2	→ $C_6H_5CH{=}CH_2$ + $CH_2{=}CHCH{=}CH_2$	53.62	± 2.03
CD/CB/CD	$C_6H_5C(CH_2)CH{=}CH_2$ + CH_3CH_3	→ C_6H_5 CH_2CH_3 + $CH_2{=}CHCH{=}CH_2$	52.70	± 1.96
	$C_6H_5C(CH_2)CH{=}CH_2$ + CH_4	→ $C_6H_5CH{=}CH_2$ + $CH_3CH{=}CH_2$	52.20	± 1.94
		Average	**52.84 ± 0.72**	
C/CB/C2/CO	$C_6H_5C(CH_3)_2CH{=}O$ + 2CH_4	→ $C_6H_5CH_3$ + $CH_3CH{=}O$ + $CH_3CH_2CH_3$	-21.33	± 2.76
	$C_6H_5C(CH_3)_2CH{=}O$ + CH_3CH_3 + CH_4	→ $C_6H_5CH_2CH_3$ + $CH_3CH{=}O$ + $CH_3CH_2CH_3$	-22.59	± 2.96
		Average	**-21.96 ± 0.98**	
CB/CO	$C_6H_5CH{=}O$	→ C_6H_6 + CO	-7.27	± 1.1
	$C_6H_5CH{=}O$ + CH_3CH_3	→ C_6H_6 + $CH_3CH_2CH{=}O$	-7.66	± 1.47
		Average	**-7.46 ± 0.27**	
	$CH_3C_6H_4CH{=}O$	→ $C_6H_5CH_3$ + CO	-15.78	± 1.3
CB/CO	$CH_3C_6H_4CH{=}O$ + CH_3CH_3	→ C_6H_6 + $CH_3CH_2CH_3$ + CO	-16.32	± 1.75
	$CH_3C_6H_4CH{=}O$ + CH_4	→ C_6H_6 + CH_3CH_3 + CO	-16.53	± 1.68
	$CH_3C_6H_4CH{=}O$ + CH_3CH_3	→ $C_6H_5CH{=}O$ + $CH_3CH_2CH_3$	-16.84	± 1.75
		Average	**-16.37 ± 0.44**	
	$C_6H_5C({=}CH_2)CH{=}O$ + CH_4	→ $C_6H_5CH_3$ + CO + $CH_2{=}CH_2$	12.48	± 1.83
CD/CB/CO	$C_6H_5C({=}CH_2)CH{=}O$ + CH_4	→ C_6H_6 + CO + $CH_2{=}CHCH_3$	11.34	± 1.86
	$C_6H_5C({=}CH_2)CH{=}O$	→ $C_6H_5CH{=}CH_2$+ CO	12.14	± 1.53
		Average	**11.98 ± 0.58**	
O/CB/CO	$C_6H_5OCH{=}O$ + CH_3CH_3	→ C_6H_5OH + $CH_3CH_2CH{=}O$	-51.68	± 1.48
	$C_6H_5OCH{=}O$	→ C_6H_5OH + CO	-51.30	± 1.11
		Average	**-51.49 ± 0.26**	
	$C_6H_5C({=}CH_2)CH_3$ + 2CH_3CH_3	→ C_6H_6 + $CH_3CH_2CH_2CH{=}CH_2$	28.37	± 2.07
	$C_6H_5C({=}CH_2)CH_3$ + CH_4	→ C_6H_6 + $CH_2{=}CHCH_2CH_3$	29.82	± 1.67
CD/C/CB	$C_6H_5C({=}CH_2)CH_3$ + CH_4	→ $C_6H_5CH{=}CH_2$ + CH_3CH_3	29.01	± 1.72
	$C_6H_5C({=}CH_2)CH_3$ + CH_4	→ $C_6H_5CH_3$ + $CH_3CH{=}CH_2$	29.01	± 1.67
	$C_6H_5C({=}CH_2)CH_3$ + CH_3CH_3	→ $C_6H_5CH{=}CH_2$ + $CH_3CH_2CH_3$	29.02	± 1.79
		Average	**29.05 ± 0.51**	
	$C_6H_5C({=}CH_2)OH$ + CH_3CH_3	→ $C_6H_5CH_2CH_3$ + $CH_2{=}CHOH$	-8.49	± 3.89
CD/CB/O	$C_6H_5C({=}CH_2)OH$ + CH_3CH_3	→ $C_6H_5CH{=}CH_2$ + CH_3CH_2OH	-8.71	± 2.05
	$C_6H_5C({=}CH_2)OH$ + CH_4	→ $C_6H_6CH_3$ + $CH_2{=}CHOH$	-7.23	± 3.79
		Average	**-8.14 ± 0.79**	

[b]reported errors for each of the standard species (see App. A) + estimated error due to the method (see App. C).

Derived radicals (loss of hydrogen R—H) from stable species (called parent species) are calculated in this chapter. The isodesmic working reactions, along with the $\Delta_f H^0_{298}$ values of these radical, and reported in Tables 4.4.

Several previous papers on thermochemistry of oxygenated hydrocarbons [114, 115, 116] report good agreement between enthalpies of stable molecules and radicals calculated with Density Functional theory and those calculated with high level methods. To validate our results we compare in the next chapter enthalpies of these oxygenated hydrocarbons and radicals using the G3MP2B3 composite method. The calculated data are used to estimate corresponding R—

H bond dissociation enthalpies (BDE) at 298 K. The first column of Table 4.4 lists the BDE group identifying the parent and radical.

Table 4.4: Calculated $\Delta_f H^0_{298}$ of the target species

Bond Energy Name	Reactions Series	$\Delta_f H^0_{298}$ kcal mol^{-1}	Error Limit[b]
	Furan	-8.29	
FURANJ1 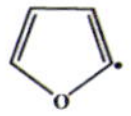	***Furanj1*** + $CH_3CH_3 \rightarrow$ Furan + $CH_3CH_2\bullet$	61.15	± 1.68
	Furanj1 + $C_6H_6 \rightarrow$ Furan + $C_6H_5\bullet$	60.86	± 4.18
	Furanj1 + $CH_2CH_2 \rightarrow$ Furan + $CH_2CH\bullet$	61.11	± 2.16
	Furanj1 + $CH_3CH{=}O \rightarrow$ Furan-CH=O + $CH_3\bullet$	63.59	± 2.61
	Average	**61.67 ± 1.28**	
FURANJ2 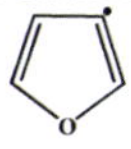	***Furanj2*** + $CH_3CH_3 \rightarrow$ Furan + $CH_3CH_2\bullet$	61.27	± 1.68
	Furanj2 + $C_6H_6 \rightarrow$ Furan + $C_6H_5\bullet$	60.99	± 4.18
	Furanj2 + $CH_2CH_2 \rightarrow$ Furan + $CH_2CH\bullet$	61.24	± 3.16
	Furanj2 + $CH_3CH{=}O \rightarrow$ Furan-CH=O + $CH_3\bullet$	63.72	± 2.61
	Average	**61.80 ± 1.28**	
	Y(C₆H₆)=O + $2CH_4 \rightarrow CH_2{=}CHCH{=}O + CH_2{=}CH_2 + CH_3CH{=}CH_2$	-4.80	± 2.35
	Y(C₆H₆)=O + $CH_2CH_2 \rightarrow C_6H_6 + CH_3CH{=}O$	-4.79	± 1.45
Y(C6J)*O	**Average**	**-4.80 ± 0.007**	
	***Y(C₆H₅•)=O*** + $2CH_4 \rightarrow CH_2{=}CHCH{=}O + CH_2CH_2 + CH_2{=}CHCH_2\bullet$	13.43	± 3.33
	Y(C₆H₅•)=O + $CH_2CH_2 \rightarrow$ Y(C_6H_6)=O + $CH_2CH\bullet$	12.51	± 2.32
	Y(C₆H₅•)=O + $CH_3CH_3 \rightarrow$ Y(C_6H_6)=O + $CH_3CH_2\bullet$	12.55	± 1.84
	Y(C₆H₅•)=O + $CH_4 \rightarrow C_6H_6 + CH_3O\bullet$	12.61	± 2.41
	Average	**12.77 ± 0.44**	
Y(C5)COJ*O 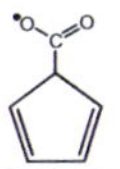	***Y(C₅H₅)C(OH)=O*** + $2CH_4 \rightarrow$ Y(C_5H_6) + $CH_3OH + CH_3CH{=}O$	-53.54	± 2.67
	Y(C₅H₅)C(OH)=O + $CH_3CH_3 \rightarrow$ Y(C_5H_6) + $CH_3CH_2C(=O)OH$	-52.08	± 2.43
	Average	**-52.81 ± 1.03**	
	Y(C₅H₅)C(O•)=O	**5.1**	Ref. 126
	C₆H₅CH=O	**-7.46 ± 0.27**	
PhCJ*O	***C₆H₅C•=O*** + $CH_4 \rightarrow C_6H_5CH{=}O + CH_3\bullet$	30.21	± 2.84
	C₆H₅C•=O + $CH_4 \rightarrow C_6H_6 + CH_3C\bullet{=}O$	28.46	± 3.09
	C₆H₅C•=O + $CH_4 \rightarrow C_6H_6 + CO + CH_3\bullet$	31.72	± 2.74
	C₆H₅C•=O + $CH_3CH_3 \rightarrow C_6H_5CH{=}O + CH_3CH_2\bullet$	31.11	± 3.09
	Average	**30.75 ± 1.42**	
	Y(C₅H₅O)CH=O + $CH_4 \rightarrow$ Y(C_5H_6O) + $CH_3CH{=}O$	-26.34	± 3.93
	Y(C₅H₅O)CH=O $\rightarrow$ Y(C_5H_6O) + CO	-22.42	± 3.44
Y(C5O)CJ*O	***Y(C₅H₅O)CH=O*** + $CH_4 \rightarrow$ Y(C_5H_5)OCH=O	-27.04	± 3.78
	Average	**-25.27 ± 2.5**	
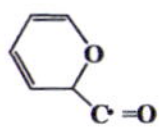	***Y(C₅H₅O)C•=O*** + $CH_4 \rightarrow$ Y(C_5H_6O) + $CH_3C\bullet{=}O$	10.10	± 2.95
	Y(C₅H₅O)C•=O + $H_2 \rightarrow$ Y(C_5H_6O) + $CH\bullet{=}O$	9.12	± 2.07
	Y ***Y(C₅H₅O)C•=O*** + $CH_3CH{=}O \rightarrow$ Y(C_5H_5O)CH=O + $CH_3C\bullet{=}O$	11.39	± 5.50
	Y(C₅H₅O)C•=O + $CH_3CH_3 \rightarrow$ Y(C_5H_5O)CH=O + $CH_3CH_2\bullet$	11.63	± 6.07
	Average	**10.56 ± 1.17**	

[b]reported errors for each of the standard species (Appendix A) + estimated error due to the method (Appendix C).

Table 4.4 (con't): Calculated $\Delta_f H^0_{298}$ of the target species

Bond Energy Name	Reactions Series	$\Delta_f H^0_{298}$ kcal mol^{-1}	Error Limit[b]
Y(C5)OCJ*O	$Y(C_5H_5)OCH{=}O + CH_4 \rightarrow C_5H_6 + CH_3OCH{=}O$	-34.96	± 2.48
	$Y(C_5H_5)OCH{=}O + CH_3CH_3 \rightarrow C_5H_6 + CO + CH_3CH_2OH$	-34.16	± 1.61
	$Y(C_5H_5)OCH{=}O + CH_4 \rightarrow C_5H_6 + CO + CH_3OH$	-34.88	± 1.49
	Average	**-34.67 ± 0.44**	
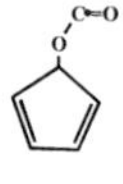	$Y(C_5H_5)OC{\bullet}{=}O + CH_4 \rightarrow Y(C_5H_5{\bullet}) + Y(C5)OCH{=}O$	8.19	± 2.46
	$Y(C_5H_5)OC{\bullet}{=}O + CH_4 \rightarrow Y(C_5H_5{\bullet}) + CH_3OH + CO$	9.13	± 2.19
	$Y(C_5H_5)OC{\bullet}{=}O + CH_3CH_3 \rightarrow Y(C_5H_5{\bullet}) + CO + CH_3CH_2OH$	9.84	± 2.34
	$Y(C_5H_5)OC{\bullet}{=}O + CH_2CH_2 \rightarrow Y(C_5H_5{\bullet}) + CO + CH_3CH{=}O$	10.98	± 2.55
	Average	**9.54 ± 1.17**	
COCJ*O	$CH_3OCH{=}O + CH_3CH_3 \rightarrow CH_3OCH_2CH_3 + CH_2{=}O$	-83.63	± 2.23
	$CH_3OCH{=}O + CH_4 \rightarrow CH_3OCH_3 + CH_2{=}O$	-84.45	± 2.14
	$CH_3OCH{=}O \rightarrow CH_3OH + CO$	-82.09	± 1.95
	Average	**-83.39 ± 1.19**	
	$CH_3OC{\bullet}{=}O + CH_3CH_3 \rightarrow CH_3OCH{=}O + CH_3CH_2{\bullet}$	-41.31	± 3.28
	$CH_3OC{\bullet}{=}O + CH_4 \rightarrow CH_3OH + CH_3C{\bullet}{=}O$	-42.59	± 2.23
	$CH_3OC{\bullet}{=}O + CH_4 \rightarrow CH_3OCH{=}O + CH_3{\bullet}$	-42.20	-
	Average	**-42.03 ± 0.65**	
C*CJC*O	$CH_2{=}CHCH{=}O$	**-18.65 ± 0.95**	
	$CH_2{=}C{\bullet}CH{=}O + CH_3CH_3 \rightarrow CH_2{=}CHCH{=}O + CH_3CH_2{\bullet}$	41.99	± 2.48
	$CH_2{=}C{\bullet}CH{=}O + CH_2CH_2 \rightarrow CH_2{=}CHCH{=}O + CH_2CH{\bullet}$	41.96	± 2.96
	$CH_2{=}C{\bullet}CH{=}O + CH_4 \rightarrow CH_2{=}CH{\bullet} + CH_3CH{=}O$	42.12	± 2.33
	$CH_2{=}C{\bullet}CH{=}O + CH_3CH{=}O \rightarrow CH_2{=}CHCH{=}O + CH_3C{\bullet}{=}O$	41.76	± 2.91
	$CH_2{=}C{\bullet}CH{=}O + CH_4 \rightarrow CH_2{=}CHCH_3 + CH{\bullet}{=}O$	42.03	± 1.14
	Average	**41.97 ± 0.13**	
C*CCJ*O	$CH_2{=}CHC{\bullet}{=}O + CH_4 \rightarrow CH_2CHCH{=}O + CH_3{\bullet}$	19.7	± 1.70
	$CH_2{=}CHC{\bullet}{=}O + CH_3CH_3 \rightarrow CH_2CHCH{=}O + CH_3CH_2{\bullet}$	20.61	± 2.05
	$CH_2{=}CHC{\bullet}{=}O + CH_3CH{=}O \rightarrow CH_2{=}CHCH{=}O + CH_3C{\bullet}{=}O$	20.37	± 3.25
	Average	**20.22 ± 0.47**	
VCJC*OC#C	$CH_2{=}CHCH(CH{=}O)C{\equiv}CH$	**45.23 ± 0.22**	
	$CH_2{=}CHC{\bullet}(CH{=}O)C{\equiv}CH + CH_3CH_3 \rightarrow CH_3CH_2{\bullet} + CH_2CHCH(CHO)C{\equiv}CH$	62.68	± 1.85
	$CH_2{=}CHC{\bullet}(CH{=}O)C{\equiv}CH + CH_3CH_3 \rightarrow CH_2{=}O + CH{\equiv}C{\bullet} + C_5H_8$	62.80	± 3.46
	$CH_2{=}CHC{\bullet}(CH{=}O)C{\equiv}CH + CH_4 \rightarrow CO + CH{\equiv}C{\bullet} + CH_2{=}CHCH_2CH_3$	62.94	± 3.32
	Average	**62.80 ± 0.13**	
O*CCJ(OH)2	$CH(OH)_2CH{=}O$	**-119.08 ± 0.44**	
	$C{\bullet}(OH)_2CH{=}O + CH_3CH_3 \rightarrow CH_3CH_2OH + CO + CH_2{\bullet}OH$	-96.15	± 3.13
	$C{\bullet}(OH)_2CH{=}O + CH_4 \rightarrow CH_3OH + CO + CH_2{\bullet}OH$	-96.79	± 3.31
	$C{\bullet}(OH)_2CH{=}O + CH_4 \rightarrow CH(OH)_2CH{=}O + CH_3{\bullet}$	-96.25	± 2.75
	$C{\bullet}(OH)_2CH{=}O + CH_4 \rightarrow CH_3OH + CH_2{=}O + CH{\bullet}{=}O$	-93.93	± 2.27
	Average	**-95.78 ± 1.26**	
VCJ(C*O)2	$CH_2{=}CHCH(CH{=}O)CH{=}O$	**-39.56 ± 0.78**	
	$CH_2{=}CHC{\bullet}(CH{=}O)CH{=}O + CH_3CH_3 \rightarrow CH{\bullet}{=}O + CH_2{=}O + C_5H_8$	-19.25	± 2.10
	$CH_2{=}CHC{\bullet}(CH{=}O)CH{=}O + CH_3CH_3 \rightarrow CH_3CH_2{\bullet} + CH_2CHCH(CH{=}O)CHO$	-19.62	± 2.96
	$CH_2{=}CHC{\bullet}(CH{=}O)CH{=}O + CH_4 \rightarrow CH{\bullet}{=}O + CO + CH_2{=}CHCH_2CH_3$	-18.11	± 1.97
	$CH_2{=}CHC{\bullet}(CH{=}O)CH{=}O + CH_2CH_2 \rightarrow CH{\bullet}{=}O + CO + CH_2CHCH_2CHCH_2$	-20.24	± 2.12
	Average	**-18.99 ± 0.78**	

[b]reported errors for each of the standard species (Appendix A) + estimated error due to the method (Appendix C).

4.2.2 Entropy, Heat Capacity and Hindered Rotation Contribution to Thermodynamic Parameters of the Target Species

Contributions to S^0_{298} and $Cp_{298}(T)$ from translation, scaled vibrations frequencies, and external rotations (TVR) of each species are calculated. TVR represents the sum of these contributions the frequencies representing internal rotor torsions are not included in TVR. Instead, a more exact contribution from hindered rotations is calculated.

Contribution to S^0_{298} and $Cp_{298}(T)$ from internal rotations (IR) are calculated by using the technique explained in chapter 2. Potential barriers for internal rotations about the C—C, C—OH, C—O, C—CH=O, C_d—C, C_d—C_d, C_d—O, C_d—OH, C_d—CH=O, C_b—C, C_b—C_d, C_b—O, C_b—CH=O and O—CH(=O) bonds in the stable molecules are determined at the B3LYP/6-31G(d,p) calculation level. A lower basis set, 6-31G(d,p) instead of 6-311G(d,p) is used, because the difference in accuracy between the two basis sets is negligible whereas the computation time difference is important. The same level was used for hindered rotation of radicals around RC•—OH, C—C•=O, RC•—C=O, C—C(O•)=O, C_d—C•R, RC_d•—CH=O, C_d—C•=O, C_b—C•=O and RO—C•(=O). Contributions for entropy and heat capacity from the internal rotors are estimated using these potentials.

The total entropy and heat capacities (TVR + IR) are given in Table D1 and D2 of Appendix D. These values are used to determine entropy and heat capacities values of the new GA groups.

4.2.3 Development of new Groups for GA

In this work, we develop a consistent set of oxygenated peroxy-hydrocarbon and acetylene-alcohol groups derived from the thermodynamic properties data of a set of hydroperoxides, peroxides, ethers and alcohols determined in previous chapters.

All values are intrinsic, that is the entropy component excludes contributions to symmetry and the number of optical isomer in the parent molecule.

4.2.3.1 Enthalpy of Formation

The enthalpy of a group is determined using the calculated heat of formation of the stable molecule containing that group, and using literature data for group additivity values for the remaining groups in the molecules. As mentioned above, the GA nomenclature means that (CD) is a carbon with a double bond (e.g. in CH_2=CH_2 both carbons here are CD) and (CB) is related to the carbons contains in benzene.

The following example describes the calculation of $\Delta_f H^0_{298}$ group additivity value for the O/CB/CO group (eq. 4.1):

$$\Delta_f H^0_{298}(C_6H_5OCH{=}O) = \Delta_f H^0_{298}(CB/H) \times 5 + \Delta_f H^0_{298}(CB/O) + \Delta_f H^0_{298}(CO/H/O) + \Delta_f H^0_{298}(\mathbf{O/CB/CO}) \quad (4.1)$$

The calculated heat of formation of $C_6H_5OCH{=}O$ is -51.49 kcal mol^{-1} is given in Table 4.3. The literature values of $\Delta_f H^0_{298}$ of CB/H, CB/O and CO/H/O groups are given in Table 4.5. We obtain then:

$$\Delta_f H^0_{298}\ (\mathbf{O/CB/CO}) = (-51.49) - (3.3 \times 5) - (-0.9) - (-32.1) = -34.99 \text{ kcal mol}^{-1} \quad (4.2)$$

The calculated enthalpies for the groups developed in this work are summarized in Table 4.7 below.

4.2.3.2 Entropy

The entropy, S^0_{298}, of each group, is derived from the calculated entropy of stable molecules. As noted before, the entropy depends on the species structure, vibration frequencies, and internal rotors.

S^0_{298} of O/CB/CO calculation is determined as follows (eq. 4.3):

$$S^0_{298}(C_6H_5OCH{=}O) = (S^0_{298}\ CB/H) \times 5 + S^0_{298}(CB/O) + S^0_{298}(CO/H/O) + S^0_{298}(O/CB/CO) + R\ln(OI) - R\ln(\sigma) \quad (4.3)$$

Where $R = 1.987$ cal mol^{-1} K^{-1}, OI denotes number of optical isomers, and σ is the symmetry number. The calculated entropies for the developed groups are summarized in Table 4.7.

4.2.3.3 Heat Capacities

The heat capacity values are also determined using the above oxygenated and non-oxygenated hydrocarbons according to equation 4.1. The results are given is Table 4.7 for a temperature range from 300 K to 1500 K. Heat capacity $Cp_{298}(T)$ of O/CB/CO is for example determined as follows:

$$Cp_{298}(T)(C_6H_5OCH{=}O) = (Cp_{298}(T)\ CB/H) \times 5 + Cp_{298}(T)(CB/O) + Cp_{298}(T)(CO/H/O) + Cp_{298}(T)(\mathbf{O/CB/CO}) \quad (4.4)$$

Table 4.5: Reference Group Values

Groups	ΔH^0_{f298} (kcal mol^{-1})	S^0_{298} cal (mol K)$^{-1}$	Cp (T) cal (mol K)$^{-1}$						
			300 K	400 K	500 K	600 K	800 K	1000 K	1500 K
CD/H2[a]	6.26	27.61	5.10	6.36	7.51	8.50	10.07	11.27	13.19
CD/C/H[b]	8.59	7.97	4.16	5.03	5.81	6.50	7.65	8.45	9.62
C/CD/H3[b]	-10.20	30.41	6.19	7.84	9.40	10.79	13.02	14.77	17.58
C/C/H3[b]	-10.20	30.41	6.19	7.84	9.40	10.79	13.02	14.77	17.58
CD/CD/H[b]	6.74	6.38	4.46	5.79	6.75	7.42	8.35	9.11	10.09
O/CD/H[c]	-46.30	25.34	4.68	6.02	7.02	7.72	8.22	8.38	8.59
O/C/H[b]	-37.90	29.07	4.30	4.50	4.82	5.23	6.02	6.61	7.44
C_d/H/O[b]	8.60	6.20	4.75	6.46	7.64	8.35	9.10	9.56	10.46
CB/O[b]	-0.90	-10.20	3.90	5.30	6.20	6.60	6.90	6.90	7.07
CB/H[b]	3.30	11.53	3.24	4.40	5.46	6.30	7.54	8.41	9.73
CB/C[b]	5.51	-7.69	2.67	3.14	3.68	4.15	4.96	5.44	5.98
CB/CD[d]	5.69	-7.80	3.59	3.97	4.38	4.72	5.28	5.61	5.75
CD/CO/H[e]	4.32	6.38	4.46	5.79	6.75	7.42	8.35	9.11	10.09
CO/H/O[b]	-32.10	34.90	7.03	7.87	8.82	9.68	11.20	12.20	-
CO/C/H[b]	-29.10	34.90	7.03	7.87	8.82	9.68	11.20	12.20	-
CD/CO/O[e]	**5.13**	-14.60	4.40	5.37	5.93	6.18	6.50	6.62	6.72
CT/H[d]	26.93	24.70	5.28	5.99	6.49	6.87	7.47	7.96	8.85
CT/C[b]	27.55	6.35	3.13	3.48	3.81	4.09	4.60	4.92	6.35
CT/CD[e]	28.20	6.43	2.57	3.54	3.50	4.92	5.34	5.50	5.80

[a]Chen and Bozzelli [127]; [b]Benson [2] also used by Holmes, Turecek [128] and Cohen [129]; [c]Li zhu, Chen and Bozzelli [130]; [d]Stein and Fahr [131]; [e]based on unpublished data of Bozzelli.

4.2.4 Thermodynamic Values of new Groups

The thermochemical properties of all groups are estimated the way described above and listed below in Table 4.7. Each groups is estimated from only one molecule system except for the calculation of O/CD/O, O/CT/O, CT/O and O/C/CT groups which are based on 2 to 5 molecules. The detailed calculations for these four groups are given in Table 4.6. All groups are derived from thermodynamic data determined by DFT calculations and isodesmic reactions.

The group value of CT/O (Table 4.6) is determined from ethynyl alcohols, ethynol (HC≡COH) and methyl ethynol ($CH_3C{\equiv}COH$). Results for both species are discussed in Chapter 3. The enthalpy value for CT/O is calculated to be 31.79 kcal mol^{-1} based on HC≡COH and 30.39 kcal mol^{-1} based on $CH_3C{\equiv}COH$. The recommended value of CT/O is taken as the average, 30.79 kcal mol^{-1} and given in Table 4.7.

The value of O/C/CT group is derived from enthalpy values for $HC{\equiv}COCH_3$ and $CH_3C{\equiv}COCH_3$ calculated in Chapter 3. O/C/CT derived from $HC{\equiv}COCH_3$ and $CH_3C{\equiv}COCH_3$ are -21.94 and -21.89 kcal mol^{-1}, respectively. Enthalpy, ΔH^0_{f298} of the O/C/CT group is the average, -21.91 kcal mol^{-1} (Table 4.7).

Table 4.6: Thermodynamic Data Calculations for New Groups Using Group Additivity

Groups	ΔH^0_{f298} (kcal mol^{-1})	S^0_{298} cal (mol K)$^{-1}$	Cp (T) cal (mol K)$^{-1}$						
			300 K	400 K	500 K	600 K	800 K	1000 K	1500 K
						$CH\equiv COH$			
CT/O	31.79	2.92	2.39	3.12	3.39	3.55	3.65	3.74	4.08
						$CH_3C\equiv COH$			
	30.39	-2.23	1.16	1.87	2.14	2.3	2.41	2.52	1.89
						$CH\equiv COCH_3$			
O/C/CT	-21.94	12.14	2.55	2.43	2.71	3.04	3.72	4.15	4.89
						$CH_3C\equiv COCH_3$			
	-21.89	7.88	2.52	2.28	2.53	2.84	3.51	3.95	3.82
						$CH\equiv COOH$			
O/CT/O	0.43	9.72	1.55	1.8	2.19	2.39	3.01	3.3	3.96
						$CH_3C\equiv COOH$			
	-1.98	3.38	0.66	0.81	1.15	1.31	1.88	2.16	1.81
						trans-$CH_3CH=CHOOH$			
O/CD/O	-11.13	10.27	3.14	2.95	3.03	3.15	3.74	4.07	4.23
						Cis-$CH_3CH=CHOOH$			
	-12.35	11.21	2.86	2.96	3.17	3.3	3.71	3.83	3.65
						$(CH_3)_2C=CHOOH$			
	-13.03	5.26	1.11	1.05	1.29	1.57	2.33	2.73	3
						$CH_2=C(CH_3)OOH$			
	-9.76	7.70	4.95	5.68	5.97	5.9	5.63	5.32	4.92
						$CH_3CH=C(CH_3)OOH$			
	-10.12	7.15	3.96	4.29	4.69	4.84	4.95	4.96	5.04

The O/CT/O group value is derived from the enthalpy values of ethynyl hydroperoxides: $HC\equiv COOH$ and $CH_3C\equiv COOH$. The ΔH^0_{f298} mean value for O/CT/O is -0.77 kcal mol^{-1}. This value allows us to evaluate the two ethynyl peroxides considered above in this work (Chapter 3). With the calculated O/CT/O value, we can estimate enthalpy of $HC\equiv COOCH_3$, 41.87 kcal mol^{-1} and $CH_3C\equiv COOCH_3$ 32.27 kcal mol^{-1}. These values are in good agreement with the DFT calculation results (41.23 ± 0.70 and 29.51± 0.27 kcal mol^{-1}, respectively).

The O/CD/O group is determined using five hydroperoxides calculated in Chapter 3: *trans*-$CH_3CH=CHOOH$, *cis*- $CH_3CH=CHOOH$, $(CH_3)_2C=CHOOH$, $CH_3CH=C(CH_3)OOH$ and $CH_2=C(CH_3)OOH$. Enthalpy of the O/CD/O group is calculated to be -11.27 kcal mol^{-1} taken as the average results from the five hydroperoxides, where the maximum deviation is 1.4 kcal mol^{-1}. This O/CD/O group is used to estimate the vinyl peroxide species by using group additivity, and comparing them to the DFT calculations.

The entropy of the O/CD/O group is calculated according to equation 4.3 as well, and use of the thermochemical data of the six hydroperoxides listed in Table 4.6. The heat capacity group values are also determined using the hydroperoxide molecules. The average values are given in Table 4.7 for a temperature range from 300 K to 1500K.

Table 4.7: Thermodynamic data calculation for new groups using Group Additivity

Groups	ΔH^0_{f298} (kcal mol^{-1})	S^0_{298} cal (mol K)$^{-1}$	Cp (T) cal (mol K)$^{-1}$ 300 K	400 K	500 K	600 K	800 K	1000 K	1500 K
CT/O[a]	30.79	0.35	1.78	2.50	2.77	2.93	3.03	3.13	2.99
O/C/CT[a]	-21.91	10.01	2.54	2.36	2.62	2.94	3.62	4.05	4.36
O/CT/O[a]	-0.77	6.55	1.11	1.31	1.67	1.85	2.45	2.73	2.89
O/CD/O[b]	-11.27	8.53	3.20	3.39	3.63	3.75	4.07	4.18	4.17
O/CB/O	-1.98	10.7	2.85	2.92	3.02	3.09	3.48	3.66	3.23
cis-correction	-1.22	0.94	-0.28	0.01	0.14	0.15	-0.03	-0.24	-0.58
C/CO/O2/H	-14.18	-14.14	6.04	9.48	11.37	12.16	12.38	12.39	-
C/CD/CO/CO/H	3.79	-9.95	5.55	8.06	9.63	10.76	11.93	12.85	-
C/C/CO/O2	-15.47	-38.54	6.01	8.88	10.43	10.98	10.77	10.31	-
C/CD2/H/O	-4.15	-12.57	2.35	4.8	5.92	6.28	6	5.67	5.88
C/CD/CO/H2	-6.37	-1.79	9.08	11.65	13.87	15.74	18.5	20.61	-
C/CD2/CO/H	-1.43	-8.81	4.74	6.23	7.36	8.17	9.01	9.63	-
C/CD2/O2	-15.20	-39.25	7.83	11.51	13.29	13.62	12.55	11.34	9.74
C/CD/CO/CT/H	5.0	-25.82	-0.9	1.94	3.9	5.38	7.2	8.6	-
CD/O2	12.41	-9.22	3.56	3.37	3.64	4.69	5.35	5.76	6.31
C/CB/H2/O	-22.6	7.62	7.03	8.98	10.32	11.17	12.14	12.87	14.01
C/CB/C/O2	-15.53	-41.24	8.89	11.88	12.81	12.59	11.23	10.06	8.55
C/CB/C2/CO	5.53	-37.15	7.22	8.34	8.61	8.49	7.89	7.39	-
CD/CB/CO	12.76	-15.48	7.21	8.76	9.45	9.62	9.42	9.11	8.13
CD/CB/O	7.98	-14.25	3.52	4.23	4.8	5.5	5.65	5.85	6.0
CD/CD/CO	7.37	-13.16	5.03	6.57	7.55	8.12	8.56	8.57	8.06
CD/CD/O	9	-12.92	3.85	4.66	5.31	6.09	6.49	6.45	6.37
CD/CO/CT	7.02	-22.74	2.29	3.12	4.25	3.51	3.91	4.35	4.63
CB/CO	4.35	-6.56	5.39	6.64	7.40	7.76	8.03	8.14	7.66
CO/CD/H	-29.23	35.10	7.55	8.82	9.95	10.95	12.39	13.13	14.1
CO/CB/H[c]	-29.23	31.46	4.58	5.54	6.62	7.67	9.38	10.47	12.2
O/CB/CO	-34.99	7.64	3.84	5.25	5.88	6.16	6.11	6.97	-
O/CD2	-35.87	11.18	2.91	3.24	3.61	3.97	4.39	4.33	4.35
CD/CB/H	6.66	5.47	4.01	5.66	6.84	7.62	8.61	9.23	10.13
C/CD2/C/H	-1.52	-9.60	4.89	6.46	7.61	8.38	9.3	9.87	10.59
	-1.10[d]	-13.03[d]	5.28[d]	6.54[d]	7.67[d]	8.48[d]	9.45[d]	10.18[d]	11.24[d]
CD/C/CB	10.80	-14.28	3.15	4.41	5.14	5.51	5.82	5.86	5.87
CD/CB/CD	11.35	-15.78	5.52	6.56	6.95	7.04	6.9	6.51	6.06

[a] ΔH^0_{f298}, $Cp^0(T)$ and S^0_{f298} are the average value derived from two species (see text). [b]average value derived from five hydroperoxides (see text). [c] ΔH^0_{f298} and Cp are based on CO/CD/H; [d]from Green et al.

Calculated values for the entropy group of the O/CD/O, show a range from 10.0 to 5.26 cal (mol K)$^{-1}$. Both *trans*-$CH_3CH{=}CHOOH$ and *cis*-$CH_3CH{=}CHOOH$ have S^0_{298} group values near 10 cal mol^{-1} K^{-1} (± 0.66). The two methyl substituted groups (3-fold symmetric, rotation barrier of 1.9 kcal mol^{-1}) on the peroxide carbon, $CH_3CH{=}C(CH_3)OOH$ and $CH_2{=}C(CH_3)OOH$ provide O/CD/O group values for entropy, S^0_{298} of 7.15 and 7.70 cal mol^{-1} K^{-1}, respectively. The

O/CD/O entropy value of $(CH_3)_2C{=}CHOOH$ which contains two methyl rotors on the non peroxide double-bond carbon has the lowest value, 5.26 cal mol^{-1} K^{-1}.

The recommended value of S^0_{298} for group O/CD/O is determined as the average value of all six hydroperoxides, 8.53 (±1.5) cal (mol K)$^{-1}$. This data suggests that group additivity for entropy may not hold where methyl and peroxide groups are both interacting with the unsaturated bond. It is possible that the CD/C/O group values for vinyl alcohols need re-evaluation.

cis correction (*cis* CH_3/OOH): A *cis* correction for ΔH^0_{f298} is assigned by Benson for the *cis*-olefins as 1 kcal mol^{-1}. In this work, we estimate the value of the (*cis* (CH_3/OOH)) correction for use in group additivity. The *cis* correction is determined by comparing enthalpies of the *cis* form of $CH_3CH{=}CHOOH$ to the *trans* form. The value given in Table 4.7 is the difference between group O/CD/O (*trans*) and O/CD/O (*cis*). The ΔH^0_{f298} value derived from $CH_3CH{=}CHOOH$ for *cis* is 1.22 kcal mol^{-1}, which is similar to that of Benson for methyl groups. S^0_{f298} and $Cp_{298}(T)$ values are also determined.

Green and sumathi have reported values for the C/CD2/C/H group [132] as derived from G2 calculation on $(CH_2{=}CH)_2CHCH_2CH_3$. The enthalpy they estimated for this group is found to be -1.10 kcal mol^{-1}, which is in good agreement with our value, -1.52 kcal mol^{-1}. The entropy and heat capacity differences between the two studies are about 3.4 and 0.65 cal mol^{-1} K^{-1} respectively. The thermodynamic values of Green and Sumathi are listed in Table 4.7 for comparison with our calculated values.

4.2.5 C—H Bond Energies and Hydrogen Bond Increment (HBI) Values.

A method to estimate thermochemical properties for radicals from the corresponding properties of the parent and of derivation of hydrogen bond increment (HBI) groups, is described by Lay et al. [25] and Sun and Bozzelli [133]. The method uses the bond energy (298.K) for loss of a hydrogen on the central atom for the enthalpy term, the difference between the radical and the parent for the heat capacity (*Cp(T)*) term and the intrinsic entropy difference for the S^0_{298} term. Thirteen new (HBI) increments for estimation of radical species and bond dissociation enthalpy (BDE) at 298 K have been developed in this work.

Enthalpy, entropy, and heat capacities, are based on the values of the parent molecule and the corresponding radical and derived as follows:

A hydrogen atom bond increment (HBI) group for ΔH^0_{f298} reflects the enthalpy change due to loss of a H atom form a stable parent molecule in the form of the R—H bond energy (BE), which is defined as the enthalpy of reaction for a simple carbon—hydrogen dissociation reaction.

$$\Delta H^{0}_{f298}(\text{BE}) = \Delta H^{0}_{f298}(\text{radical}) + \Delta H^{0}_{f298}(\text{H}) - \Delta H^{0}_{f298}(\text{parent}) \quad (4.6)$$

As an Example, from CH_2=CHCH=O $\rightarrow$ CH_2=CHC•=O + H, we get

$$\text{BE(O=C—H)} = \Delta H^{0}_{rxn,298} = 52.1 + (20.22) - (-18.65) = 90.97 \text{ kcal mol}^{-1}$$

Heat capacity, $Cp_{298}(T)$, is determined more directly, as it is a simple difference in the corresponding $Cp_{298}(T)$ properties:

$$Cp_{298}(T)\ (\text{radical}) = Cp_{298}(T)\ (\text{parent}) - \text{HBI}\ Cp_{298}(T)\ (\text{radical}) \quad (4.7)$$

The effects for changes in symmetry between the radical and parents are included in evaluation of the entropy of each species.

The Entropy (298 K) HBI group, S^{0}_{298}, can be written as

$$\text{HBI}\ S^{0}_{298}\ (\text{radical}) = [\,S^{0}_{298}\ (\text{radical})\, S^{0}_{298} + R\ \ln\sigma_{\text{radical}}] - [\,S^{0}_{298}\ (\text{parent}) + R\ \ln\sigma_{\text{parent}}] \quad (4.8)$$

σ is the symmetry. Electronic degeneracy (R ln(2)) of the radical electronic state is included in HBI group. The entropy value also includes correction for spin degeneracy of the electron and loss of the optical isomer, when appropriate (example: loss of H in ROOH in hydroperoxides).

The following notation is used to describe the structures and bond sites: * denotes a double bond, **J** denotes a radical site on a central atom, # means a triple bond, **Ph** is the phenyl (C_6H_5—), **V** is a vinyl moiety (CH_2=CH—) and **Y** designates the start of a cyclic structure.

Thermochemical properties for the HBI groups are listed in Table 4.8. An example is loss of a hydrogen atom from the secondary vinylic carbon in CH_2=CHCH=O, represented by C*CJC*O (CH_2=C•CH=O). The calculated BDE (Table 4.8) are given in Table 4.4 with the corresponding parent and radical.

The unpaired electrons on several of the carbonyl radicals (CJ*O) in this study have the possibility for resonance overlap with adjacent unsaturated carbons. Two of these resonance systems are the C*CCJ*O group (dissociation of CH_2=CHCH=O to CH_2=CHC•=O + H) and the PhCJ*O group (C6H5C(=O)—H). One involves a vinyl and the second has a phenyl ring bonded to the CJ*O. The bond dissociation energies are near identical at 90.97 and 90.31 kcal mol^{-1}. A third system, the Y(C5O)CJ*O radical from Y(C_5H_5O)CH(=O)—H (Y(OC=CC=CCH-)CH=O) has a slightly lower bond energy, 88.11 kcal mol^{-1}, implying that the ring oxygen in this structure involves more overlap and results in a more stabilized (resonant) structure with the carbonyl radical relative to the olefin or phenyl ring. These bond energies are similar to those of formaldehyde and acetaldehyde. Comparison of C*CCJ*O,

PhCJ*O and Y(C5O)CJ*O to the more common carbonyl radicals HC•=O (in H_2C=O) at 90.20 kcal mol^{-1}, and CCJ*O (from CH_3CH=O to CH_3C•=O) determined by Gutman [134] at 89.0 kcal mol^{-1} and more recently by Lee and Bozzelli [115] 89.82 kcal mol^{-1} shows good agreement.

Table 4.8: Bond dissociation Energy calculation[a]

Species	$\Delta_f H^0_{298}$ kcal mol^{-1}	S^0_{298} cal (mol K)$^{-1}$	Cp (T) in cal (mol K)$^{-1}$						
			300 K	400 K	500 K	600 K	800 K	1000 K	1500 K
C*CJC*O	112.22	1.76	2.58	3.55	4.25	4.74	5.29	5.6	6.01
O*CCJ(OH)2	75.4	5.69	0.53	1.04	1.4	1.57	1.79	1.98	2.5
RCJC*OC#C	69.67	-0.88	0.62	1.19	1.72	2.19	2.98	3.6	4.6
RCJ(C*O)2	72.37	15.22	5.02	5.11	5.12	5.14	5.19	5.24	5.31
FURANJ2	122.06	0.21	0.52	1.21	1.82	2.31	3.06	3.63	4.56
FURANJ1	122.19	0.04	0.35	1.04	1.67	2.18	2.96	3.56	4.52
C*CCJ*O	90.97	2.58	3.2	4.04	4.66	5.11	5.65	5.95	6.29
PhCJ*O	90.31	0.9	-0.17	0.32	0.94	1.54	2.56	3.33	4.46
Y(C6J)*O	68.84	2.28	1.38	1.57	1.89	2.24	2.97	3.6	4.61
Y(C5)COJ*O	110.01	1.34	0.64	1.88	2.98	3.87	5.19	6.01	6.96
Y(C5O)CJ*O	88.11	0.27	-0.4	0.58	1.53	2.32	3.38	4.03	4.88
Y(C5)OCJ*O	96.31	-0.94	1.73	1.94	1.83	1.72	1.79	2.21	3.44
COCJ*O	93.47	-2.28	0.23	1.03	1.91	2.72	4.06	4.76	5.65

The values determined from the density function calculations in this study suggest that there is little resonance overlap of the unpaired electrons on the carbonyl radicals with the adjacent unsaturated vinyl or phenyl groups. The slight increase in the bond energy by ca. 1 kcal mol^{-1} may even indicate withdrawal of the electron from the stabilized carbonyl to the unsaturated sites being a de-stabilizing effect. Higher level calculations or calculations with a larger basis set may show better configuration interaction calculations and define the most appropriate; but are beyond the scope of this study.

There are several other formyl radical groups in this study. Two BDE have the CJ*O group attached to an ether oxygen link, where the other ether bond strength is to a sp^3 carbon. Our data show that the ether linked carbonyl radical, RO—CJ*O bond increases significantly, 5-7 kcal mol^{-1}, relative to normal aldehydes [115]. The COCJ*O bond energy is 95.05 in the dissociation of CH_3OCH=O to CH_3OC•=O, whereas Y(C5)OCJ*O) (Y(C=CC=CCH—)OC•=O) from Y(C_5H_5)OC—H=O (Y(C=CC=CCH—)OCH=O) is even higher at 96.31 kcal mol^{-1}.

The secondary vinylic carbon-hydrogen bond in 2-propenal-2-yl radical, C*CJC*O, resulting from the dissociation of CH_2=C(CH=O)—H has a bond strength of 112.72 kcal mol^{-1}. This BDE is similar to the primary vinyl bond energy of ethylene C=C—H, given by Lineberger et al. [135] to be 111.2 kcal mol^{-1}, and 7 kcal mol^{-1} stronger than the secondary vinyl C—H bond in propene [136] 105.6 kcal mol^{-1}; the 2-propenyl radical is estimated in this study (see text above and Table 4.4).

The carbonyl group appears to withdraw electrons from the adjacent vinyl carbons and therefore increase the dissociation energy of the C_d—H bond. The methyl C—H bond strength in acetaldehyde, $CH_3CH{=}O$ to $CH_2{\bullet}CH{=}O$ + H of 96.4 kcal mol^{-1} [115] shows resonance stabilization of the methyl radical with the carbonyl of ca. 4.5 kcal-mol^{-1} relative to the 101 value of ethane.

C—H bond energy in the CH_3 group bonded to the carbonyl further decreases when replacing one hydrogen on the methyl by an OH, where the dissociation energy for $CH_2(OH)CH{=}O$ to $CH{\bullet}(OH)CH{=}O$ + H is 91 kcal mol^{-1}.

We note a surprising large decrease of the dihydroxy carbon C—H bond energy in $CH(OH)_2CH{=}O$ to $C{\bullet}(OH)_2CH{=}O$ + H. We have looked at the structure of the corresponding radical, $C{\bullet}(OH)_2CH{=}O$, as given by the density functional calculations shown in Figure 4.2. The geometry of the radical, which is fully optimized, shows a significant shortening of the C—C and the two C—O bond distances. That suggests resonance in the two C—O and C—C bonds. The similar bond length in the two C—O bond (1.33 Å) also suggests O—C—O resonance. The coordinates of the molecule $CH(OH)_2CH{=}O$ and the radical $C{\bullet}(OH)_2CH{=}O$ are given in the Appendix B. This decrease noted above may be explained by the electronegative OH groups and a resulting conjugation. The derived bond dissociation energy, (O*CCJ(OH)2) has a value of 75.4 kcal mol^{-1}, which appears abnormally low.

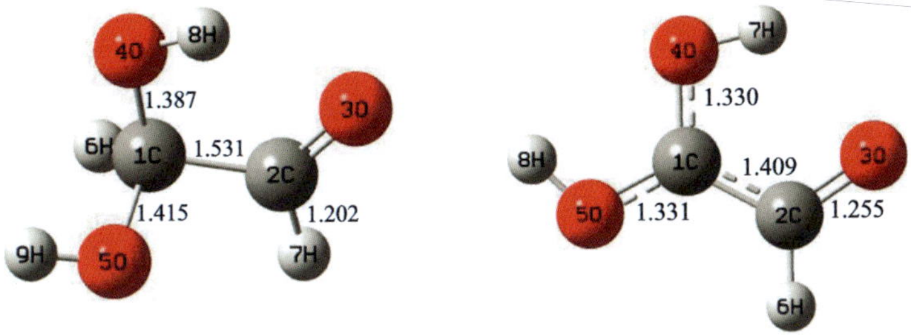

Fig. 4.2: Dissociation of $CH(OH)_2CH{=}O$ to $C{\bullet}(OH)_2CH{=}O$

The Y(C5)COJ*O acid radical which results from the dissociation of $Y(C_5H_5)C(OH){=}O$ (Y(C=CC=CCH—)C(OH)=O to Y(C=CC=CCH-)C(O•)=O), in which the carboxyl group is attached to an a sp^3 carbon has a BE of 110.01 kcal mol^{-1}. The bond dissociation energy value of benzoic acid reported by Blanksby [137] at 111 kcal mol^{-1} is slightly higher. A second comparison acid O—H bond occurs in the dissociation of prop-2-enoic acid, $CH_2{=}CHC$(O—H)=O to $CH_2{=}CHC(O{\bullet}){=}O$, where we find 110.14 kcal mol^{-1}, in agreement with the three values above.

The cyclic group Y(C6J)*O, where the carbonyl group is adjacent to the sp^3 radical, is derived from the dissociation of $Y(C_6H_6)=O$ to $Y(C_6H_5\bullet)=O$ (Y(C=CC=CC•C-)=O) + H at 69.75 kcal mol^{-1}. VCJC*OC#C derived from $CH_2=CHCH(CH=O)C\equiv CH$ to $CH_2=CHC\bullet(CH=O)C\equiv CH$ + H, is 69.67 kcal mol^{-1}; both have a similar bond energy about 69.7 kcal mol^{-1} due to the double allylic resonance and the resonance with the carbonyl group.

The VCJ(C*O)2 group resulting from $CH_2=CHCH(CH=O)CH=O$ to $CH_2=CHC\bullet(CH=O)CH=O$ + H has resonance with two carbonyl groups C*O, and results in a low bond energy 72.7 kcal mol^{-1}.

Furanj1 (the radical is on the carbon next to oxygen) and Furanj2 (the radical is on the second carbon next to oxygen) have very high BDE's, 122 kcal mol^{-1}.

4.3 Conclusion

Thermodynamic properties, $\Delta_f H^0_{298}$, S^0_{298} and $Cp_{298}(T)$, bond energies, barriers to internal rotations, and groups for group additivity of a series of oxygenated and non oxygenated species (28 stable species and 12 radicals) are reported. Group additivity parameters have been determined for 26 groups and for 13 hydrogen bond increment groups. Bond energies have been determined for 13 C—H or O—H bonds in unsaturated oxygenated hydrocarbons.

5. Validation of DFT with the G3MP2B3 Method

It is important and of value to try to validate the accuracy of the Density Functional method in order to use DFT to estimate thermochemistry in large oxygenated hydrocarbons, where high level ab initio calculations may not be possible. Enthalpies ($\Delta_f H^0_{298}$) are calculated for the previous set of twenty seven oxygenated and non-oxygenate, unsaturated hydrocarbons and twelve radicals at the G3MP2B3 level of theory and compared with the commonly used B3LYP/6-311G(d,p) Density Functional Theory (DFT) method. Standard enthalpies of formation are determined from the calculated enthalpy of reaction ($\Delta H^0_{rxn,298}$) using isodesmic work reactions with reference species that have accurately known $\Delta_f H^0_{298}$ values. The deviation between G3MP2B3 and B3LYP methods is less than ± 0.5 kcal mol^{-1} for nine species, eighteen other species differs by less than ± 1 kcal mol^{-1}, and eleven species differ by about 1.5 kcal mol^{-1}. Among all species, eleven radicals are derived from the above-oxygenated hydrocarbons that show good agreement between G3MP2B3 and B3LYP methods. G3 calculations have been performed to validate enthalpy values for which a discrepancy of more than 2.5 kcal mol^{-1} between the G3MP3B3 and the DFT calculations are found. Surprisingly, the G3 calculations support the DFT calculations in these several cases.

5.1 Introduction

Application of computational chemistry to estimation of thermochemical or kinetic properties is moving to larger and larger species including biochemicals, large aromatic intermediates in soot formation, carbon nanotubes and in many other systems. It is of value to try to validate the accuracy of the commonly used Density Functional Theory methods (DFT) for these systems for which higher accurate methods need long computation time or are not possible [138], [139]. DFT may be one of the few applicable calculation methods for these large molecules systems.

In the present chapter, we test the use of the common Density Functional Theory (B3LYP/6-311G(d,p) by comparing enthalpy results ($\Delta_f H^0_{298}$) obtained for the unsaturated oxy-hydrocarbons studied in chapter 4 [29] with the results obtained by the higher-level G3MP2B3 method. The results are utilized to validate or improve the oxy-hydrocarbon groups and radical increments previously developed by B3LYP methods [29]. The groups can be used to estimate thermochemical properties for stable species and radicals on larger unsaturated oxygenated hydrocarbons.

5.2 Results and Discussion

5.2.1 Methodology

All calculations are performed using the Gaussian 98 program suite [64, 65]. The method used in this study is the modified Gaussian-3 (G3) theory referred as G3MP2B3 [48], [19] combined with work reactions that are designed to optimize cancellation of errors in the calculations. Results calculated with DFT at the B3LYP/6-311G(d,p) level are compared with those of the higher-level G3MP2B3 results of this study. Comparisons of these different levels provide a calibration of the B3LYP/6-311G(d,p) values (with similar working reactions). The G3MP2B3 (G3MP2//B3LYP/6-31G(d)) uses geometries and ZPVE scaled by 0.96 from B3LYP/6-31G(d) calculations.

Janoschek and Rossi [140] have calculated energies, harmonic vibration frequencies, moment of inertia, and thermochemical properties, on a set of 32 selected free radicals at the G3MP2B3 level of theory. They compared their calculated values to literature data and show a mean absolute deviation between calculated and experimental enthalpies values of 0.9 kcal mol^{-1}, which was close to the average experimental uncertainty of ± 0.85 kcal mol^{-1}.

Fortunately, determination of geometries and frequencies with B3LYP calculations are known to provide good results [141, 142, 143, 144, 145]. Part of the success of the G3MP2B3 methods is a result of the accuracy of the geometries and vibration frequency data.

5.2.2 Enthalpy of Formation

In the previous Chapter, the thermodynamic properties of fourteen unsaturated C_2-C_6 stable species with an oxygenated containing functional group (alcohol, peroxide, and aldehyde), and in addition, the hydrocarbon 3 methyl 1,4 pentadiene (CH_2=$CHCH(CH_3)CH$=CH_2) were determined using DFT method. To validate these data, standard enthalpies of formation for a number of oxygenated and non-oxygenated hydrocarbons (stable molecules and radicals) are calculated, using G3MP2B3. The work reactions used to determine $\Delta_f H^0_{298}$ on the stable and radical species are illustrated in appendix E (Tables E 1, E 2, and E 3). Literature enthalpy values for standard reference species in the work reactions are listed in appendix A.

The error resulting from the G3MP2B3 work reaction computation is estimated by comparing the calculated enthalpy of reaction for a series of work reactions with $\Delta H^0_{rxn,298}$ for each reaction using thermochemical data from the literature.

The reaction enthalpies ($\Delta H^0_{rxn,298}$) for a series of work reactions, for which each species in the reaction has a literature value, are determined and compared to the experimental enthalpies of reaction $\Delta H^0_{rxn,298}$ (see Table C2 in appendix C). The difference $\Delta H^0_{rxn,298}$(cal) - $\Delta H^0_{rxn,298}$(exp) on these work reactions provides a reasonable evaluation of the error of the calculation method. The mean absolute average from the 15 calculated difference values for G3MP2B3 computation method is 0.31 kcal mol^{-1}. This value is considered as the error due to the calculation method for the molecules represented in this study.

Table 5.1 compares the B3LYP/6-311G(d,p) and G3MP2B3 enthalpy values for 27 unsaturated, stable hydrocarbons (oxygenated / non-oxygenated, cyclic / linear). The results obtained with DFT are, overall, in good agreement with G3MP2B3 values. The difference between the two methods is less than ± 1.0 kcal mol^{-1} for sixteen species. Nine other species have DFT enthalpy values that differ from the G3MP2-B3 of this study by ca. ± 1.5 kcal mol^{-1}. Two of these molecules, CH_2=CHCH(CH=O)C≡CH and CH_2=CHCH(CH_3)CH=CH_2, show deviation of ± 1.6 kcal mol^{-1}. The recommended $\Delta_f H^0_{298}$ values are obtained by averaging the results of all working reactions at G3MP2B3 calculation level.

TABLE 5.1: Comparison of $\Delta_f H^0_{298}$ values of stable molecules: G3MP2B vs. B3LYP

Species		$\Delta_f H^0_{298}$ in kcal mol^{-1}		
Formula	Name of the species	**B3LYP/ 6-311G(d,p)**[a]	**G3MP2/ B3LYP6-31G(d,p)**	$(\Delta_f H^0_{298})_{DFT} - (\Delta_f H^0_{298})_{G3MP2B3}$
CH_2=CHCH(CH=O)CH=O	3-butenal, 2-formyl	-39.56 ± 0.78	-40.76 ± 0.90	1.2
CH_2=CHCH(CH=O)C≡CH	1-penten, 3-formyl-4-yn	45.23 ± 0.22	43.57 ± 0.11	1.66
$CH(OH)_2$CH=O	Ethanal, 2-diol	-119.08 ± 0.44	-118.55 ± 0.09	-0.53
$CH_3C(OH)_2$CH=O	Propanal, 2-diol	-130.57 ± 1.31	-131.64 ± 1.69	1.07
CH_2=CHCH(CH=O)CH=CH_2	1,4-pentadiene, 3-formyl	-0.83 ± 0.85	-0.79 ± 2.09	-0.04
CH_2=CHC$(OH)_2$CH=CH_2	1,4-pentadiene, 3-diol	-61.30 ± 0.41	-60.68 ± 0.91	-0.62
CH_2=C$(OH)_2$	Ethene, 1-diol	-73.95 ± 0.65	-73.89 ± 0.27	-0.06
CH_2=CHCH_2CH=O	3-Butenal	-20.62 ± 0.74	-20.32 ± 1.08	-0.3
CH_2=CHCH(CH_3)CH=CH_2	1,4penten, 3-methyl	17.98 ± 0.20	16.3 ± 0.28	1.68
CH_2=C(CH=O)CH=CH_2	Formyl butadiene	-2.56 ± 0.55	-2.00 ± 0.41	-0.56
CH_2=C(CH=O)C≡CH	1-buten, 2-formyl, 3-yn	39.18 ± 1.20	39.01 ± 0.41	0.17
CH_2=CHCH(OH)CH=CH_2	1,4pentadiene, 3-ol	-12.35 ± 0.78	-12.99 ± 1.11	0.64
CH_2=CHOCH=CH_2	vinyl ether	-6.15 ± 0.31	-5.23 ± 0.29	-0.92
CH_2=CHCH=O	Propenal	-18.65 ± 0.95	-17.76 ± 0.41	-0.89
CH_3OCH=O	Methyl formate	-83.39 ± 1.19	-82.18 ± 0.42	-1.21
$C_6H_5C(OH)_2CH_3$	Ethylbenzene, 2-diol	-79.52 ± 0.81	-79.98 ± 1.8	0.46
$C_6H_5C(CH_2$)CH=CH_2	Vinyl Styrene	52.84 ± 0.72	51.47± 0.56	1.37
C_6H_5CH=O	Benzaldehyde	-7.46 ± 0.27	-8.79 ± 0.99	1.33
$CH_3C_6H_4$CH=O	Para-methyl Benzaldehyde,	-16.37 ± 0.44	-16.2 ± 0.37	-0.17
$C_6H_5C(=CH_2)$CH=O	Formyl benzaldehyde	11.98 ± 0.58	10.61 ± 0.52	1.37
C_6H_5OCH=O	Benzyl formate	-51.49 ± 0.26	-52.54 ± 1.42	1.05
$C_6H_5C(=CH_2)CH_3$	Methyl styrene	29.05 ± 0.51	27.80 ± 0.43	1.25
$C_6H_5C(=CH_2)$OH	1-phenylethanal	-8.14 ± 0.79	-7.35 ± 0.12	-0.79
$Y(C_6H_6)$=O	1,4-cyclohexadienone	-4.5 ± 1.14	-4.88 ± 0.78	0.38
$Y(C_5H_5)$C(OH)=O	2,4-cyclopentadiene formic acid	-52.81 ± 1.0	-53.84 ± 1.0	1.03
$Y(C_5H_5O)$CH=O	2H-pyran, 2-formyl	-25.27 ± 2.5	-25.25 ± 0.51	0.02
$Y(C_5H_5)$OCH=O	cyclopentadiene formate	-34.67± 0.44	-34.49 ± 0.92	-0.11

[a]Chapter 4;

The enthalpy of formation of methylformate, CH_3OCH=O, is calculated at -82.2 kcal mol^{-1} by the G3MP2B3 method and at -83.4 kcal mol^{-1} with the B3LYP/6-311G(d,p) method. There are previous calculations on methylformate by Sumathi and Green [146] using CBS-Q with spin-orbit and bond additivity corrections; they reported a value at -85.87 kcal mol^{-1}.

The NIST Standard Reference Database (2003) reports the experimental enthalpy for $CH_3OCH{=}O$ determined by Hine and Klueppet [147] to be -86.6 kcal mol^{-1} and a value of -84.9 kcal mol^{-1} by Hall and Baldt [148]. The B3LYP value is 2.4 kcal mol^{-1} higher than that of Sumathi and Green, 3.2 and 1.5 kcal mol^{-1} higher than Hine and Klueppet and Hall and Baldt's experimental values, respectively. Our calculated G3MP2B3 values show larger deviations than the B3LYP calculations (see Table 5.1).

We have further evaluated the enthalpy of $CH_3OCH{=}O$ by using the G3 [149, 150] method and the same work reactions. We find -82.3 kcal mol^{-1} which supports our G3MP2B3 calculated value. In addition we use an extra reaction for methylformate analysis:

$$CH_3OCH{=}O + CH_2{=}CHCH_3 \rightarrow CH_3OCH{=}CH_2 + CH_3CH{=}O$$

The enthapy of $CH_3OCH{=}O$ from this reaction is -81, -83.3 and -82.8 kcal mol^{-1} by B3LYP, G3MP2B3, and G3 respectively. We recommend the G3 value of -82.3 ± 0.49 kcal mol^{-1} for $CH_3OCH{=}O$ (Table 5.3).

Table 5.2 lists $\Delta_f H^0_{298}$ data for twelve unsaturated oxygenated radical species, linear or cyclic, estimated by DFT and G3MP2B3 level. Good agreement between the two calculation methods is observed where $\Delta_f H^0_{298}$ values being below ± 0.5 kcal mol^{-1} for nine radicals and below ± 1.0 kcal mol^{-1} for two radicals.

The enthalpy of formation of the radical resulting from the dissociation of $Y(C_6H_6){=}O$ to $Y(C_6H_5\bullet){=}O$ + H is not 2,4-cyclohexadienone-yl-6 (A), but 2,5-cyclohexadienone-yl-4 (B).

Table 5.2: Comparison of $\Delta_f H^0_{298}$ values for Radicals: G3MP2B3 vs. B3LYP

Species	$\Delta_f H^0_{298}$ in kcal mol^{-1}		
	B3LYP/6-311G(d,p)[a]	**G3MP2//B3LYP/6-31G(d,p)**	$(\Delta_f H^0_{298})_{DFT} - (\Delta_f H^0_{298})_{G3MP2B3}$
Furan•1	61.67 ± 1.28	60.88 ± 1.27	0.79
Furan•2	61.80 ± 1.28	60.96 ± 1.27	0.84
$Y(C_6H_5\bullet){=}O$[b]	12.77 ± 0.44	15.35 ± 0.22	-2.58
$Y(C_5H_5)C(O\bullet){=}O$[b]	5.1[c]	5.43 ± 0.73	-0.33
$C_6H_5C\bullet{=}O$	30.75 ± 1.42	30.47 ± 0.86	0.28
$Y(C_5H_5O)C\bullet{=}O$	10.56 ± 1.17	11.02 ± 0.38	-0.46
$Y(C_5H_5)OC\bullet{=}O$	9.54 ± 1.17	9.78 ± 0.19	-0.24
$CH_3OC\bullet{=}O$	-42.02 ± 0.69	-41.53 ± 0.57	-0.49
$CH_2{=}C\bullet CH{=}O$	41.97 ± 0.13	42.47 ± 0.42	-0.5
$CH_2{=}CHC\bullet(CH{=}O)C{\equiv}CH$	62.80 ± 0.13	62.53 ± 1.7	0.27
$C\bullet(OH)_2CH{=}O$	-95.78 ± 1.26	-95.56 ± 0.26	-0.22
$CH_2{=}CHC\bullet(CH{=}O)CH{=}O$	-18.99 ± 0.78	-18.70 ± 0.44	-0.29

[a]Chapter 4; [b]determined with G3 (see text and Table 5.3) as well; [c]Ref.32

The structures obtained from both methods (B3LYP and G3MP2B3) indicate that the radical site in the para position (B), is the favored resonance structure, as shown in Fig. 5.1.

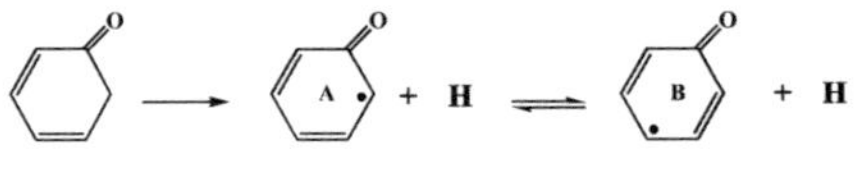

Figure 5.1: 2,5-cyclohexadienone-yl-4 (B) resulting from dissociation of a weak (doubly allylic) C—H bond on 2,4-cyclohexadienone

Enthalpy results for Y(C_6H_5•)=O listed in Table 5.2 show a difference of 2.6 kcal mol^{-1} between the two calculation methods, 12.77 ± 0.44 with DFT and 15.35 ± 0.22 kcal mol^{-1} for G3MP2B3. Geometries are fully optimized and the final structures are the same in both methods. To further investigate the enthalpy value we have performed calculations on this Y(C_6H_5•)=O radical with Gaussian-3 (G3) [149, 150]. The enthalpy value for Y(C_6H_5•)=O estimated with G3 is 13.43 ± 0.29 kcal mol^{-1} (Table 5.3) which is close to the B3LYP value (0.66 kcal mol^{-1} higher) and nearly 2 kcal mol^{-1} lower than the G3MP2B3 value. We recommend the G3 enthalpy of Y(C_6H_5•)=O, 13.43 ± 0.29 kcal mol^{-1} here.

Table 5.3: Comparison of $\Delta_f H^0_{298}$ values: Cyclic Radicals

Species	$\Delta_f H^0_{298}$ in kcal mol^{-1}		
	Y(C_6H_5•)=O	Y(C_5H_5)C(O•)=O	CH_3OCH=O
B3LYP/6-311g(d,p)	12.77 ± 0.44[a]	5.1[b]	-83.39 ± 1.19
G3MP2B3//B3LYP/6-31g(d,p)	15.35 ± 0.22	5.78 ± 0.60	-82.18 ± 0.42
G3	13.43 ± 0.29	5.16 ± 1.12	-82.32 ± 0.49
Recommended value	**13.43 ± 0.29**	**5.1 ± 1.12**	**-82.3 ± 0.49**

[a]Chapter 4; [b]Ref. 32 determined at B3LYP/6-31g(d,p)

A somewhat similar radical Y(C_5H_6)C(O•)=O with resonance overlap, is also calculated with G3 and it can also be compared with the previously reported B3LYP density functional value from Hadad's group [151]. The result obtained with G3MP2B3, 5.78 ± 0.60 kcal mol^{-1}, agrees reasonably with the G3 value, 5.16 ± 1.12 kcal mol^{-1}, and with the value, 5.1 kcal mol^{-1} by Hadad at al. (Table 5.3). The density functional value is 6.35 kcal mol^{-1}. We recommend the G3 value, 5.2 ± 1.12 kcal mol^{-1}, for the enthalpy of formation of Y(C_5H_6)C(O•)=O.

5.2.3 Bond Energies

Reported C—H bond energies are derived from the G3MP2B3 calculations and compared with C—H bond energies from B3LYP calculation values previously reported in chapter 4. The bond energies from the two calculation methods show good agreement for all radicals – parent pairs listed in Table 5.4, with the exception of Y(C_5H_5)C(OH)=O / Y(C_5H_5)C(O•)=O and Y(C_6H_6)=O / Y(C_6H_5•)=O.

There are two furan radical sites called Furanj1 and Furanj2, which differ in the position of the radical site with respect to oxygen. In Furanj1, the radical site is adjacent to the oxygen and in Furanj2 the radical site is on the secondary olefinic carbons. The strongest C—H bond in the set (Table 5.4) results from these furan C—H bonds at 121.3 kcal mol^{-1} for Furanj1 and Furanj2. The density functional values are in good agreement and only slightly higher (0.8 kcal-mol^{-1}) at 122.1 and 122.2 kcal mol^{-1} respectively.

The acid O—H dissociation energy from $Y(C_5H_5)C(OH)$=O (forming $Y(C_5H_5)C(O\bullet)$=O + H) is 111.4 kcal mol^{-1} with G3MP2B3 and 110.1 kcal mol^{-1} with G3. This compares with a value derived by the research group of Hadad et al. [32] at the B3LYP/6-311G(d) level of 110.01 kcal mol^{-1}. We recommend the G3 value of 110.1 kcal mol^{-1}. These values are somewhat lower than the value of 112 kcal mol^{-1} recommended in the review (Accounts of Chemical Research) by Blanksby and Ellison [152].

Table 5.4: C—H Bond energies and comparison between B3LYP and G3MP2B3 values

Reactions	C—H bond energy in kcal mol^{-1}	
	B3LYP/6-311g(d,p)[a]	G3MP2B3/ B3LYP6-31g(d,p)
Furan → Furanj1 + H	122.06	121.27
Furan → Furanj2 + H	122.19	121.35
CH_2=CHCH=O → CH_2=C•CH=O + H	112.72	112.33
$Y(C_5H_5)C(OH)$=O → $Y(C_5H_5)C(O\bullet)$=O + H	110.01	111.37 / **110.1**[b]
$Y(C_5H_5)OCH$=O → $Y(C_5H_5)OC\bullet$=O + H	96.31	96.37
CH_3OCH=O → $CH_3OC\bullet$=O + H	95.05	95.54
C_6H_5CH=O → $C_6H_5C\bullet$=O + H	90.31	91.36
$Y(C_5H_5O)CH$=O → $Y(C_5H_5O)C\bullet$=O + H	87.93	88.37
$CH(OH)_2CH$=O → $C\bullet(OH)_2CH$=O + H	75.4	75.09
CH_2=CHCH(CH=O)CH=O → CH_2=CHC•(CH=O)CH=O + H	72.67	74.16
$Y(C_6H_6)$=O → $Y(C_6H_5\bullet)$=O + H	69.67	72.33 / **70.3**[b]
CH_2=CHCH(CH=O)C≡CH → CH_2=CHC•(CH=O)C≡CH + H	69.67	71.06

[a]Chapter 4; [b]the G3 method - recommended value.

This study includes four carbonyl groups with a radical on the carbon site (substituted formyl radicals). Two of these carbonyl radical systems are esters and have the carbonyl groups bonded to an ether oxygen atom link, $Y(C_5H_5)OC\bullet$=O from the dissociation of cyclopentadiene formate $Y(C_5H_5)OCH$=O, and $CH_3OC\bullet$=O from the methyl formate CH_3OCH=O. These carbonyl radicals show similar bond energies, 96.3 and 95.0 kcal mol^{-1} respectively with the difference between the G3MP2B3 and B3LYP calculations for the respective bond strengths being less than 0.5 kcal mol^{-1}. Benzaldehyde, C_6H_5CH=O, which substituted carbonyl radical is bonded to a benzene carbon has a C—H energy dissociation at 91.4 kcal mol^{-1} (C_6H_5CH=O to $C_6H_5C\bullet$=O + H). This is 5 kcal mol^{-1} lower than the above ester carbonyl systems. There is one carbonyl bonded to a sp^3 carbon $Y(C_5H_5O)CH$=O where the bond energy to $Y(C_5H_5O)C\bullet$=O + H is ca. 8 kcal mol^{-1} lower than that in the ether link (ester) carbonyls, 88.4 kcal mol^{-1}. This 88 to 90 kcal mol^{-1} range is typical of C—H bonds in aldehyde groups of alkanes such as acetaldehyde [115, 152].

In the dissociation of ethanal,2-diol, $CH(OH)_2CH{=}O$ to $C\bullet(OH)_2CH{=}O$ + H, We observe a very small, (only 75.4 kcal mol^{-1}) RC—H energy in the dihydroxy C—H bond. This can be explained with resonance of the carbon radical with a carbonyl group. We have calculated this energy with the G3 method to verify the data and find a similar value, 75.1 kcal mol^{-1}. We have not observed this effect with species containing only one hydroxy (OH) and the carbonyl and look to study on simple dihydroxy alkanes to further, understand this effect of a second OH.

To further study the carbon – hydrogen bonds in hydroxyl carbonyls system we have evaluated the C—H bond energy in $CH_2(OH)CH{=}O$ to $CH\bullet(OH)CH{=}O$ + H, which have only one α-OH group. We used B3LYP, G3MP2B3 and G3 calculations (see Table 5.4). The C—H energy is some 7 kcal mol^{-1} higher than in dihydroxyl species but is still quite low (82.6 kcal mol^{-1}). If we compare the stabilizing effects of the carbonyl group on $CH_3CH{=}O \rightarrow C\bullet H_2CH{=}O + H$ which is reported to be 94 kcal mol^{-1} [115], and the hydroxyl stabilizing effect on $CH_3CH_2OH \rightarrow CH_3C\bullet HOH + H$ (95 kcal mol^{-1}) [152], we see that both adjacent carbonyl and hydroxyl groups provide ca. 6 kcal mol^{-1} stabilization. The combined effect of the two groups, leads to some 18 kcal mol^{-1} stabilization, which is significantly more stabilization than we expected.

The dissociation of the sp^3 C—H bond in $CH_2{=}CHCH(CH{=}O)CH{=}O$ (to $CH_2{=}CHC\bullet(CH{=}O)CH{=}O$ + H) has resonance with two carbonyl groups and one vinyl group. This C—H bond energy is very low, 74.2 kcal mol^{-1} with G3MP2B3 and the B3LYP value is even lower at 72.7 kcal mol^{-1}.

The C—H bond from dissociation of $CH_2{=}CHCH(CH{=}O)C{\equiv}CH$ to $CH_2{=}CHC\bullet(CH{=}O)C{\equiv}CH$ + H is 71.1 kcal mol^{-1} at G3MP2B3. This is similar to the respective bond energy in $CH_2{=}CHC({-}H)(CH{=}O)CH{=}O$ above. The low bond energy is due to the multiple resonance stabilization. The B3LYP value is similar, at 69.7 kcal mol^{-1}.

The C—H bond in 1,4-cyclohexadienone, $Y(C_6H_6){=}O$ to $Y(C_6H_5\bullet){=}O$ + H, is 73.3 kcal mol^{-1} which is even lower than the double allylic C—H bond in 1, 4 cyclohexadiene, which is 75 to 76 kcal-mol^{-1} . The value from density functional analysis is lower at 69.7. We used G3 for this system, which yielded a value of 70.3 kcal mol^{-1}. We recommend the G3 value for $\Delta_f H^0_{298}$ of $Y(C_6H_5\bullet){=}O$ (13.4 kcal mol^{-1}), which results in the C—H bond energy of 70.3 kcal mol^{-1}.

The good agreement between the enthalpy of formation values for the parent molecules, the radicals and for the bond energies, where both the parent and radical enthalpy were determined from work reactions shows the good precision and suggests that errors in the calculations are significantly cancelled by the use of work reactions. The work reaction method appears to work reasonably well for the B3LYP/6-311 Density Functional Theory and suggests this method may be useful for the calculation of enthalpy values for larger unsaturated oxygenated hydrocarbons.

5.2.4 Group Additivity

In this work, we have performed G3MP2B3 enthalpy calculations on 27 unsaturated oxygenated hydrocarbons, and we have utilized these data to derive a consistent set of oxygen-hydrocarbon GA groups, which were used to validate the groups estimated by DFT (B3LYP) calculations.

Table 5.5: Thermodynamic data calculation for new groups using Group Additivity

New groups	From B3LYP calculations[a]	From G3MP2B3 calculations	New groups	From B3LYP calculations[a]	From G3MP2B3 calculations
C/CO/O2/H	-14.18	-13.65	CO/CD/H	-29.23	-28.34
C/CD/CO/CO/H	3.79	2.59	C/CD2/C/H	-1.52	-3.2
C/C/CO/O2	-15.47	-16.54	O/CD2	-35.87	-34.95
C/CD2/H/O	-4.15	-4.79	C/CB/C/O2	-15.53	-15.99
C/CD/CO/H2	-6.37	-6.07	CD/CB/CD	11.35	9.98
C/CD2/CO/H	-1.43	-1.39	O/CB/CO	-34.99	-36.04
C/CD2/O2	-15.20	-14.58	CD/C/CB	10.80	9.55
C/CD/CO/CT/H	5.0	3.34	CB/CO	4.35	4.52
CD/O2	12.41	12.45	CD/CB/CO	12.76	11.39
CD/CD/CO	7.37	7.93	CD/CB/O	7.98	8.77
CD/CO/CT	7.02	6.85			

[a]Chapter 4

We note an overall good agreement between B3LYP and G3MP2B3 results as shown in Table 5.5. Determination of thermochemical properties of groups is described in chapter 4. We recommend the DFT values. Entropy and heat capacities for these groups are reported in chapter 4 using B3LYP structures, internal rotor potentials and frequencies.

5.3 Conclusion

Enthalpies for 27 oxygenated and non-oxygenated stable species and 12 radicals are calculated at the G3MP2B3 level of calculation and compared with enthalpies obtained at the B3LYP/6-311G(d,p) Density Functional Theory level. The overall results show good agreement between the G3MP2B3 and DFT values with one exception, the radical Y(C_5H_5•)=O (2,5-cyclohexadienone-yl-4). For this radical the B3LYP enthalpy value deviates by -2.58 kcal mol^{-1} from the G3MP2B3 value. The G3 data support the (B3LYP/6-311G(d,p)) values in the two cases where large difference between the DFT and G3MP2B3 method occurred. The data reported in this study suggest that DFT calculations combined with work reactions provide a reasonable method for the determination of enthalpies of formation within a standard deviation of ± 0.43 kcal-mol^{-1} relative to G3MP2B3.

6. The Phenyl System

The phenyl radical is formed in combustion and oxidation systems by abstraction of a hydrogen atom from benzene by a number of active radical species. Unimolecular reaction of the energized and stabilized PhOO• radical lead to chain branching reactions and new products that need to be included in detailed combustion models. Analysis and determination of thermochemical parameters, transition state structures and kinetic parameters of important reaction paths of the phenyl radical + O_2 reaction system are reported in this chapter.

Enthalpies are evaluated using ab initio (G3MP2B3 and G3) and Density Functional Theory (DFT) methods, group additivity (GA) and literature data. The ab initio and DFT methods are combined with isodesmic reaction analysis to improve the calculated enthalpy values of stable species (reactants, intermediate adducts and products).

Entropies and heat capacities are calculated using DFT and group additivity and compared to literature data when available.

High Pressure limit kinetic parameters are obtained from canonical Transition State Theory calculations. Multifrequency Quantum Rice-Ramsperger-Kassel (QRRK) analysis is used to calculate k(E) data and master equation analysis is applied to evaluate fall-off in this chemically activated reaction system.

The Phenyl + O_2 association results in a chemically activated phenyl-peroxy radical $[PhOO\bullet]^{\#}$ with 49 kcal mol^{-1} excess energy relative to stabilized PhOO•. This energized adduct can dissociate to phenoxy + O, react back to phenyl + O_2, or react to form a hydroperoxy-phenyl radical. All mentioned reaction channels have barriers that are at or below the energy of the reactants. The terminal oxygen of the peroxy radical can undergo an internal addition reaction to one of several unsaturated carbon sites of the phenyl ring. In contrast to the first mentioned channels, the internal addition paths will proceed via tight transition states. One of these ring addition pathways forms a bicyclic di-oxiranyl radical. This channel has a low-lying barrier of ~ 20 kcal mol^{-1} below the entrance channel, which suggests that it could be of considerable importance. This study will show that intramolecular addition channels can further react through several paths undergoing ring opening (unsaturated + carbonyl moieties) as well as cyclopentadienyl radical + CO_2. There are at least two reactions pathways that can lead to ortho-quinone and H atom. Kinetic parameters for important product channels are reported and a reduced mechanism is proposed.

6.1 Introduction

The use and importance of aromatic compounds in fuels sharply contrasts the limited kinetic data available in the literature, regarding their combustion kinetics and reaction pathways. A number of experimental and modelling studies on benzene [153, 154, 155, 156, 157, 158], toluene [159, 160] and phenol [161] oxidation exist in the literature, but it would still be helpful to have more data on initial product and species concentration profiles to understand or evaluate important reaction paths and to validate detailed mechanisms. The above studies show that phenyl and phenoxy radicals are key intermediates in the gas phase thermal oxidation of aromatics. The formation of the phenyl radical usually involves abstraction of a strong (111 to 114 kcal mol^{-1}) aromatic—H bond by the radical pool. These abstraction reactions are often endothermic and usually involve a 6 – 8 kcal mol^{-1} barrier above the endothermicity; but they still occur readily under moderate or high temperature combustion or pyrolysis conditions. The phenoxy radical in aromatic oxidation can result from an exothermic process involving several steps, (i) formation of phenol by OH addition to the aromatic ring with subsequent H or R elimination from the addition site [162]; (ii) the phenoxy radical is then easily formed via abstraction of the weak (ca. 86 kcal mol^{-1}) phenolic hydrogen atom.

These reaction processes of OH addition / H atom elimination to yield phenol, and abstraction to form a phenyl radical, are two major routes to initiate decomposition of the aromatic ring and the subsequent reactions in combustion and oxidation systems [163].

The thermal decomposition of the phenoxy radical [153, 161, 164, 165, 166, 33] (Figure 6.1) shows an interesting mechanism in which the bicyclo [3,1,0] hexenone radical is formed by intramolecular addition in the resonant form of phenoxy (bond formation across the ring by addition of the radical site to a Π bond in the ring). This bicyclic radical then undergoes a beta scission reaction to cleave a cyclic CO—C bond and form a cyclopentadienyl formyl radical. This RC•=O radical then undergoes a further beta scission to eliminate carbon monoxide and form the resonance stabilized cyclopentadienyl radical. In addition to loss of the phenolic H atom, phenoxy can isomerize with a 70 kcal mol^{-1} barrier to 2,4 cyclo-hexadienyl ketone (keto – enol isomerization), which can undergo ring opening via cleavage of a weak – doubly allylic carbon – carbon bond to a di-radical [167].

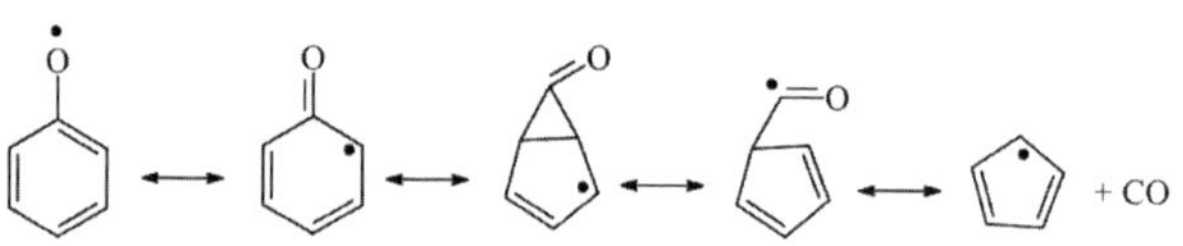

Figure 6.1: Phenoxy Degradation Reaction Pathway

Formation of the highly reactive phenyl radical in combustion systems is important. This occurs even at moderate temperatures in several downstream zones of an incinerator. The phenyl radicals will rapidly react with molecular oxygen in the combustion environment to form an energized (49 kcal mol^{-1}) phenyl-peroxy adduct (chemical activation). The energized peroxy adduct can undergo intramolecular reaction(s) before it is stabilized, resulting in a new intermediates and products with loss of the aromatic ring. These reaction paths, products and kinetics of the phenyl radical reactions with O_2 are important to understanding and modeling the oxidation chemistry of many aromatic systems and need to be analysed as a function of temperature and pressure.

Frank et al. [168] studied the reactions of phenyl radicals with molecular and atomic oxygen in shock tube experiments in the temperature range of 1000 K to 1800 K and pressures from 1.3 to 2.5 bar. For the initiation reaction $C_6H_5 + O_2$, two products channels were reported: $C_6H_5 + O_2 \rightarrow C_6H_5O + O$ with the apparent activation energy for the formation of C_6H_5O of 6 kcal mol^{-1}, and a fast channel producing H-atoms, which they assigned as $C_6H_5 + O_2 \rightarrow C_6H_4O_2 + H$.

Lovell et al [153] reported flow reactor data on benzene combustion at three oxygen concentrations corresponding to rich, lean and stoichiometric conditions. The experiments showed that the initial oxygen concentration significantly effects fuel consumption rates along with formation of carbon monoxide and cyclopentadiene. They suggested that phenoxy is formed and phenoxy degradation is the major reaction path for both carbon monoxide formation and benzene consumption. Lovell et al. [153] also reported experimental data on the impact of NO_2 addition on benzene oxidation. The benzene oxidation rate doubled in the presence of NO_2 and they suggested additional reaction pathways, including the association of hydroxyl radical with cyclopentadienyl radical.

Mebel and Lin [169] report enthalpy values for phenyl peroxy and a number of isomers using HF/6-31 level of ab initio calculations. They report a well depth of the phenyl peroxy adduct relative to reactants phenyl radical + O_2 of 41 kcal mol^{-1}. They later performed higher-level calculations [33] on the vinyl radical + O_2 reaction system, and found a well depth of 46 kcal mol^{-1}. The vinyl reaction with O_2 is considered similar to phenyl + O_2.

A number of researchers have recently published models for benzene or toluene oxidation; having made a number of modifications to reaction rate constant modifications to better fit experimental data. But phenyl radical reactions were not analyzed or treated in detail. Bittker [170] published a mechanism, which was based upon previously published reaction paths for ignition delay times and benzene and toluene loss profiles. Bittker employed a sensitivity code to determine the important reactions and then optimized the agreement to the data by adjusting rate constants. Davis et al. [171] modified the mechanism (phenol reactions) proposed by Brezinsky et al. to fit better flame speed data. Tan and Frank [172] published a benzene oxidation model and used it to explain H and O atom production in their shock tube data. Emdee et al. [173] report an updated mechanism for toluene oxidation based on Brezinsky's model [153, 174, 175]. They indicated that the branching reaction: $C_6H_5CH_3 + O_2 \rightarrow C_6H_5CH_2 + HO_2$ is of major importance. In Emdee et al.'s toluene-mechanism, the reactions of resonance stabilized benzyl radicals control the reaction at early times. None of these studies included phenyl peroxy radical reactions, however.

Several recent publications suggest that new reaction paths are needed to model aromatic oxidation and combustion [156, 154, 175, 176] Zhang et al. [156] published an elementary reaction model for high-temperature benzene combustion under fuel rich conditions - near sooting environment. They report that the flame speeds for benzene could not be matched by their model and suggested that some important reaction paths may be missing. Shandross et al. [154] reported data from molecular beam experiments on benzene flames and showed that current models strongly over predicted destruction of phenol at high temperatures. Shandross et al. modified the phenol reaction chemistry of Emdee and Brezinsky, of Lindstedt and Skevis [155] and of Zhang and McKinnon to obtain improved model comparisons with experiments.

The research groups of Glarborg [176] and of Tester et al. [177] have recently published a benzene mechanism indicating that the products of the reaction phenyl radical with oxygen are uncertain and need to be more clearly identified.

Hadad and coworkers [32, 178,] have studied the unimolecular decomposition pathways of phenyl-peroxy using B3LYP Density Functional Theory calculations by calculating Gibbs energies of reactants, transition states and products. The association reaction results in a chemically activated phenyl-peroxy radical, which can dissociate to a phenoxy radical + O atom or undergo intramolecular addition of the peroxy moiety to the ipso carbon. This intramolecular addition product further reacts via several steps to ring opening and ring expansion products, which are much lower in energies and are chemically activated species. Hadad et al. reported that the oxy-pinoxy radical is an important highly stable intermediate. They have performed an extensive study on the reactions of this oxy-pinoxy radical with O_2 based on its stability. They omitted the evaluation of unimolecular beta scission reactions to ring opening, however, and these show that this radical, which is formed with ~ 90 kcal mol of excess energy, will dissociate with a barrier of less than 30 kcal mol^{-1} and will not be stabilized in thermal or even under atmospheric conditions.

The present study calculates thermochemical properties of intermediates, transition states and products important to the degradation of the aromatic ring in the phenyl radical + O_2 reaction system. Kinetic parameters are developed for the important elementary reaction paths through each channel as a function of temperature and pressure. The calculation is done via a bimolecular chemical activation and master equation analysis for fall-off.

Our approach starts with and extends a mechanism initially proposed by Carpenter [175] who reported a computational study showing several very important new reaction pathways for the reaction of vinyl and phenyl radicals with molecular oxygen using semi-empirical calculations. Mebel et al. [33] extended the work of Carpenter on vinyl with Density Functional Theory calculations, B3LYP/6-311g(d,p), for structures and energies as well as G2M calculations to verify the energetics.

In this study enthalpies, entropies and heat capacities for species in the decomposition of the first ring are evaluated using ab initio (G3MP2B3 and G3) calculations in addition to DFT calculations (B3LYP/6-311G(d,p)), group additivity (GA) and by comparisons to the vinyl + O_2 data of Mebel et al. [180]. We also include comparisons with the reported Gibbs energies for the phenyl + O_2 system published by Hadad et al. [32] We show that the vinyl radical is a good model for phenyl, hence high level calculations on the smaller vinyl system can be used to calibrate ab initio and DFT calculations on the phenyl system. Isodesmic reaction analysis is used to estimate enthalpy values of the intermediates and well-depths of the adducts.

The nomenclature in this work: **Ph** represents phenyl, **Y(A)** represents a cyclic structure (e.g. Y(C5) is cyclopentadiene), **D** is a double bond (CDO is C=O), A• or **AJ** represent a radical site on the structure. In some abbreviated nomenclature the double bond symbol (= or D) is omitted in the name, e.g. CH_2=CHCH=CHC•=O is named C4C•DO. There are often several resonant structures for a given species.

6.2 Results and Discussion

Optimized geometries, harmonic vibrational frequencies and moments of inertia calculated at the B3LYP/6–311G(d,p) level of Density Functional Theory are reported in Appendix B.

6.2.1 Enthalpy of Formation

Standard enthalpies of formation of reactants, transition states and products are determined using:

i. The total energies obtained from B3LYP/6-311G(d,p), G3 and G3MP2B3 calculation methods combined with isodesmic work reactions.

ii. The B3LYP/6-311+G(d.p) energies and Gibbs energies reported in Fadden et al. study [32].

iii. The G2M [181, 182, 183, 184] calculations reported by Tokmakov et al. [180]

iiii. The group additivity (GA) method.

Comparisons of density function parameters with higher level calculation G2M(RCC,MP2) on the vinyl + O_2 system from Mebel et al. [180] are also performed for the important transition state of phenyl peroxy reaction to Phenoxy + O atom.

Total energies obtained at B3LYP/6-311G(d,p), G3 and G3MP2B3 calculation level are used with isodesmic work reactions having bond conservation, when possible, for the calculation of the standard enthalpies. The total energies obtained with DFT are corrected by zero-point vibration energies (ZPVE), which are scaled by 0.97 as recommended by Scott and Radom [120].

Absolute enthalpies of formation of the target species (stable species and radicals) are estimated using calculated $\Delta H^0_{rxn,298}$ values for each reaction and evaluated literature thermodynamic properties for the reference species in the reactions. Three to five work reactions are used to evaluate each species to minimize possible errors from use of only one work reaction and to provide more confidence in the determined $\Delta_f H^0_{298}$.

The Group additivity (GA) method, with a number of new developed groups (see chapter 4) for unsaturated oxygenated hydrocarbons [29] is also used to estimate enthalpies values of species.

Evaluated enthalpy values for standard reference species used in the work reactions are listed in Appendix A along with literature references.

6.2.1.1 Enthalpy of Formation of the Reference Species

Examination of the working reactions used to determine $\Delta_f H^0_{298}$ for the target species reveals that accurate enthalpies of several stable species: Y($C_2H_4O_2$), CH_2=CHCH=CHCH=O, Y(C_3H_4O)=O, Y(C_6H_8O Y(C_5H_6O), C_6H_5OH and C_6H_5OOH; and for the radicals: CH•=CHCH=CHCH=CH_2, O=CHC•=O, CH_2=CHC•=O, Y(C_6H_7•), and •OC(=O)CH=CH_2 are needed to be used as reference. Enthalpy values for these species could not be found in the literature and were calculated using DFT, G3MP2B3 and G3 methods. Values were also compared to GA values. CH_2=CHCH=O is calculated in this work using the G3 method only, DFT and G3MP2B3 values were obtained in previous chapters [29, 30].

The enthalpy of formation of Y(C_3H_4O)=O and Y(C_6H_8O), listed in Table 6.1, are determined in order to deduce two cyclic groups, CY/C3O/DO and CY/C6O/DE/13 for use in group additivity. These cyclic groups are then used for group additivity comparisons to computational chemistry values of Y(C5•)Y(C3O)DO and Y(C6•O)DO radicals in the phenyl + O_2 reaction system. The isodesmic reactions used to estimate the enthalpy of these reference species are listed in Table 6.1. All work reactions for these species show good agreement among the DFT, *ab initio* methods and GA results.

Table 6.1: Calculated $\Delta_f H^0_{298}$ of species used in reference reactions and comparison with GA

Reactions Series (kcal mol^{-1})	$\Delta_f H^0_{298}$	Error limit[a]	Therm[b]
$C_6H_5OH + CH_2{=}CHOOH \rightarrow CH_2{=}CHOH + C_6H_5OOH$	-23.96	± 2.21	-23.03[d]
$C_6H_5OH + CH_3CH_2OOH \rightarrow CH_3CH_2OH + C_6H_5OOH$	-23.82	± 2.28	
$C_6H_5OH + CH_3OCH_3 \rightarrow CH_3OH + C_6H_5OCH_3$	-24.41	± 1.72	-23
Average	**-24.06 ± 0.28**		
$C_6H_5OOH + CH_2{=}CH_2 \rightarrow CH_2{=}CHOOH + C_6H_6$	-2.15	± 1.61	
$C_6H_5OOH + CH_3CH_3 \rightarrow CH_3CH_2OOH + C_6H_6$	-2.76	± 2.11	
$C_6H_5OOH + CH_3OCH_3 \rightarrow CH_3OOH + C_6H_5OCH_3$	-3.13	± 3.26	
Average	**-2.68 ± 0.49**		-3

[a]reported errors for each standard species (see Appendix A) and estimate error due to the method (see Appendix C); [b]values are from group additivity unless noted otherwise; [d]reference [185]

Table 6.1 (con't): Calculated $\Delta_f H^0_{298}$ species used in Reference Reactions; Comparison with GA

	Reactions Series	$\Delta_f H^0_{298}$ (kcal mol^{-1})			
		B3LYP	**G3MP2B3**	**G3**	**GA**[a]
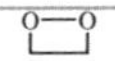 *Y(C2O2)*	$Y(C_2H_4O_2) + CH_4 \rightarrow Y(CH_2O_2) + CH_3CH_3$	10.35	11.33	11.30	11.30
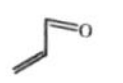	$CH_2{=}CHCH{=}O + CH_3CH_3 \rightarrow CH_2{=}CHCH_2CH_3 + CH_2{=}O$			-18.66	
	$CH_2{=}CHCH{=}O + CH_2{=}CH_2 \rightarrow CH_2CHCHCH_2 + CH_2{=}O$			-17.88	
	$CH_2{=}CHCH{=}O + CH_4 \rightarrow CH_2CHCH_3 + CH_2{=}O$			-17.14	
CDCCDO	Average	**-18.65 ± 0.95**[b]	**-17.76 ± 0.78**[c]	**-17.89 ± 0.76**	**-18.65**
	$CH_2CHCHCHCH{=}O + CH_2{=}CH_2 \rightarrow C_4H_6 + CH_2{=}CHCH{=}O$	-6.84	-5.31	-5.49	
	$CH_2CHCHCHCH{=}O + CH_4 \rightarrow CH_2CHCHCH_2 + CH_3CH{=}O$	-5.79	-4.09	-4.67	
	$CH_2CHCHCHCH{=}O + CH_3\bullet \rightarrow C_4H_5CHCHCH\bullet + CH_3CH{=}O$	-5.21	-4.49	-5.10	**-5.17**
C4CDO	Average	**-5.95 ± 0.91**	**-4.63 ± 0.62**	**-5.09 ± 0.41**	
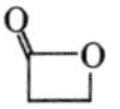	$Y(C_3H_4O){=}O \rightarrow CO_2 + CH_2{=}CH_2$		-65.45	-65.35	
	$Y(C_3H_4O){=}O \rightarrow 2CO + CH_4$		-65.72	-67.51	
	$Y(C_3H_4O){=}O + CH_3CH_3 \rightarrow CO + CH_2{=}CH_2 + CH_3OCH_3$		-65.55	-67.54	
Y(C3O)DO	Average		**-65.57 ± 0.13**	**-66.8 ±1.25**	**d**
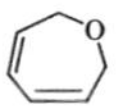	$Y(C_6H_8O) + 2H_2 \rightarrow Y(C_2H_4O) + 2CH_2{=}CH_2$		-1.64	-	
	$Y(C_6H_8O) + CH_4 \rightarrow CH_2{=}CHCH{=}CHCH{=}O + CH_3CH_3$		-2.75	-2.31	
	$Y(C_6H_8O) \rightarrow Y(C_5H_6) + CH_2{=}O$		-0.84	-1.22	
Y(C6O)	$Y(C_6H_8O) + H_2 \rightarrow CH_2{=}CHCH{=}O + CH_2{=}CHCH_3$		-3.33	-2.10	
	$Y(C_6H_8O) + CH_4 \rightarrow 2CH_2{=}CH_2 + CH_2{=}CHCH{=}O$		-1.07	-1.64	
	Average		**-1.93 ± 1.07**	**-1.83 ± 0.48**	**d**
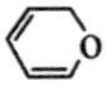	$Y(C_5H_6O) + 2CH_4 \rightarrow CH_2{=}O + 2CH_2{=}CHCH_3$	-2.14	-2.25	-3.49	
	$Y(C_5H_6O) + CH_4 \rightarrow Y(C_5H_6) + CH_3CH{=}O$	-2.34	-4.15	-4.32	
Y(C5O)	$Y(C_5H_6O) + CH_4 \rightarrow CH_2{=}CHCH{=}O + CH_3CH{=}CH_2$	-1.21	-3.50	-4.11	
	Average	**-1.9 ± 0.6**	**-3.3 ± 0.96**	**-3.97 ± 0.43**	**-4.79**
	$CH\bullet CHCHCHCHCH_2 + CH_3CH_3 \rightarrow C_6H_8 + CH_3CH_2\bullet$	100.43	104.00	100.42	
	$CH\bullet CHCHCHCHCH_2 + CH_2{=}CH_2 \rightarrow C_6H_8 + CH_2{=}CH\bullet$	100.39	104.42	101.10	
C•C5	$CH\bullet CHCHCHCHCH_2 + CH_4 \rightarrow C_6H_8 + CH_3\bullet$	99.53	104.82	101.97	
	Average	**100.12 ± 0.51**	**104.41 ± 0.41**	**101.1 ± 0.77**	**98.58**
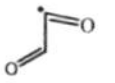	$O{=}CHC\bullet{=}O + CH_2{=}CH_2 \rightarrow O{=}CHCH{=}O + CH_2{=}CH\bullet$	-11.58	-11.41	-10.63	
	$O{=}CHC\bullet{=}O + CH_3CH{=}O \rightarrow O{=}CHCH{=}O + CH_3C\bullet{=}O$	-11.78	-10.89	-10.30	
ODCC•DO	$O{=}CHC\bullet{=}O + CH_3CH_3 \rightarrow O{=}CHCH{=}O + CH_3CH_2\bullet$	-11.54	-11.83	-11.31	
	Average	**-11.63 ± 0.13**	**-11.38 ± 0.47**	**-10.75 ± 0.51**	**-11.73**
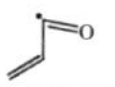	$CH_2{=}CHC\bullet{=}O + CH_4 \rightarrow CH_2CHCH{=}O + CH_3\bullet$	19.76	19.90	21.76	
	$CH_2{=}CHC\bullet{=}O + CH_3CH_3 \rightarrow CH_2CHCH{=}O + CH_3CH_2\bullet$	20.65	19.09	20.76	
	$CH_2{=}CHC\bullet{=}O + CH_3CH{=}O \rightarrow CH_2{=}CHCH{=}O + CH_3C\bullet{=}O$	20.42	20.02	21.78	
CDCC•DO	$CH_2{=}CHC\bullet{=}O + CH_2CH_2 \rightarrow CH_2{=}CHCH{=}O + CH_2CH\bullet$	20.62	19.50	21.45	
	Average	**20.36 ± 0.41**	**19.63 ± 0.42**	**21.44 ± 0.47**	**20.22**
	$Y(C_6\bullet H_7) + CH_2{=}CH_2 \rightarrow Y(C_6H_8) + CH_2{=}CH\bullet$	47.96	51.51	49.91	
	$Y(C_6\bullet H_7) + CH_3CHCH_2 \rightarrow Y(C_6H_8) + CH_2CHCH_2\bullet$	50.10	51.29	49.22	
Y(C6•)	$Y(C_6\bullet H_7) + CH_3CH_3 \rightarrow Y(C_6H_8) + CH_3CH_2\bullet$	47.99	51.09	50.71	
	Average	**48.68 ± 1.22**	**51.29 ± 0.21**	**49.94 ± 0.74**	**49.93**
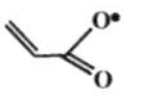	$\bullet OC({=}O)CH{=}CH_2 + H_2 \rightarrow CH\bullet{=}CH_2 + HOCH({=}O)$	-26.29			
	$\bullet OC({=}O)CH{=}CH_2 + CH_2CH_2 \rightarrow HOC(O)CHCH_2 + CH_2CH\bullet$	-26.58			
•OC(DO)CDC	$\bullet OC({=}O)CH{=}CH_2 + CH_3CH_3 \rightarrow HOC({=}O)CHCH_2 + C_2H_5\bullet$	-26.54			
	Average	**-26.47 ± 0.16**			**-27.72**

[a]values are from group additivity; [b]chapter 3; [c]Chapter 5; [d]Y(C3O)DO is used to develop the cyclic group CY/C3ODO and Y(C6O) to develop the cyclic group, CY/C6O/DE/13 for use in group additivity (see Table 6.6).

6.2.1.2 Enthalpy of Formation of the Target Species

The enthalpy of formation of twenty target radicals and three stable molecules in reactions are listed in Table 6.2. Enthalpy values derived from Gibbs energies and supplemental material data of Fadden et al. [32] are also listed. The recent enthalpy values from Lin research group [180] are given as well.

The $\Delta_f H^0_{298}$ values for many species show good precision through the different work reactions and B3LYP enthalpies are in good agreement with enthalpy values from G3 and G3MP2B3, when available.

Most of the enthalpies of formation that we have calculated from the supplemental information by Hadad et al. [32] are in agreement with the B3LYP, G3 and G3MP2B3 values reported across in Table 6.2, but there are a few exceptions. As an example, a deviation of about 8 kcal mol^{-1} is observed for Y(C5)OC•DO, Y(C5O)C•DO and *p*-ODY(C6)O• while the enthalpie data by Lin et al. [176] agree with our calculated B3LYP values.

The DFT method appears to fail for the radical Y(C5•)Y(C3O)DO. At the B3LYP level, an enthalpy of -2.27 kcal mol^{-1} was found which differs significantly from the G3 and G3MP2B3 calculations that yield values around -11 kcal mol^{-1}. The DFT optimized geometry does not appear to be responsible for the enthalpy value deviation, as the B3LYP geometry is alike the G3MP2B3 and G3 optimized structures. We notice, however, that our B3LYP enthalpy value and that of the Hadad et al. are similar. We recommend the G3 value, -12.4 kcal mol^{-1}, which is supported by the G2M value of Lin (-14 kcal mol^{-1}). Surprisingly, the DFT enthalpy value of the stable *o*-quinone species deviates by ca. 7 kcal mol^{-1} from G3 and G3MP2B3. This deviation can also not be related to geometry differences. Again, we recommend the G3 value (-24.4 kcal mol^{-1}) as it is supported by the G2M value of Lin et al. (-24.9 kcal mol^{-1}).

The comparison of the G2M enthalpies reported by Lin's group with our calculation methods (B3LYP, G3MP2B3 and G3) shows for four species an absolute deviation of 4 to 7 kcal mol^{-1}. All other calculated enthalpies differences are less than 3.5 kcal mol^{-1} and therefore in good agreement with the G2M values of Lin et al. We recommend the enthalpy values resulting from G3 calculations but we point out that the calculated G3 values support most of the B3LYP calculations.

Table 6.2: Calculated $\Delta_f H^0_{298}$ for Species

	Reactions Series	$\Delta_f H^0_{298}$ (kcal mol^{-1})					
		B3LYP	**G3MP2B3**	**G3**	**GA**[a]	**Hadad**[b]	**Lin**[c]
 **Ph•**	$C_6H_5\bullet + CH_3OH \rightarrow CH_3\bullet + C_6H_5OH$	81.55	82.67	82.59			
	$C_6H_5\bullet + CH_3CH_3 \rightarrow CH_3CH_2\bullet + C_6H_6$	81.45	82.85	81.87			
	$C_6H_5\bullet + CH_3OCH_3 \rightarrow CH_3\bullet + C_6H_5OCH_3$	81.22	83.20	-			
	$C_6H_5\bullet + CH_2{=}CH_2 \rightarrow CH_2{=}CH\bullet + C_6H_6$	81.24	83.27	82.55			
	Average	**81.4 ± 0.16**	**82.99 ± 0.28**	**82.34 ± 0.40**	**81**	**81**	-

[a]values from group additivity method; [b]ref. [32]; [c]Ref. [180]

Table 6.2: Calculated $\Delta_f H^0_{298}$ for Species

	Reactions Series	$\Delta_f H^0_{298}$ (kcal mol^{-1})					
		B3LYP	**G3MP2B3**	**G3**	**GA**[a]	**Hadad**[b]	**Lin**[c]
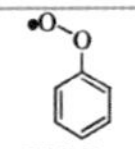 **PhOO•**	$C_6H_5OO\bullet + CH_3CH_2OH \rightarrow C_6H_5OH + CH_3CH_2OO\bullet$	31.73	32.10	32.07			
	$C_6H_5OO\bullet + CH_3OOH \rightarrow C_6H_5OOH + CH_3OO\bullet$	31.50	32.69	32.51			
	$C_6H_5OO\bullet + CH_2{=}CH_2 \rightarrow C_6H_5OOH + CH_2{=}CH\bullet$	31.80	32.40	31.86			
	$C_6H_5OO\bullet + CH_2{=}CHOOH \rightarrow C_6H_5OOH + CH_2{=}CHOO\bullet$	31.21	32.77	32.02			
	Average	**31.56 ± 0.26**	**32.49 ± 0.30**	**32.12 ± 0.28**	**31.52**	**36.11**	**34.7**
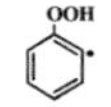 **Ph•OOH**	$C_6H_4\bullet OOH + CH_2{=}CH_2 \rightarrow C_6H_5OOH + CH_2{=}CH\bullet$	60.46	62.58	62.02			
	$C_6H_4\bullet OOH + C_6H_6 \rightarrow C_6H_5OOH + C_6H_5\bullet$	60.22	60.31	60.47			
	Average	**60.34 ± 0.17**	**61.44 ± 1.6**	**61.24 ± 1.1**	**58.22**	**-**	**-**
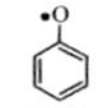 **PhO•**	$C_6H_5O\bullet + CH_4 \rightarrow C_6H_6 + CH_3O\bullet$	12.61	14.51	12.18			
	$C_6H_5O\bullet + CH_3OH \rightarrow C_6H_5OH + CH_3O\bullet$	14.84	15.30	13.28			
	$C_6H_5O\bullet + CH_2{=}CH_2 \rightarrow C_6H_5OH + CH_2{=}CH\bullet$	10.60	15.52	12.45			
	Average	**12.68 ± 2.12**	**15.11 ± 0.53**	**12.64 ± 0.57**	**13.17**	**12.07**	**12.35**
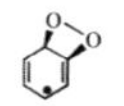 *Y(C6•)Y(C2O2)*	$Y(C_6H_5\bullet)Y(C2O2) + CH_4 \rightarrow Y(C_6H_7\bullet) + CH_2O_2$	78.15	82.37	81.98			
	$Y(C_6H_5\bullet)Y(C2O2) \rightarrow Y(C6\bullet O)DO$	79.31	76.33	74.51			
	$C_6H_5OO\bullet \rightarrow Y(C_6H_5\bullet)Y(C2O2)$	79.35	76.88	74.65			
	$Y(C_6H_5\bullet)Y(C2O2) \rightarrow Y(C_5H_5\bullet) + CO_2$	80.69	76.18	74.96			
	$Y(C_6H_5\bullet)Y(C2O2) + CH_3CH_3 \rightarrow Y(C_6H_7\bullet) + C_2H_4O_2$	79.74	82.37	81.45			
	Average	**79.44 ± 0.91**	**78.83 ± 3.24**	**77.51 ± 3.84**	**77.38**	**80**	**72.7**
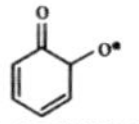 *O-ODY(C6)O•*	$O\text{-}O{=}Y(C_6H_5)O\bullet + H_2 \rightarrow CH_2{=}CHCH{=}CH\bullet + O{=}CHCH{=}O$	14.46	15.42				
	$O\text{-}O{=}Y(C_6H_5)O\bullet + H_2 \rightarrow CH_2{=}CHCH{=}CH_2 + O{=}CHCH\bullet{=}O$	14.68	16.57				
	$O\text{-}O{=}Y(C_6H_5)O\bullet \rightarrow Y(C_6H_5\bullet O){=}O$	17.98	17.46				
	$O\text{-}O{=}Y(C_6H_5)O\bullet \rightarrow C_6H_5OO\bullet$	18.16	18.01				
	Average	**16.32 ± 2.0**	**16.87 ± 1.13**		**15.04**	**18.7**	**15.9**
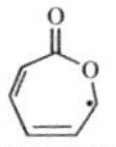 *Y(C6•O)DO*	$Y(C_6H_5\bullet O){=}O \rightarrow Y(C_5H_5\bullet) + CO_2$	-11.12	-12.58	-14.47			
	$Y(C_6H_5\bullet O){=}O + H_2 \rightarrow CH_2{=}CHCH{=}CH_2 + CH\bullet{=}O + CO$	-13.93	-12.15	-15.23			
	$C_6H_5OO\bullet \rightarrow Y(C_6H_5\bullet O){=}O$	-12.32	-11.89	-16.04			
	$Y(C_6H_5\bullet O){=}O \rightarrow Y(C_5H_5\bullet O) + CO$	-12.63	-13.14	-16.07			
	$Y(C_6H_5\bullet O){=}O + CH_4 \rightarrow Y(C_5H_6O) + CO + CH_3\bullet$	-16.76	-12.11	-13.71			
	Average	**-13.35 ± 2.15**	**-12.37 ± 0.49**	**-15.10 ± 1.02**	**-10.71**	**-11.57**	**-15.4**
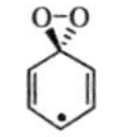 *Y(C6•)Y(CO2)*	$Y(C_6H_5\bullet)Y(CO_2) + CH_4 \rightarrow Y(C_6H_7\bullet) + Y(CH_2O_2)$	55.08	60.56	57.79			
	$C_6H_5OO\bullet \rightarrow Y(C_6H_5\bullet)Y(CO2)$	56.94	52.45	50.46			
	$Y(C_6H_5\bullet)Y(CO2) \rightarrow Y(C_5H_5\bullet) + CO_2$	58.15	51.76	49.70			
	$Y(C_6H_5\bullet)Y(CO2) \rightarrow Y(C_6H_5\bullet O){=}O$	55.91	51.96	49.06			
	Average	**56.52 ± 1.32**	**54.18 ± 4.27**	**51.75 ± 4.06**	**62.99**	**58.42**	**49.4**
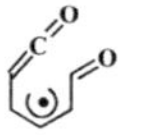 *ODC6•DO*	$O{=}C\bullet CH{=}CHCH{=}CHCH{=}O \rightarrow Y(C_5H_5\bullet) + CO_2$	3.96	7.35	6.19			
	$ODC6\bullet DO + CH_2CH_2 \rightarrow CH\bullet{=}O + CO + C_6H_8$	5.61	9.54	6.38			
	$ODC6\bullet DO + CH_3CH_3 \rightarrow CH\bullet{=}O + CH_2{=}CHCH{=}O + C_4H_6$	5.43	6.75	3.48			
	Average	**5.0 ± 0.9**	**7.88 ± 1.47**	**5.35 ± 1.62**	**8.0**	**4.83**	**4.34**
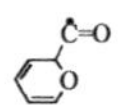 **Y(C5O)C•DO**	$Y(C_5H_5O)C\bullet{=}O + CH_3CH_3 \rightarrow CH\bullet{=}O + CH_2{=}CHCH{=}O + C_4H_6$			9.06			
	$Y(C_5H_5O)C\bullet{=}O + CH_3CH_3 \rightarrow CH_3CHCHCH_3 + CO + CH_2C\bullet CH{=}O$			9.16			
	$Y(C_5H_5O)C\bullet{=}O + H_2 \rightarrow Y(C_5H_6O) + CH\bullet{=}O$			9.43			
	$Y(C_5H_5O)C\bullet{=}O + CH_4 \rightarrow CH_3CH\bullet{=}O + CH_2{=}CHCH{=}CH_2 + CO$			9.99			
	Average	**10.56 ± 1.17**[d]	**11.03 ± 0.38**[e]	**9.41 ± 0.42**	**10.3**	**17.9**	**7.8**
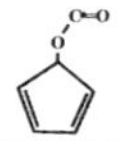 **Y(C5)OC•DO**	$Y(C_5H_5)OC\bullet{=}O + CH_4 \rightarrow CH_2{=}O + Y(C_5H_6) + CH\bullet{=}O$			7.32			
	$Y(C_5H_5)OC\bullet{=}O + CH_3CH_3 \rightarrow CH_2{=}CHOH + Y(C_5H_6) + CH\bullet{=}O$			7.68			
	$Y(C_5H_5)OC\bullet{=}O + \rightarrow Y(C_5H_5\bullet) + CO_2$			9.63			
	Average	**9.54 ± 1.17**[d]	**9.78 ± 0.19**[e]	**8.21 ± 1.24**	**10.40**	**17.2**	**8.5**
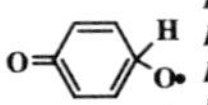 *P-ODY(C6)O•*	$P\text{-}ODY(C6)O\bullet + CH_2CH_2 \rightarrow CH_2CHC\bullet{=}O + CH_2C(CHO)CHCH_2$	9.85	7.75	10.00			
	$P\text{-}ODY(C6)O\bullet + 2H_2 \rightarrow CH_2{=}O + CH\bullet{=}O + CH_2{=}CHCH{=}CH_2$	10.92	11.20	15.66			
	$P\text{-}ODY(C6)O\bullet + H_2 \rightarrow CH_2{=}CHCH{=}O + CH_2{=}CHC\bullet{=}O$	9.14	8.33	11.49			
	$P\text{-}ODY(C6)O\bullet + H_2 \rightarrow CH_2{=}CHCH{=}O + CH\bullet{=}CHCH{=}O$	9.82	8.40	12.97			
	$P\text{-}ODY(C6)O\bullet \rightarrow$ *p*-quinone + H	10.72	11.13	15.31			
	$P\text{-}ODY(C6)O\bullet + 3H_2 \rightarrow CH_2{=}CH_2 + 2CH_2{=}O + CH_2{=}CH\bullet$	7.62	8.81	14.73			
	Average	**9.67 ± 1.2**	**9.27 ± 1.5**	**13.36 ± 2.27**	**9.02**	**18.4**	**10.3**

[a]values from group additivity method; [b]ref. [32]; [c]Ref. [180]; [d]chapter 4; [e]chapter 5

Table 6.2 (con't): Calculated $\Delta_f H^0_{298}$ for species

	Reactions Series		$\Delta_f H^0_{298}$ (kcal mol^{-1})					
			B3LYP	**G3MP2B3**	**G3**	**GA**[a]	**Hadad**[b]	**Lin**[c]
	Y(C_5H_5•)Y(C3O)=O	→ Y(C_5H_5•) + CO_2	-1.12	-11.07	-12.44			
	Y(C_6H_5•O)=O	→ Y(C_5H_5•)Y(C3O)=O	-3.35	-10.37	-13.07			
	C_6H_5OO•	→ Y(C_5H_5•)Y(C3O)=O	-3.35	-10.86	-11.67			
Y(C5•)Y(C3O)DO		**Average**	**-2.6 ± 1.28**	**-10.77 ± 0.36**	**-12.39 ± 0.70**	**-9.88**	**-2.5**	**-14**
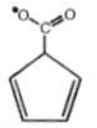	Y(C_5H_5)C(O•)=O	→ Y(C_5H_5•O) + CO_2	4.13					
	Y(C6•O)DO	→ Y(C_5H_5)C(O•)=O	6.37					
Y(C5)C(O•)DO	PhOO•	→ Y(C_5H_5)C(O•)=O	5.16					
		Average	**5.22 ± 1.12**	**5.78 ± 0.60**[c]	**5.16 ± 1.12**[d]	**2.76**	**5.1**	**9**
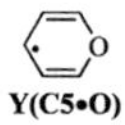	Y(C_5H_5•O) + CH_2=CH_2	→ O=CHCH=CHCH=CH_2 + CH_2=CH•	21.95	19.59	17.74			
	PhOO•	→ Y(C_5H_5•O) + CO	22.46	21.02	20.89			
Y(C5•O)	Y(C_5H_5•O)	→ O=CHCH=CHCH=CH•	21.69	20.86	17.72			
	Y(C_5H_5•O)	→ O=C•CH=CHCH=CH_2	21.86	19.80	17.83			
		Average	**21.99 ± 0.33**	**20.31 ± 0.73**	**18.54 ± 1.56**	**19.1**	**25.19**	**16.7**
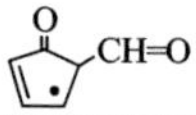	**ODY(C5•)CO**	→ O=Y(C_5H_4) + CH•=O	-10.51	-12.48	-17.07			
	ODY(C5•)CO + H_2 + CH_4	→ CH_2CHCH_2• + $CH_3C(O)CH_3$ + CO	-15.13	-13.37	-15.19			
ODY(C5•)CDO	**ODY(C5•)CO** + CH_4	→ Y(C_5H_6) + CH_2=O + CH•=O	-9.88	-12.94	-17.69			
		Average	**-11.84 ± 2.86**	**-12.93 ± 0.44**	**-16.65 ± 1.3**			
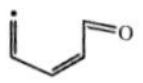	Y(C_5H_5•) + CH_2=CH_2	→ Y(C_5H_6) + CH_2=CH•	61.71	63.03	63.40			
	Y(C_5H_5•) + CH_3CH_3	→ Y(C_5H_6) + CH_3CH_2•	61.75	62.62	62.71			
	Y(C_5H_5•) + CH_4	→ Y(C_5H_6) + CH_3•	60.85	63.43	63.71			
C•C4DO	Y(C_5H_5•) + $CH_3CH_2CH_3$	→ Y(C_5H_6) + $CH_3CH_2CH_2$•	61.42	62.38	62.48			
		Average	**61.43 ± 0.41**	**62.87 ± 0.46**	**63.07 ± 0.57**	**61.59**	**66.46**	**60.25**
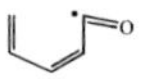	CH•CHCHCHCH=O + CH_3CH_3	→ C_4H_5CH=O + CH_3CH_2•	55.43	58.09	56.36			
	CH•CHCHCHCH=O + CH_4	→ CH•CHCHCH$_2$ + CH_3CH=O	55.21	59.11	57.34			
	CH•CHCHCHCH=O + CH_4	→ C_4H_5CH=O + CH_3•	54.54	58.91	57.35			
C4C•DO	CH•CHCHCHCH=O + CH_2=CH_2	→ C_4H_5CH=O + CH_2=CH•	55.40	58.51	57.05			
		Average	**55.15 ± 0.41**	**58.65 ± 0.45**	**57.03 ± 0.46**	**53.93**	-	-
	$CH_2CHCHCHC$•=O + CH_3CH_3	→ C_4H_5CH=O + CH_3CH_2•	33.05	33.33	33.21			
	$CH_2CHCHCHC$•=O + CH_4	→ $CH_2CHCHCH_2$ + CH_3C•=O	32.89	35.16	34.65			
CDCCDC•	$CH_2CHCHCHC$•=O + CH_2=CH_2	→ C_4H_5CH=O + CH_2=CH•	33.02	33.74	33.90			
	$CH_2CHCHCHC$•=O + CH_4	→ C_4H_5CH=O + CH_3•	32.16	34.14	34.21			
		Average	**32.78 ± 0.42**	**34.09 ± 0.78**	**33.99± 0.6**	**33.70**	-	-
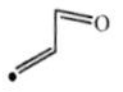	$CH_2CHCHCH$• + CH_3CH_3	→ $CH_2CHCHCH_2$ + CH_3CH_2•	87.04	87.68	86.01			
	$CH_2CHCHCH$• + CH_2=CH_2	→ $CH_2CHCHCH_2$ + CH_2=CH•	87.00	88.10	86.69			
C•DCCDO	$CH_2CHCHCH$• + CH_4	→ $CH_2CHCHCH_2$ + CH_3•	86.14	88.50	87.00			
		Average	**86.72 ± 0.5**	**88.09 ± 0.41**	**86.57 ± 0.5**	**85.10**		
	CH•=CHCH=O + CH_3CH_3	→ CH_2CHCH=O + CH_3CH_2•	41.63	41.90	41.87			
	CH•=CHCH=O + CH_2=CH_2	→ CH_2CHCH=O + CH•=CH_2	41.60	42.32	42.55			
Y(C5•)	CH•=CHCH=O + CH_3CH=CH_2	→ CH_2•CHCH$_2$ + CH_2CHCH=O	43.73	42.10	43.36			
	CH•=CHCH=O + CH_4	→ CH_2CHCH=O + CH_3•	40.74	40.74	42.86			
		Average	**41.92 ± 1.27**	**41.76 ± 0.7**	**42.66 ± 0.62**	**40.45**		
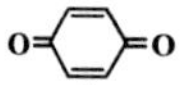	**p-quinone** + 2H_2	→ 2CH_2=CHCH=O	-40.92	-35.85	-34.67	-		
p-quinone	**p-quinone** + 2H_2	→ CH_2=CHCH=CH_2 + O=CHCH=O	-35.41	-29.64	-28.92			
	p-quinone + H_2	→ CH_2=CHCH=CH_2 + 2CO	-28.49	-27.27	-28.83			
		Average	**-34.47 ± 6.22**	**-30.92 ± 4.43**	**-30.81 ± 3.34**	**-34.57**	-	**-32.7**
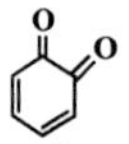	**o-quinone** + 2H_2	→ 2CH_2=CHCH=O	-33.83	-26.29	-26.71			
	o-quinone + 2CH_2=CH_2	→ 2CH_2=C(CH=O)CH=CH_2	-32.41	-26.50	-26.18			
o-quinone		**Average**	**-33.12 ± 1.0**	**-26.39 ± 0.15**	**-26.44 ± 0.37**	**-31.89**	-	**-24.9**
	**ODY(C5)** + CH_4	→ Y(C_5H_6) + CO	11.46	12.09	11.24			
	ODY(C5) + 2CH_4	→ CH_2=CHCH=CH_2 + $CH_3C(=O)CH_3$	10.20	12.99	12.48			
ODY(C5)		**Average**	**10.83 ± 0.89**	**12.54 ± 0.63**	**11.86 ± 0.87**	-	-	-

[a]values from group additivity method; [b]Ref. [32]; [c]Ref. [180]

6.2.1.3 Enthalpy of Formation of the Transition state structures

Enthalpies of transition state structures are determined with the above composite ab initio and DFT methods. Enthalpies of transition states structures are calculated as the difference between the calculated value of the TS structure and the value of the stable radical adduct(s) (adjacent product and reactant where both are a single species) at the corresponding levels of calculation. The computational methods for enthalpies are G3MP2B3 and G3 (whenever possible), and DFT. Zero-point energies (ZPVEs) and thermal corrections to 298.15 K are from DFT. Table 6.3 contains the nineteen transition state structures and the enthalpy, relative to reactant and product adducts. The calculation of $\Delta_f H^0_{298}$ (TS) is performed with the help of isodesmic reactions as it has been done for the stable species. This table provides the identification of the reaction path for each transition state structure as well. To test our method, two well known transition states structures have been calculated; TSC_2H_3 which is the TST structure of the hydrogen elimination of vinyl to acetylene CH_2=CH• → HC≡CH + H [186, 187] and TSC•DO which is the TST structure of the reaction CH•=O → CO + H [188, 186]. For these two reactions kinetic parameters are calculated and compared to the literature.

The different calculation methods result in reasonably good agreement for a given transition state structure with a few exceptions. The B3LYP enthalpy value for the intramolecular abstraction, hydrogen transfer from the phenyl ring to the peroxy oxygen radical (TSPH•OOH) differs by some 10 kcal mol^{-1} from that calculated at the G3MP2B3 level. The optimized structures are however very similar. We recommend the average value resulting from the G3MP2B3 and the B3LYP calculations as 73.6 kcal mol^{-1} because for abstraction calculations of TST structures, MP2 calculations (from G3MP2B3) are always high and DFT are low.

Table 6.3: Identification of the Reaction Path and Calculated $\Delta_f H^0_{298}$ for Transition State Structures

	Reactions Series	$\Delta_f H^0_{298}$ (kcal mol^{-1})		
		B3LYP	**G3MP2B3**	**G3**
Identification path: C_6H_5OO• → TSPH•OOH → C_6H_4•OOH				
TSPH•OOH 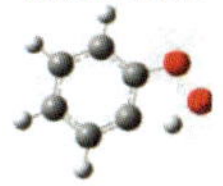	C_6H_5OO• → **TSPH•OOH**	69.92	79.30	
	TSPH•OOH → C_6H_4•OOH	67.54	77.38	
	Average	**68.73 ± 1.68**	**78.34 ± 1.36**	
Identification path: C_6H_5OO• → TSY(C6•)Y(C2O2) → Y(C6•O)DO				
TSY(C6•)Y(C2O2) 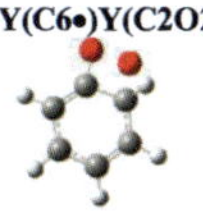	C_6H_5OO• → **TSY(C6•)Y(C2O2)**	76.13	77.23	78.01
	TSY(C6•)Y(C2O2) → Y(C6•O)DO	79.20	79.74	80.50
	TSY(C6•)Y(C2O2) → Y(C6•)Y(C2O2)	80.30	82.17	76.61
	Average	**78.54 ± 2.16**	**79.71 ± 2.47**	**78.37 ± 1.97**

[a]it is recognized that these saddle are transient species

Table 6.3 (con't): Identification of the Reaction Path and Calculated $\Delta_f H^0_{298}$ for Transition State Structures (saddle points)

	Reactions Series	$\Delta_f H^0_{298}$ (kcal mol^{-1}) B3LYP	G3MP2B3	G3
Identification path: $C_6H_5OO\bullet$ → TSY(C6•)Y(CO2) → Y(C6•O)DO				
	$C_6H_5OO\bullet$ → **TSY(C6•)Y(CO2)**	57.05	58.96	57.37
	TSY(C6•)Y(CO2) → Y(C6•O)DO	60.13	61.47	59.22
	TSY(C6•)Y(CO2) → ODC6•DO	60.13	61.79	61.91
TSY(C6•)Y(CO2)	**TSY(C6•)Y(CO2)** → Y(C6•)Y(CO2)	60.73	63.69	63.12
	Average	**59.51 ± 1.66**	**61.48 ± 1.94**	**60.41 ± 2.59**
Identification path: CH•=CHCH=CHCH=O → TSC•DC-C3DO → CH•=CHCH=O + CH≡CH				
TSC•DC-C3DO 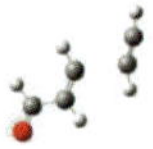	CH•=CHCH=CHCH=O → **TSC•DC-C3DO**	96.03	99.63	100.85
	TSC•DC-C3DO → CH•=CHCH=O + CH≡CH	101.73	100.09	101.59
	Average	**98.88 ± 4.03**	**99.86 ± 0.32**	**101.22 ± 0.52**
Identification path: CH_2=CHCH=CH•→ TSCDC-CDC• → HC≡CH + CH•=CH_2				
TSCDC-CDC• 	CH_2=CHCH=CH•→ **TSCDC-CDC•**	125.61	129.07	129.64
	TSCDC-CDC• → HC≡CH + CH•=CH_2	131.97	130.36	130.78
	Average	**128.79 ± 4.49**	**129.72 ± 0.91**	**130.21 ± 0.80**
Identification path: CH_2=CHC•=O → TSCDC-C•DO → CH_2=CH• + CO				
	CH_2=CHC•=O → **TSCDC-C•DO**	41.31	42.50	45.11
	TSCDC-C•DO → CH_2=CH• + CO	46.48	46.48	47.60
TSCDC-C•DO	Average	**43.89 ± 3.65**	**44.49 ± 2.81**	**46.35 ± 1.76**
Identification path: CH•=CHCH=O → TSC•DC-CDO1 → CO + CH_2=CH•				
TSC•DC-CDO1 	CH•=CHCH=O→ **TSC•DC-CDO1**	41.78	43.04	45.41
	TSC•DC-CDO1 → CO + CH_2=CH•	46.42	46.36	47.64
	Average	**44.1 ± 3.28**	**44.70 ± 2.35**	**46.52 ± 1.57**
Identification path: CH•=CHCH=O → TSC•DC-CDO2 → CH•=O + CH≡CH				
TSC•DC-CDO2 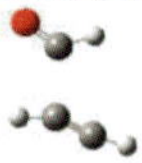	CH•=CHCH=O→ **TSC•DC-CDO2**	64.99	69.1	71.07
	TSC•DC-CDO2 → CH•=O + CH≡CH	73.12	72.56	73.4
	Average	**69.06 ± 5.75**	**70.83 ± 2.44**	**72.24 ± 1.64**
Identification path: CH•=O → TSC•DO → CO + H•				
TSC•DO 	CH•=O → **TSC•DO**	26.26	26.37	27.63
	TSC•DO → CO + H•	26.13	26.39	27.21
	Average	**26.19 ± 0.09**	**26.38 ± 0.01**	**27.42 ± 0.29**
Identification path: CH_2=CH• → TSC2H3 → CH≡CH + H•				
TSC2H3 	CH_2=CH• → **TSC2H3**	106.3	107.89	108.79
	TSC2H3 → CH≡CH + H•	109.66	108.06	108.48
	Average	**107.98 ± 2.37**	**107.97 ± 0.12**	**108.64 ± 0.22**

[a]it is recognized that these saddle are transient species

Table 6.3 (con't): Calculated $\Delta_f H^0_{298}$ for Transition State Structures (saddle points)[a]

Reactions Series		$\Delta_f H^0_{298}$ (kcal mol^{-1})	
		B3LYP	**G3MP2B3 /G3[b]**
Identification path: O=CHCH=CHCH=CHC•=O → TSODC•-C5DO → O=CHCH=CHCH=CH• + CO			
TSODC•-C5DO 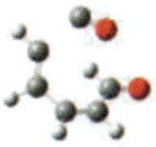	O=CHCH=CHCH=CHC•=O → **TSODC•-C5DO**	37.12	40.21
	TSODC•-C5DO → O=CHCH=CHCH=CH• + CO	37.39	39.8
	Average	**37.26 ± 0.19**	**40.0[b]± 0.29**
Identification path: •CH=CHCH=CHCH=O → TSC•C4DO → CH_2=CHCH=CHC•=O			
TSC•C4DO(abstr.) 	•CH=CHCH=CHCH=O → **TSC•C4DO**	57.91	
	TSC•C4DO → CH_2=CHCH=CHC•=O	57.92	
	Average	**57.91 ± 0.007**	
Identification path: CH_2=CHCH=CHC•=O → TSC4-C•DO → CH_2=CHCH=CH• + CO			
TSC4-C•DO 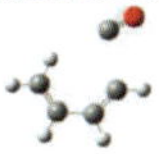	CH_2=CHCH=CHC•=O → **TSC4-C•DO**	61.04	65.17
	TSC4-C•DO → CH_2=CHCH=CH• + CO	65.00	67.53
	Average	**63.02 ± 2.8**	**66.35 ± 1.66**
Identification path: CH•=CHCH=O → TSCDCC•DO → CH_2=CHCH•=O			
TSCDCC•DO (abst.) 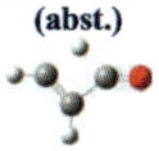	CH•=CHCH=O → **TSCDCC•DO**	66.89	71.09
	TSCDCC•DO → CH_2=CHCH•=O	66.36	70.44
	Average	**66.62 ± 0.37**	**70.76 ± 0.46**
Identification path: ODC6•DO → TSO-ODY(C6)O• → *o*-O=Y(C6)O•			
TSO-ODY(C6)O• 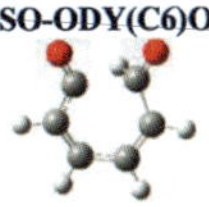	ODC6•DO → **TSO-ODY(C6)O•**	20.79	21.01
	TSO-ODY(C6)O• → *o*-O=Y(C6)O•	16.70	20.04
	Average	**18.75 ± 2.89**	**20.53 ± 0.68**
Identification path: ODC6•DO → TSODYC5•CDO → O=Y(C5•)CH=O			
TSODYC5•CDO 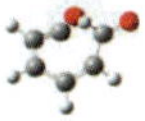	ODC6•DO → **TSODYC5•CDO**	10.39	15.79
	TSODYC5•CDO → O=Y(C5•)CH=O	-	14.78
	Average	**10.39**	**15.29 ± 0.71**
Identification path: O=Y(C5•)CH=O → TSODYC5•-CDO → O=Y(C5) + CH•=O			
TSODYC5•-CDO 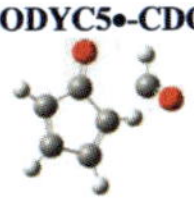	O=Y(C5•)CH=O → **TSODYC5•-CDO**	20.64	25.07
	TSODYC5•-CDO → O=Y(C5) + CH•=O	21.96	25.52
	Average	**21.30 ± 0.93**	**25.30 ± 0.32**
Identification path: *o*-O=Y(C6)O• → TSODYC6O• → *o*-quinone + H			
TSODYC6O• 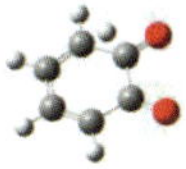	*o*-O=Y(C6)O• → **TSODYC6O•**	**26.02**	**29.97**
Identification path: Y(C5O)C•=O → TSY(C5O) –C•DO → Y(C5•O) + CO			
TSY(C5O)-C•DO 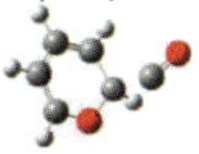	Y(C5O)C•=O → **TSY(C5O) -C•DO**	9.51	14.55
	TSY(C5O) -C•DO → Y(C5•O) + CO	12.79	15.04
	Average	**11.15 ± 2.31**	**14.79 ± 0.34**

[a]it is recognized that these saddle are transient species; [b]G3 calculation

6.2.2 Entropy and Heat Capacity

6.2.2.1 Entropies and Heat Capacities of Stable Molecules and Radicals

Entropy and heat capacities are calculated using the rigid-rotor-harmonic-oscillator approximation, based on:

i. The calculated frequencies and moments of inertia of the optimised B3LYP/6-311G(d,p) structures.

ii. The frequencies and moments of inertia given by Fadden et al [32].

iii. The Group additivity method (GA).

S^0_{298} and $Cp_{298}(T)$ for radicals and stable species are listed in Table 6.4. The statistical mechanics calculations are performed with the "SMCPS" computer program [74]. The DFT calculations are chosen because they are obtained with a larger basis set (6-311G(d,p)) than the G3MP2B3 (B3LYP/6-31G(d)) and the G3 (HF/6-31G(d)) methods.

Based on geometries and frequencies reported by Fadden et al. in their published supplemental tables [32], we recalculated S^0_{298} and $Cp_{298}(T)$ by use of the "SMCPS" code. The comparison of our calculated S^0_{298} and $Cp_{298}(T)$ values with those resulting from Fadden et al. data and with and group additivity (GA) results are given in Table 6.4.

The DFT determined S^0_{298} and $Cp_{298}(T)$ are generally in agreement with the entropies and heat capacities obtained from the vibration frequency and structure data reported by Fadden et al. For species for which no data from the literature or from GA are available (ODY(C5•)CDO and ODY(C5)), we have evaluated the entropy and heat capacities from the structures and the frequency data obtained from G3MP2B3 calculations. These values are included in the Table 6.4 for comparison.

We have seen above that the DFT method failed for the calculation of the enthalpy of formation of the Y(C5•)Y(C3O)DO radical. For this reason, the entropy and heat capacities values of this radical were verified with the G3 method. The DFT and G3 entropy values show good agreement with a difference of only 1 cal mol^{-1} K^{-1}. The comparison of heat capacities shows a higher but still reasonable discrepancy within 1.1- 2.4 cal mol^{-1} K^{-1}.

The comparison of heat capacities, $Cp_{298}(T)$ between DFT calculations and GA are all within 1- 5 cal mol^{-1} K^{-1}. Only the heat of capacities of the ODC6•DO radical differs significantly from the DFT calculations by around 9 cal mol^{-1} K^{-1}. For this radical G3MP2B3 calculations have been performed to calculate $Cp_{298}(T)$ and to compare these to the DFT and GA results. The DFT and G3MP2B3 heat capacities values show good agreement. These results indicate that all GA heat capacities values have to be re-evaluated.

Table 6.4: Calculated Thermochemical Properties of Radicals and Stable Species[a]

	Species		$\Delta_f H^0_{298}$ [b]	S^0_{298} [b]	Cp (T) cal/mole K						
					300 K	**400 K**	**500 K**	**600 K**	**800 K**	**1000 K**	**1500 K**
	Ph•	**DFT**[e]	81.40	68.84	19.39	26.03	31.58	36.01	42.42	46.80	53.18
	(1)[c](2)[d]	**Hadad**[f]	81.00	68.82	19.31	25.90	31.43	35.85	42.27	46.67	53.08
		GA	80.70	69.28	19.03	25.46	30.83	35.11	41.49	45.98	53.14
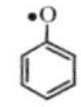	PhO•	**DFT**[e]	12.68	73.89	23.06	30.25	36.19	40.91	47.68	52.27	58.8
	(1)[c](2)[d]	**Hadad**[f]	12.07	75.21	22.99	30.13	36.05	40.76	47.54	52.14	58.80
		GA	13.17	73.49	24.04	32.05	38.27	42.91	49.33	53.56	60.27
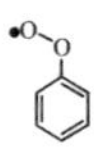	PhOO•	**DFT**[e]	31.28	77.53	24.46	32.17	38.55	43.59	50.79	55.63	62.55
	(1)[c](1)[d]		C_b—OO	5.61	2.10	2.19	2.21	2.18	2.01	1.81	1.47
		Total		**83.14**	**26.46**	**34.36**	**40.76**	**45.77**	**52.80**	**57.44**	**64.02**
		Hadad[f]	36.11	82.49	26.37	34.04	40.40	45.43	52.64	57.49	64.45
		GA	**31.52**	**86.20**	**26.11**	**33.30**	**39.14**	**43.76**	**50.51**	**55.25**	**62.30**
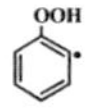	Ph•OOH	**DFT**	TVR	76.73	24.49	31.81	37.83	42.58	49.37	53.95	60.60
	(1)[c](2)[d]		Cb—OOH	6.11	1.89	1.87	1.78	1.66	1.44	1.25	0.87
			O—OH	3.38	1.41	1.44	1.47	1.49	1.50	1.49	1.41
		Total	**60.34**	**86.22**	**27.79**	**35.12**	**41.08**	**45.73**	**52.31**	**56.69**	**62.88**
		GA	**58.22**	**86.08**	**27.75**	**34.96**	**40.76**	**45.16**	**51.48**	**55.74**	**62.14**
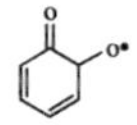	O-ODY(C6)O•	**DFT**	16.32	83.09	28.03	35.49	41.57	46.38	53.35	58.08	64.85
	(1)[c](1)[d]	**GA**	15.04	85.53	25.60	35.26	42.87	48.64	56.63	62.04	--
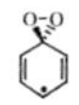	Y(C6•)Y(CO2)	**DFT**	56.52	80.10	27.32	35.27	41.62	46.55	53.51	58.16	64.82
	(1)[c](1)[d]	**Hadad**[f]	58.42	80.03	27.21	35.11	41.45	46.38	53.36	58.03	64.73
		GA	**62.99**	**79.31**	**25.87**	**34.71**	**41.84**	**47.11**	**54.63**	**59.33**	
	Y(C6•O)DO	**DFT**	-13.35	82.68	27.12	34.66	40.84	45.73	52.81	57.62	64.53
	(1)[c](1)[d]	**Hadad**[f]	-11.57	82.49	27.00	34.49	40.64	45.54	52.63	57.46	64.43
		GA	-10.71	84.61	28.43	35.18	40.67	45.05	51.79	57.27	
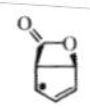	Y(C5•)Y(C3O)DO	**DFT**	-2.6	79.15	25.85	34.02	40.63	45.76	52.99	57.81	64.67
	(1)[c](1)[d]	**Hadad**[f]	-2.53	79.17	25.76	33.86	40.44	45.56	52.81	57.65	64.56
		G3	-12.39	78.09	23.92	31.62	38.17	43.42	51.01	56.14	63.57
		GA	-9.88	80.34	21.96	30.35	38.08	43.78	51.98	57.02	
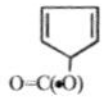	Y(C5)CO•DO	**DFT**	5.22	84.10	27.84	35.46	41.67	46.53	53.44	58.08	64.76
	(1)[c](1)[d]	**Hadad**[f]	5.1	83.42	27.43	35.08	41.33	46.23	53.20	57.88	64.63
		GA	2.76	83.19	26.77	33.95	40.06	44.86	51.71	56.23	--
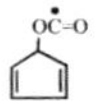	Y(C5)OC•DO	**DFT**	9.54	89.56	27.93	34.91	40.83	45.60	52.54	57.23	63.79
	(1)[c](1)[d]	**Hadad**[f]	17.25	85.26	27.49	34.87	40.99	45.85	52.86	57.62	64.50
		GA	10.40	90.83	29.22	35.98	41.89	46.27	53.31	58.00	.00
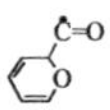	Y(C5O)C•DO	**DFT**	10.56	84.24	28.02	35.32	41.32	46.08	53.01	57.74	64.58
	(1)[c](1)[d]	**Hadad**[f]	17.94	84.33	27.93	35.18	41.15	45.92	52.85	57.60	64.48
		GA	10.30	82.86	27.72	36.42	43.25	47.90	54.53	58.67	
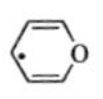	Y(C5•O)	**DFT**	21.99	71.94	20.88	27.45	32.81	37.02	43.08	47.23	53.35
	(1)[c](1)[d]	**Hadad**[f]	25.2	71.97	20.87	27.38	32.70	36.90	42.95	47.11	53.26
		GA	19.10	69.29	18.28	25.56	31.86	36.78	43.47	47.47	
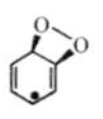	Y(C6•)Y(C2O2)	**DFT**	79.44	79.14	26.60	34.46	40.61	45.19	51.07	54.40	58.17
	(1)[c](1)[d]	**Hadad**[f]	80.36	79.12	26.55	34.53	40.99	46.02	53.17	57.95	64.75
		GA	83.30	78.73	21.57	29.21	35.64	40.62	48.08	52.36	
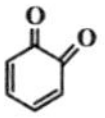	o-quinone	**DFT**	-33.12	81.30	25.96	32.70	38.25	42.69	49.14	53.51	59.70
		GA	-31.89	80.29	23.94	31.16	37.34	42.18	49.23	54.34	-

[a]Thermodynamic properties are referred to a standard state of an ideal gas at 1 atm; [b] ΔH^0_{f298} in kcal mol^{-1}; S^0_{298} in cal mol^{-1} K^{-1}; [c]optical isomers number; [d]symmetry number; [e]Density Functional Theory at B3LYP/6-311G(d,p); [e]from chapter 3; [f]based on Hadad's calculated Gibbs energies ref.[32].

Table 6.4 (con't): Calculated Thermochemical Properties of Radicals and Stable Species

	Species		$\Delta_f H^0_{298}$ [b]	S^0_{298} [b]	Cp (T) cal mol^{-1} K^{-1} 300 K	400 K	500 K	600 K	800 K	1000 K	1500 K
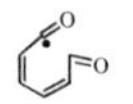	ODC6•DO (1)[c](1)[d]	**DFT**	5.0	80.75	29.15	35.70	41.17	45.63	52.27	56.91	63.70
		Hadad[e]	4.84	80.78	29.13	35.63	41.07	45.50	52.13	56.78	63.60
		G3MP2B3	7.14	80.78	29.13	35.63	41.07	45.50	52.13	56.78	63.60
		GA	8.00	83.74	37.41	44.22	49.99	54.24	60.61	65.72	-
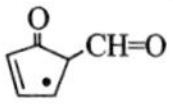	ODY(C5•)CDO (1)[c] (1)[d]	**DFT**	TVR	72.12	24.86	32.22	38.31	43.16	50.18	54.95	61.77
			IR	3.32	2.14	2.65	2.98	3.12	3.06	2.80	2.15
		Total	**-11.84**	**75.44**	**27**	**34.87**	**41.29**	**46.28**	**53.24**	**57.75**	**63.92**
		G3MP2B3	TVR	72.12	24.79	32.09	38.15	42.99	50.01	54.79	61.66
			IR	3.32	2.14	2.65	2.98	3.12	3.06	2.80	2.15
		Total	**-12.93**	**75.44**	**26.93**	**34.74**	**41.13**	**46.11**	**53.07**	**57.59**	**63.81**
		GA	**-12.36**	85.05	25.27	33.28	40.14	45.28	52.96	58.11	-
	Y(C5•) (1)[c] (2)[d]	**DFT**	61.43	71.67	19.11	24.89	29.61	33.30	38.63	42.33	47.90
		GA	62.9	66.61	18.31	24.78	30.05	34.13	39.89	43.80	49.84
	ODY(C5) (1)[c] (1)[d]	**B3LYP**	10.84	63.58	18.94	24.76	29.49	33.21	38.57	42.23	47.56
		G3MP2B3	12.54	63.55	18.86	24.64	29.35	33.06	38.43	42.10	47.47
	C•C4DO (1)[c] (2)[d]	**DFT**	55.15	79.91	24.18	29.51	34.09	37.85	43.55	47.59	53.60
		GA	53.93	81.86	25.84	31.80	36.35	39.79	44.69	48.24	53.03
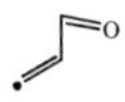	C•DCCDO (1)[c] (2)[d]	**DFT**	TVR	63.92	14.08	17.15	19.76	21.90	25.18	27.52	31.05
			IR	3.32	2.14	2.65	2.98	3.12	3.06	2.80	2.15
		Total	**41.92**	**67.24**	**16.22**	**19.80**	**22.74**	**25.02**	**28.24**	**30.32**	**33.20**
		GA	40.45	69.10	16.92	20.07	22.85	24.95	27.99	30.02	32.85
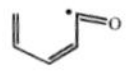	C4C•DO (1)[c] (2)[d]	**DFT**	TVR	74.81	21.97	27.20	31.73	35.48	41.18	45.26	51.37
			IR	3.63	2.39	2.98	3.35	3.45	3.24	2.85	2.08
		Total	**32.78**	**78.44**	**24.36**	**30.18**	**35.08**	**38.93**	**44.42**	**48.11**	**53.45**
		G3MP2B3	TVR	74.98	22.09	27.36	31.91	35.67	41.37	45.43	51.49
			IR	3.63	2.39	2.98	3.35	3.45	3.24	2.85	2.08
			33.35	**78.61**	**24.48**	**30.34**	**35.26**	**39.12**	**44.61**	**48.28**	**53.57**
	CDCCDC• (1)[c] (2)[d]	**DFT**	86.72	67.41	18.08	22.46	26.15	29.17	33.76	37.11	42.32
		GA	85.10	67.99	18.93	23.55	27.16	29.92	34.02	37.27	42.03

[a]Thermodynamic properties are referred to a standard state of an ideal gas at 1 atm; [b] ΔH^0_{f298} in kcal mol^{-1}; S^0_{298} in cal mol^{-1} K^{-1}; [c]optical isomers number; [d]symmetry number; [e]Density Functional Theory at B3LYP/6-311G(d,p); [e]from chapter 3; [f]based on Hadad's calculated Gibbs energies ref.[32].

Table 6.5: Thermochemical Properties[a] of $Y(C_3H_4O)$=O and $Y(C_6H_8O)$ to use for Group Additivity

	Species		$\Delta_f H^0_{298}$ [b]	S^0_{298} [b]	Cp (T) cal mol^{-1} K^{-1} 300 K	400 K	500 K	600 K	800 K	1000 K	1500 K
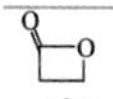	$Y(C_3H_4O)$=O (1)[c](1)[d]	G3	-66.80	68.73	15.46	19.87	23.93	27.38	32.67	36.43	42.04
	$Y(C_6H_8O)$ (1)[c] (1)[d]	G3MP2B3	-1.93	80.97	25.70	34.00	41.31	47.37	56.56	63.06	72.69
	New ring groups derived from Y(C3O)DO and Y(C6O) for GA										
GA	CY/C3O/DO		23.79	30.89	-5.74	-5.69	-5.07	-4.41	-3.73	-3.26	.00
	CY/C6O/DE/13		-2.05	22.34	-4.14	-4.54	-4.49	-4.19	-3.34	-2.37	-1.4

[a]Thermodynamic properties are referred to a standard state of an ideal gas at 1 atm; [b] ΔH^0_{f298} in kcal mol^{-1}; S^0_{298} in cal mol^{-1} K^{-1}; [c]optical isomers number; [d]symmetry number.

The entropy and heat capacity of $Y(C_3H_4O)$=O and $Y(C_6H_8O)$ are calculated in this work (and summarized in Table 6.5) to be used to determine GA values of two unknown cyclic groups, CY/C3O/DO and CY/C6/DE/13. These two cyclic groups are needed to estimate the entropy and heat capacities of the radicals Y(C5•)Y(C3O)DO and Y(C6•O)DO according to the method developed by Benson [20].

6.2.2.2 Entropies and Heat Capacities of Transition State Structures

The entropies and heat capacities of transition state structures are calculated and reported in this section. S^0_{298} and $Cp_{298}(T)$ are obtained with the SMCPS program for which the imaginary frequency is not considered. These entropies and heat capacities along with the corresponding enthalpies of formation were then converted to the NASA polynome format. The "ThermKin" code used to determine the kinetic parameters of the elementary reactions needs these NASA polynomials as input. Entropies and heat capacities were calculated from the Density Functional Theory calculations and compared to those obtained from G3MP2B3 and G3 methods. The entropies and heat capacities values of the transition state structures from the different calculation methods are listed in Table 6.6 along with the identification of the reaction path. Overall, the calculated entropies and heat capacities show good agreement among the methods used.

Table 6.6: Calculated Thermochemical Properties of the Transition State Structures[a]

Species		$\Delta_f H^0_{298}$ [b]	S^0_{f298}	Cp (T) cal/mole K						
				300 K	**400 K**	**500 K**	**600 K**	**800 K**	**1000 K**	**1500 K**
Identification path: $C_6H_5OO•$ → TSPH•OOH → $C_6H_4•OOH$										
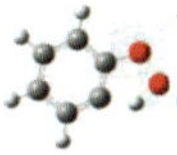 TSPH•OOH	DFT	68.71	78.07	26.29	34.04	40.37	45.34	52.38	57.02	63.48
(1)[c](2)[d]	G3MP2B3	78.52	78.14	26.29	33.99	40.28	45.24	52.28	56.92	63.41
Identification path: $C_6H_5OO•$ → TSY(C6•)Y(C2O2) → Y(C6•O)DO										
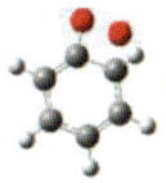 TSY(C6•)Y(C2O2)	DFT	78.54	67.68	24.30	32.10	38.42	43.37	50.39	55.10	61.83
(1)[c](2)[d]	G3MP2B3	79.72	67.64	24.18	31.93	38.24	43.19	50.23	54.95	61.73
	G3	78.37	67.98	23.59	30.92	37.09	42.03	49.18	54.02	61.06
Identification path: $C_6H_5OO•$ → TSY(C6•)Y(CO2) → Y(C6•O)DO										
 TSY(C6•)Y(CO2)	DFT	59.51	69.19	24.92	32.48	38.67	43.53	50.46	55.11	61.80
(1)[c] (2)[d]	G3MP2B3	61.48	69.27	24.84	32.35	38.51	43.37	50.30	54.97	61.70
	G3	60.41	68.15	23.90	31.23	37.37	42.26	49.30	54.07	61.04
Identification path: O=CHCH=CHCH=CHC•=O → TSODC•-C5DO → O=CHCH=CHCH=CH• + CO										
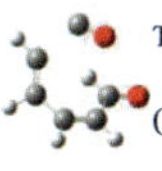 TSODC•-C5DO	DFT	37.26	84.88	30.69	36.68	41.72	45.79	51.84	56.07	62.31
(1)[c](2)[d]	G3	40	81.57	29.13	34.87	39.82	43.92	50.13	54.55	61.24

[a]Thermodynamic properties are referred to a standard state of an ideal gas at 1 atm;
[b] ΔH^0_{f298} in kcal mol^{-1}; S^0_{298} in cal mol^{-1} K^{-1};
[c]optical isomer number; [d]symmetry number; [e]Density Function Theory at B3LYP/6-311G(d,p);

Table 6.6 (con't): Calculated Thermochemical Properties of Transition State Structures[a]

Species		$\Delta_f H^0_{298}$ [b]	S^0_{f298}	Cp (T) cal/mole K 300 K	400 K	500 K	600 K	800 K	1000 K	1500 K
Identification path: ODC6•DO → TSO-ODY(C6)O• → *o*-O=Y(C6)O•										
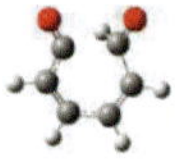 TSO-ODY(C6)O•	DFT	18.75	71.73	26.36	33.30	39.05	43.68	50.49	55.17	61.91
(1)[c](2)[d]	G3MP2B3	20.53	71.86	26.33	33.17	38.87	43.48	50.29	54.99	61.79
Identification path: ODC6•DO → TSODYC5•CDO → O=Y(C5•)CH=O										
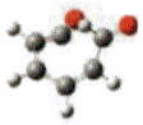 TSODYC5•CDO	DFT	10.39	73.88	27.01	33.67	39.22	43.70	50.34	54.96	61.71
(1)[c](2)[d]	G3MP2B3	15.29	73.64	26.98	33.73	39.33	43.83	50.49	55.10	61.81
Identification path: O=Y(C5•)CH=O → TSODYC5•-CDO → O=Y(C5) + CH•=O										
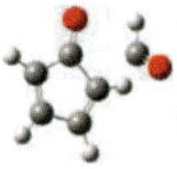 TSODYC5•-CDO	DFT	21.20	81.33	28.73	35.33	40.71	45.00	51.31	55.68	62.09
(1)[c](2)[d]	G3MP2B3	25.30	81.32	28.68	35.20	40.55	44.83	51.14	55.53	61.98
Identification path: C_6H_5OO• → CH•=CHCH=CHCH=O → TSC•DC-C3DO → CH•=CHCH=O + CH≡CH										
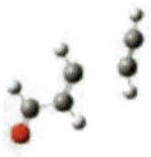 TSC•DC-C3DO	DFT[e]	98.91	82.00	24.95	29.66	33.44	36.48	41.12	44.54	49.91
(1)[c](2)[d]	G3MP2B3	99.86	82.88	25.08	29.70	33.42	36.42	41.02	44.43	49.81
Identification path: *o*-O=Y(C6)O• → TSODYC6O• → *o*-quinone + H										
 TSO-ODY(C6)O•	DFT	26.02	70.24	25.63	33.08	39.20	44.06	51.05	55.73	62.30
(1)[c](2)[d]	G3MP2B3	29.97	70.03	25.46	32.86	38.98	43.85	50.87	55.58	62.20
Identification path: •CH=CHCH=CHCH=O → TSC•C4DO → CH_2=CHCH=CHC•=O										
 TSC•C4DO(abstr) (1)[c](2)[d]	DFT	57.94	75.21	22.98	28.63	33.35	37.17	42.81	46.70	52.33
Identification path: CH_2=CHCH=CHC•=O → TSC4-C•DO → CH2=CHCH=CH• + CO										
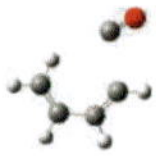 TSC4-C•DO	DFT	63.02	85.04	26.08	31.23	35.48	38.80	43.59	46.88	51.80
(1)[c](2)[d]	G3MP2B3	66.35	85.55	26.30	31.10	35.32	38.63	43.42	46.72	51.69
Identification path: CH•=CHCH=O → TSCDCC•DO → CH_2=CHCH•=O										
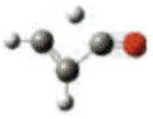 TSCDCC•DO	DFT	66.62	65.08	14.88	18.05	20.74	22.93	26.20	28.45	31.67
(abst.) (1)[c](2)[d]	G3MP2B3	70.76	65.14	14.84	17.99	20.67	22.86	26.13	28.38	31.62
Identification path: CH_2=CHCH=CH•→ TSCDC-CDC• → HC≡CH + CH•=CH_2										
 TSCDC-CDC•	DFT	128.80	70.40	19.33	22.89	25.68	27.93	31.44	34.17	38.72
(1)[c](2)[d]	G3MP2B3	129.72	77.27	21.50	24.97	27.70	29.90	33.37	36.07	40.62
	G3	130.20	76.29	21.26	24.70	27.48	29.71	33.16	35.81	40.33

[a]Thermodynamic properties are referred to a standard state of an ideal gas at 1 atm;
[b] ΔH^0_{f298} in kcal mol^{-1}; S^0_{f298} in cal mol^{-1} K^{-1};
[c]optical isomer number; [d]symmetry number; [e]Density Function Theory at B3LYP/6-311G(d,p);

Table 6.6 (con't): Calculated Thermochemical Properties of Transition State Structures[a]

Species		$\Delta_f H^0_{298}$ [b]	S^0_{f298}	Cp (T) cal/mole K						
				300 K	**400 K**	**500 K**	**600 K**	**800 K**	**1000 K**	**1500 K**
Identification path: CH_2=CHC•=O → TSCDC-C•DO → CH_2=CH• + CO										
 TSCDC-C•DO	DFT	43.9	74.04	17.87	19.99	21.83	23.41	25.98	27.97	31.18
(1)[c](2)[d]	G3MP2B3	44.5	75.51	17.99	20.04	21.84	23.39	25.93	27.92	31.13
	G3	46.36	70.04	16.62	18.89	20.87	22.55	25.25	27.31	30.69
Identification path: CH•=CHCH=O → TSC•DC-CDO1 → CO + CH_2=CH•										
 TSC•DC-CDO1	DFT	44.10	73.55	17.84	19.98	21.83	23.41	25.98	27.97	31.18
(1)[c](2)[d]	G3MP2B3	44.70	75.12	18.00	20.05	21.84	23.39	25.93	27.92	31.13
	G3	46.53	69.86	16.57	18.84	20.84	22.53	25.23	27.30	30.68
Identification path: CH•=CHCH=O → TSC•DC-CDO2 → CH•=O + CH≡CH										
 TSC•DC-CDO2	DFT	69.06	68.96	17.16	19.33	21.02	22.42	24.69	26.48	29.41
(1)[c](2)[d]	G3MP2B3	70.83	69.72	17.33	19.42	21.06	22.42	24.65	26.42	29.36
	G3	72.24	73.43	19.14	21.24	22.88	24.23	26.43	28.16	31.11
Identification path: CH•=O → TSC•DO → CO + H•										
 TSC•DO	DFT	26.19	56.77	9.72	9.86	10.01	10.19	10.55	10.86	11.34
(1)[c](1)[d]	G3MP2B3	26.38	57.00	9.75	9.88	10.02	10.20	10.56	10.87	11.34
	G3	27.42	54.85	9.06	9.40	9.67	9.91	10.35	10.70	11.24
Identification path: CH_2=CH• → TSC2H3 → CH≡CH + H•										
 TSC2H3	DFT	107.98	56.57	12.81	14.51	15.68	16.57	17.91	18.97	20.81
(1)[c](2)[d]	G3MP2B3	107.97	56.91	13.08	14.70	15.82	16.66	17.96	18.99	20.81
	G3	108.64	56.20	12.60	14.36	15.61	16.54	17.91	18.94	20.74
Identification path: CH_3O• → TSCH3O → CH_2=O + H•										
 TSCH3O	DFT	26.36	56.34	11.53	12.85	14.13	15.31	17.26	18.73	20.95
(1)[c](3)[d]	G3MP2B3	26.22	56.12	11.36	12.71	14.01	15.19	17.15	18.63	20.88
	G3	26.94	55.19	10.58	12.10	13.52	14.77	16.76	18.25	20.56

[a]Thermodynamic properties are referred to a standard state of an ideal gas at 1 atm;
[b] ΔH^0_{f298} in kcal mol^{-1}; S^0_{f298} in cal mol^{-1} K^{-1};
[c]optical isomer number; [d]symmetry number; [e]Density Function Theory at B3LYP/6-311G(d,p);

6.2.3 Comparison of the Phenyl System to the Vinyl System

It is interesting to compare the thermochemistry between the phenyl + O_2 and vinyl + O_2 systems and corresponding hydroperoxides. The close similitude outlined below, allows calculations of the vinyl system to serve as comparison or reference to the larger phenyl system. The analogy in bond energies and well depths also suggests a corresponding similarity in some of the initial reactions processes of these R• + O_2 reactions. The properties analogy between the two systems allows the smaller vinyl + O_2 system to serve to calibrate several of the reaction paths and kinetic barriers. The comparison suggests that the vinyl + O_2 system can be used as model system for which differences between high and lower level calculation methods may be applied to improve calculation values on the larger phenyl + O_2 system.

In this study, the results of the high level calculations on vinyl + O_2 from Mebel and Lin [33] are used to determine a more accurate transition state for phenoxy radical + O atom path. It also illustrates that the density functional method combined with isodesmic reactions provides very good results when compared with the G2M calculations of Mebel and Lin and other high level calculations of Mebel and Kislov [189].

6.2.3.1 Comparison of the Enthalpies

In a previous chapter enthalpies of formation have been determined for CH_2=CHOO• and C_6H_5OO• using DFT and isodesmic reactions. The $\Delta_f H^0_{298}$ value obtained for CH_2=CHOO• is 24.34 ± 0.42 kcal mol^{-1} and shows good agreement with that of Mebel and Lin value, 24.45 [33] kcal mol^{-1} determined at G2M(RCC,MP2) level. The calculated enthalpy for C_6H_5OO• has the same order of magnitude at 31.28 ± 0.48 kcal mol^{-1}.

6.2.3.2 Comparison of the Bond Strengths

Table 6.7 provides a summary of data to compare the relative bond energies R—OOH, RO—OH and ROO—H of vinyl and phenyl hydroperoxides species as well as the corresponding bonds in vinyl and phenyl peroxy radicals. The bond energies R—H in ethylene and ethenol are compared to those in benzene and phenol. All three applied methods (B3LYP/6-311G(d,p), G3MP2B3 and G3) lead to the conclusion that the corresponding bond energies in the phenyl and vinyl peroxy and hydroperoxide systems are very similar.

Table 6.7: Comparison Vinyl vs Phenyl systems: Bond energies calculated at B3LYP/6-311g(d,p), G3MP2B3 and G3 level.

		B3LYP /6-311G(d,p)	G3MP2B3	G3
C_d—OOH	CH_2=CHOOH	84.5	84.71	84.67
C_b—OOH	C_6H_5OOH	87.33	88.41	88.07
C_dO—OH	CH_2=CHOOH	23.03	22.11	21.47
C_bO—OH	C_6H_5OOH	24.32	26.3	24.08
C_dOO—H	CH_2=CHOOH	86.07	85.53	86.1
C_bOO—H	C_6H_5OOH	86.16	85.72	86.7
C_d—OO•	CH_2=CHOO•	47.28	48.03	47.42
C_b—OO•	C_6H_5OO•	50.22	51.54	50.02
C_dO—O•	CH_2=CHOO•	39.65	39.27	38.06
C_bO—O•	C_6H_5OO•	40.85	43.27	40.07
C_d—H	CH_2=CH_2	111.17		
C_b—H	C_6H_6	113.3		
C_dO—H	CH_2=CHOH	87.23		
C_bO—H	C_6H_5OH	87.81		

Examination of Table 6.7 shows that the DFT method works well for these peroxide systems. Overall, the bond energies (BE) among DFT, G3MP2B3 and G3 are in good agreement demonstrating that DFT provides very good relative energies. A systematic comparison of the vinyl to the phenyl compounds shows that the bond energy magnitudes are very close. The comparison of the C_d—X bond energy, with X = OOH or OO•, with C_b—X, shows that overall BE in the vinyl species are lower than in the phenyl species by ca. 3 kcal mol^{-1}. This difference could be explained by the resonance contained in the benzene ring. The comparison between phenyl and vinyl goes even further, the C_d—H bond of the ethylene and the C_b—H of the benzene are 111.17 and 113.29 respectively, showing similar to the peroxide systems a small BE difference.

Table 6.7 demonstrates that the O—H bond energy is independent of the system. The calculated O—H bond is 86.16 kcal mol^{-1} for C_6H_5OO—H and it is 86.07 kcal mol^{-1} for CH_2=COO—H. This result is confirmed by the O—H bond energy of phenol and vinyl alcohol which are found to be 87.81 and 87.23 kcal mol^{-1} respectively.

The O—O bond in C_6H_5O—OH is calculated to be 24.32 kcal mol^{-1}, while CH_2=CHO—OH bond is 23.03 kcal mol^{-1} showing a difference of 1.3 kcal mol^{-1}. This difference is found again in the peroxy systems as the C_6H_5O—O• bond energy is 40.85 kcal mol^{-1} while CH_2=CHO—O• bond is 39.65 kcal mol^{-1}.

6.2.3.3 Similarities between the Potential Energy Surface of C_6H_5OO• and CH_2=CHOO•

A comparison of the calculated potential energy surface (PES) for the dissociation reaction of phenylperoxy (C_6H_5OO•) to phenoxy + oxygen atom (C_6H_5O• + O) versus the O—O bond distance against the corresponding vinylperoxy system (CH_2=CHOO• → CH_2=CHO• + O) was performed. The O—O distance was scanned at steps of 0.1 Å (or less) for both systems allowing the optimization of the structure for each O—O distance [190, 191, 192, 193].

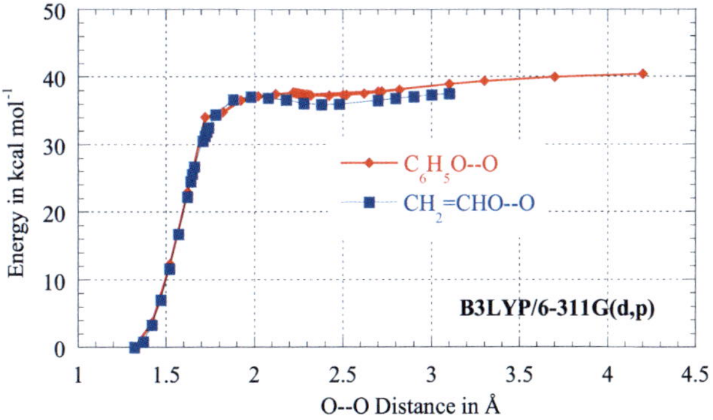

Figure 6.2: Energy diagram for the RO_2• → RO• + O dissociation with R = C_6H_5 and C_2H_3

This optimization was performed at the B3LYP/6-311G(d,p) level and the PES are illustrated in Figure 6.2. The resulting energies for the vinylperoxy radical (CH_2=CHOO•) reaction to vinoxy + oxygen atom (CH_2=CHO• + O) are very close to those reported by Mebel et al.

The C_6H_5OO• potential diagram of Figure 6.2 illustrates that the Density Functional Theory calculations illustrate the same pattern for both reaction systems, that there is a very small or no barrier in addition to the reaction of endothermicity, for the dissociation of phenyl-peroxy (PhOO•) to phenoxy + oxygen atom (PhO• + O). The DFT calculations are also similar to those determined by Mebel and Lin [33] and we conclude this dissociation is somewhat lower than the barrier previously estimated [194, 195, 196].

6.2.4 Potential energy Diagram and Kinetic Parameters

6.2.4.1 High-Pressure Limit A Factor (A_∞) and Rate Constant (k_∞) Determination

The reactions for which thermochemical properties of transition states are calculated by ab initio or Density Functional Theory methods, k_∞'s are fit by three parameters A_∞, n, and E_a over the temperature range of 298–2000 K, $k_\infty = A_\infty(T)T^n \exp(E_a/RT)$. Entropy differences between the reactants and transition state structures are used to determine the pre-exponential factor, A, via canonical Transition State Theory [197].

$$A = (k_B T/h) \exp(\Delta S^{\#}/R), \qquad E_a = \Delta H^{\#}$$

where h is the Planck's constant and k_B is the Boltzmann constant. A proper treatment of the internal rotors for S and $C_p(T)$ is important for the pre-exponential factor, because these internal rotors (and the corresponding entropy) contributions are often lost in cyclic transition state structures.

Corrections for H-atom tunnelling are applied for the intramolecular hydrogen atom transfer reactions of the transition state structures using Wigner 2nd order correction [198]. In this study the rate constants of three of our calculated transition states, TSPH•OOH and TSC•C4DO and TSCDCC•DO identified in Table 6.3, are corrected for H-atom tunnelling.

Unimolecular dissociation and isomerization reactions of chemically activated and stabilized adduct resulting from addition or combination reactions are analyzed by constructing potential energy diagrams. Some high-pressure rate constants for each channel are obtained from literature or referenced estimation techniques. Kinetics parameters for uni-molecular and bimolecular (chemical activation) reactions are then calculated using multifrequency QRRK analysis for $k(E)$ [199, 200, 63].

6.2.4.2 Important initial Reaction Paths and Transition States of the Chemically Activated PhOO• Adduct

Figures 6.3 – 6.8 describe the pathways and energetics relevant for the reaction of phenyl radicals with molecular oxygen. The phenyl-peroxy (PhOO•) is formed with nearly 50 kcal mol^{-1} of excess of energy and there are several forward reaction channels that require less energy for this chemically activated adduct to react to. The names and structures of the adduct/transition state/product have previously been described in Table 6.3.

Five reactions paths are available to the energized phenyl-peroxy adduct $[PhOO\bullet]^{\#}$, resulting from the $C_6H_5\bullet + O_2$ association and are evaluated as the primary paths for forward reaction; they are illustrated in Figure 6.3.

Three paths are relatively direct reactions:

i. Formation of phenyl-peroxy, PhOO• (stabilization), with a well depth at 298K of almost 50 kcal mol^{-1}. The activated [PhOO•]* radical is formed with no barrier and has a loose transition state making for a high prexponential factor for the reverse dissociation. Formation of the stabilized PhOO• radical is very important at lower temperatures and atmospheric and higher pressures.

ii. dissociation of the initial PhO—O• adduct to phenoxy radical + O atom (PhO• + O) with a barrier of 8 kcal mol^{-1} below the entrance channel. This channel has an intermediate loose transition state structure with cleavage of the RO—O bond but also formation of the Ph•O near double bond.

iii. Formation of ortho-phenyl-hydroperoxide radical (Ph•OOH) via intramolecular abstraction (H-shift) from $C_6H_5OO\bullet$ to $C_6H_4\bullet OOH$. This path has not been considered in previous studies of this system. This channel has a barrier of 12 kcal mol^{-1} below the entrance channel; but it has also a relative tight transition state structure due to loss of the PH—OO• rotor and ring strain.

The remaining two low barrier forward reactions are more complex involving addition of the peroxy radical site to the ring at ipso or ortho positions.

iv. The ipso addition is of key importance. This addition leads to a dioxirane cyclohexadienyl radical with a barrier of about 20 kcal mol^{-1} below the entrance channel. The transition state is tight. This dioxyrane cyclohexadienyl can react back to the peroxy adduct or to a seven member ring oxypinoxy radical that subsequently undergoes ring opening to form highly unsaturated oxy-hydrocarbon products.

Addition of the terminal oxygen of phenyl peroxy to the ortho position of the aromatic ring has a barrier that is only 2 - 3 kcal mol^{-1} lower than the reverse reaction back to phenyl radical plus O_2. Because its transition state structure is tight, we do not expect this channel to be important.

v. A sixth important reaction channel that is calculated to have a high reaction rate is the reverse reaction. The dissociation of the peroxy adduct back to phenyl + O_2 is due to the loose transition state (this is a non-reaction but it re-forms the reactants and they can re-react).

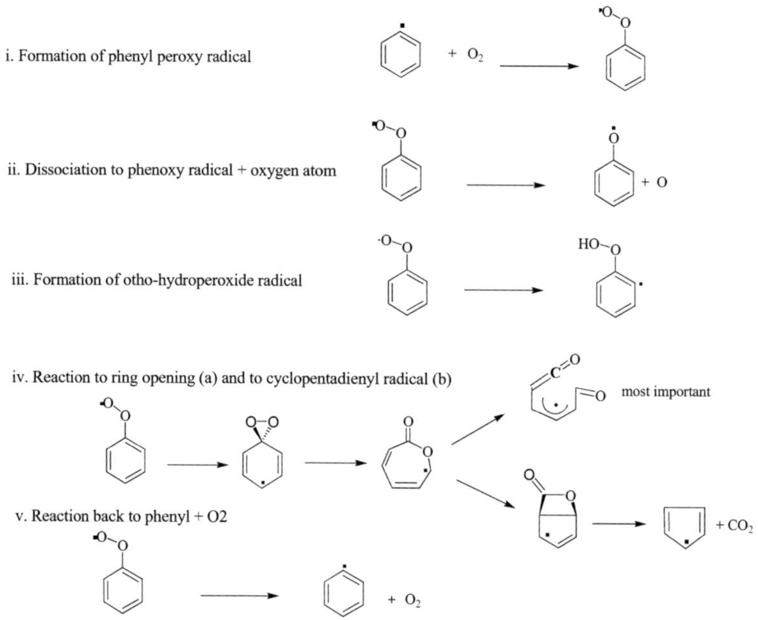

Figure 6.3: Important reactions channels

Adduct formation and reverse reaction

Figure 6.4 illustrates the phenyl + O_2 association that results in a chemically activated phenyl-peroxy radical ($C_6H_5OO\bullet$) with a 49 kcal mol^{-1} well depth. The [PhOO•]$^{\#}$ adduct is formed with no barrier and a relatively loose transition state. This loose transition state results in a high pre-exponential factor for the reverse reaction which is the dissociation back to phenyl + O_2. This is an important path at low pressures where stabilization or partial stabilization is slow.

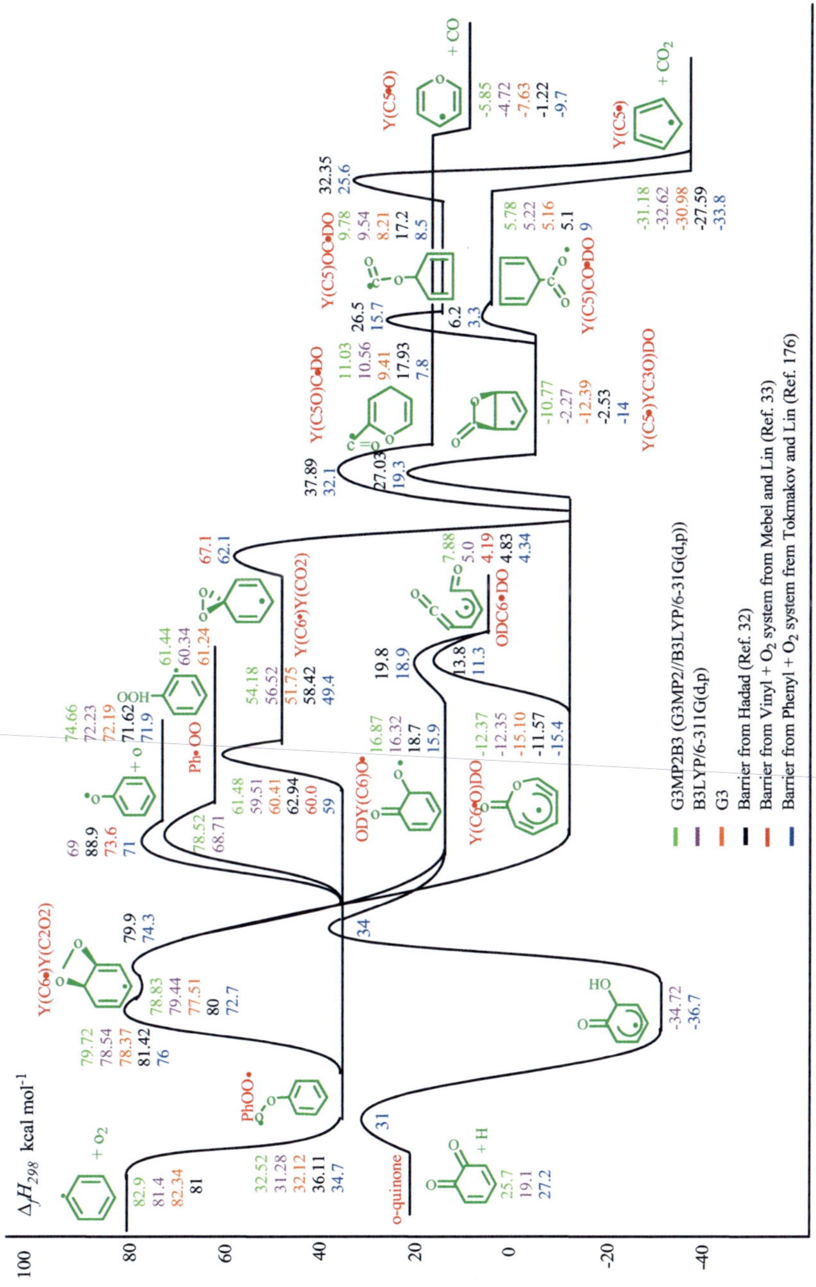

Figure 6.4: Potential diagram of phenyl + O_2

It is also an important path at high temperatures. The transition state for this peroxy radical adduct formation was initially published by Hadad et al. [32] and the canonical transition state analysis results in pre-exponential factors of ~ 2E+13 sec^{-1} at 298 K to 3.5E+15 sec^{-1} at 2000 K with the corresponding reverse reactions 2E+15 sec^{-1} to 5E+16 sec^{-1}. While this transition state is reported to be slightly below the energy of the reactants, we treat it as it occurs at the energy of the reactants. The reverse reaction has a small barrier, 13 kcal mol^{-1}, and needs to be considered as forward reaction as well.

Dissociation to phenoxy radical plus O atom

This reaction is endothermic by 41 kcal mol^{-1} relative to the stabilized phenyl-peroxy radical and has only a small or no barrier (less than 2 kcal mol^{-1}). The transition state for this dissociation reported by Fadden et al. shows a high pressure limit rate constants of 9.8E+12 to 2.4E+14 sec^{-1} from 300 K to 2000 K. This is reasonable for a system undergoing some structure rearrangement to form the near double bond in the phenoxy, in which the radical density is mostly within the ring. The energy barrier and the potential curve for the formation of the phenoxy radical + O through the cleavage of the O—O bond discussed previously is compared with the recent work of Tokmakov et al. [180]. Figure 6.5 illustrates the potential of this dissociation to PhO• + O and shows good agreement between the high calculation level method, G2M used by Tokmakov et al. and our calculations performed with DFT at B3LYP/6-311G(d,p) level.

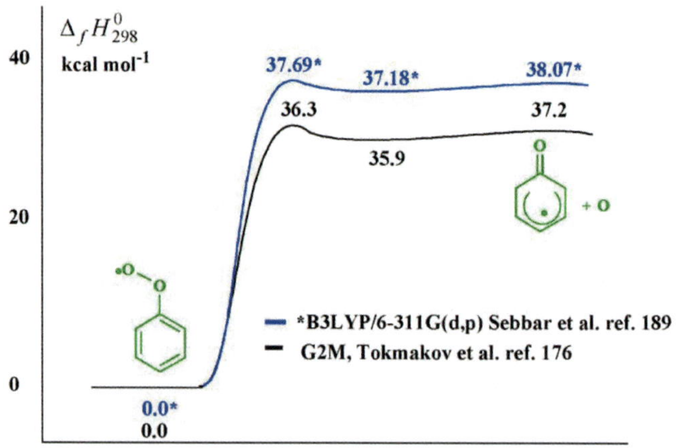

Figure 6.5: Potential Energy curve of the cleavage of the O—O bond in C_6H_5OO

Intramolecular abstraction of a phenyl H atom by the peroxy radical to form a hydroperoxide phenyl radical

This reaction has not been considered in any previous analysis of this reaction system. Abstraction from the ortho position via a five membered ring structure is found to have a barrier that is below the entrance channel and thus could be an important channel. TSPH•OOH represents the structure of this intramolecular abstraction (H-shift) or isomerization from C_6H_5OO• to C_6H_4•OOH. The H atom is in a bridge structure shifting from the carbon to the radical site. The reaction is calculated to be endothermic by 29 kcal mol^{-1} relatively to the peroxy adduct. The combination of the ring strain and the small reverse barrier brings the forward barrier to 42 kcal mol^{-1} (at the G3MP2 level) but this is still below the initial energy of $[C_6H_5OO\bullet]^{\#}$ at 49 kcal mol^{-1}. The reverse reaction has a barrier of only 13 kcal mol^{-1}. There are several plausible reaction products for the Ph •OOH adduct. It may react back to the PhOO• over a low barrier, dissociate to a diradical (Ph •O •) + OH with an 18 kcal mol^{-1} barrier [201], or the phenyl radical could attack the weak oxygen bounds in the hydroperoxide to form either a bicyclic epoxide + OH or an ortho hydroxyl phenoxy radical. This latter product channel is highly exothermic and may likely react to ortho-quinone + H atom.

The imaginary frequency of this transition state structure is equal to -2183 cm^{-1} at UB3LYP/6-311G(d,p) (DFT) and -2132 cm^{-1} at B3LYP/6-31G(d) (G3MP2B3). The leaving C—H bonds are 1.39 and 1.42 Å at B3LYP and G3MP2B3 respectively. The forming O—H bonds are 1.18 and 1.17 Å at B3LYP and G3MP2B3, respectively.

Intramolecular addition of the peroxy radical to the ortho site of the aromatic ring, 4-member ring formation

The TSY(C6•)Y(C2O2) transition state structure results from the intramolecular addition of the peroxy (PhOO•) to an ortho carbon site on the phenyl ring, forming a four member ring in which the forming O—C bond is 1.73 Å. This structure is tight and the barrier of 48 kcal mol^{-1} makes this channel relatively unfavourable if compared to the lower barrier channels discussed below and the above ortho H-abstraction path. If formed, the bicyclic dioxetane will isomerize to the oxypinoxy radical (YC6•O)DO with some 92 kcal mol^{-1} lower energy or to an ortho-quinone radical (ODY(C6)O•). Both products will undergo ring opening to form a linear unsaturated C_6H_5•O_2 (ODC6•DO) radical or to form an ortho-quinone and a hydrogen atom.

Intramolecular addition of the peroxy radical to the carbon of the peroxy group (ipso addition), 3-member ring formation

Addition of the PhOO• radical to the ipso carbon of the phenyl ring has a low barrier (20 kcal mol^{-1} below the entrance channel) and a tight transition state. This path is important and the addition results in a dioxirane cyclohexadienyl radical (Y(C6•)Y(CO2). This bicyclic cyclohexadienyl-dioxirane adduct has resonance between the ortho and para positions on the ring. The formation path has a barrier of 29 kcal mol^{-1} relatively to the stabilized peroxy radical. This dioxirane cyclohexadienyl radical forms a shallow well on the PES, less than 10 kcal mol^{-1}, and both forward and reverse reactions (out of the well) are fast. The forward reaction forms an oxypinoxy radical (Y(C6•O)DO) adduct, which is 65 kcal mol^{-1} lower in energy than Y(C6•)Y(CO2) and 77 kcal mol^{-1} below the transition state barrier. Our chemical activation calculations indicate that approximately 80% of this adduct further dissociates to new products before it is stabilized. These channels are of lower importance. The paths are described in separate discussion and analysis of this Y(C_6•)Y(CO_2) radicals' reaction paths (see below).

Intramolecular H atom transfer reactions considered and evaluated as not important

The following reactions are considered as not being important paths for the chemical activation of the unimolecular dissociation of the key stabilized adducts.

The abstraction of a meta or para hydrogen from the phenyl ring by the peroxy radical involves significant strain. These meta- or para- abstractions result in loss of aromaticity in the phenyl ring due to the strain from ring puckering, while the ring in the ortho-position is not significantly deformed. Calculations of this study indicate that the barrier energies for abstraction form meta and para positions are well above 100 kcal mol^{-1}.

Frank et al. [168] reported the formation of *p*-quinone + hydrogen atom via a transition state with a barrier that was estimated at 90 kcal mol^{-1} and in which the O—O bridge is in para-position (Figure 6.6). Hadad et al. [32] reported a higher barrier of 127 kcal mol^{-1} for this para-position transition state adduct. From G2M calculations, Tokmakov et al. [180] report a high barrier for this pathway as well (123 kcal mol^{-1}). Hadad et al. evaluated the transition state structure and determined the barrier to be around 138 kcal mol^{-1}. The O—O bridge was found in the meta-position to form Y(C5O•) + CO. The third isomer is the ortho-position transition state structure. This structure was calculated in this work by different computational methods and by Hadad et al. It was found to be at the lowest energy (around 81 kcal mol^{-1}).

Reactions of the oxy-pinoxy radical Y(C6•O)DO

Overall, we find little (ca. 10%) of this radical formed as stable radical; but it is a major intermediate in the chemical activated process, in which it is formed in a highly energized state and subsequent unimolecular reaction to products controlling a significant fraction of the reaction product. The Y(C6•O)DO 7-ring member can undergo ring opening to form the non-cyclic l-OCD6•DO. This ring opening channel has a pre-exponential factor in the range of 1 to 5E+12, which is five to ten times larger than any other exit channel for this reactant and has a barrier of only 20 kcal mol^{-1}. This barrier is relatively insignificant compared to energy needed by the adduct to be formed at ca. 80 kcal mol^{-1}.

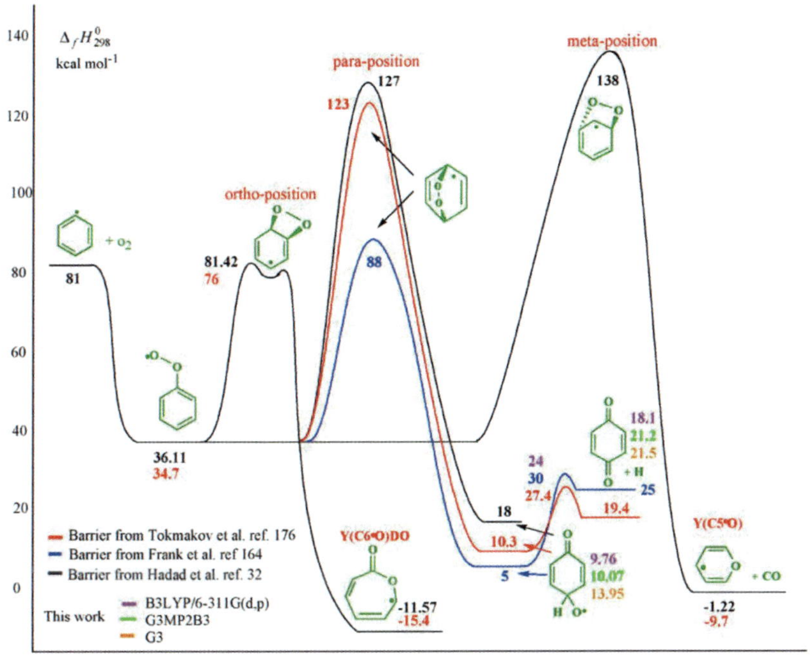

Figure 6.6: Possible transition states for PhOO• given by the literature

Y(C6•O)DO can also react to Y(C5•)Y(C3O)DO and to Y(C5O)C•DO with slightly higher barriers at 27 and 37 kcal mol^{-1}, respectively, and smaller pre-exponential factors than for the ring opening channel. These reactions lead to products considered to be important, cyclopentadienyl and carbon dioxide (Y(C5•) + CO_2) and oxiranyl and carbon monoxide (Y(C5•O) + CO), respectively.

Reaction of the linear ring opening product (from oxypinoxyl radical)

The ODC6•DO (O=C=C-C-C•-C-C=O) radical resulting from the Y(C6•O)DO ring-opening will undergo dissociation and unimolecular izomerization (formation of different ring structures). Dissociation (elimination reactions) will form acetylene, vinyl, formyl and carbonyl moieties with pre-exponential factors in the range of 1E+13 to 1E+14, and low barriers of 6 to 10 kcal mol^{-1} (Figure 6.7). As noted, ODC6•DO has several resonant structures accessible via internal rotations or electron re-arrangement barriers with relatively low (4 – 8 kcal mol^{-1}) barriers. We have chosen to illustrate the initial decomposition paths to carbon monoxide and acetylene moieties as opposed to ketenyl because of the lower barriers and the more favourable entropic factors.

ODC6•DO reacts primarily through three reaction paths:

(i) A five-member ring formation to form cyclopenten- yl ketone formaldehyde (ODY(C5) + HC•O) through a low energy path not considered previously [201].

(ii) A six-ring closing to form *o*-quinone + H [201].

(iii) A series of elimination reactions to form two carbon monoxides, two acetylenes and H atom (we term this unzipping).

The reaction of ODC6•DO to form a five member ring ODY(C5•)CDO radical has the lowest barrier of only 15 kcal mol^{-1}. This ODY(C5•)CDO radical can beta scission (eliminate) a formyl radical with a 37 kcal mol^{-1} barrier to form cyclopentadiene-one (ODY(C5)). It is important to note that resonantly stabilized radicals of cyclopentadiene and cyclopentadiene-ones are formed in the oxidation steps of aromatic systems. Their thermochemistry, reaction paths and kinetics all need to be analyzed for the inclusion of aromatics in combustion mechanisms.

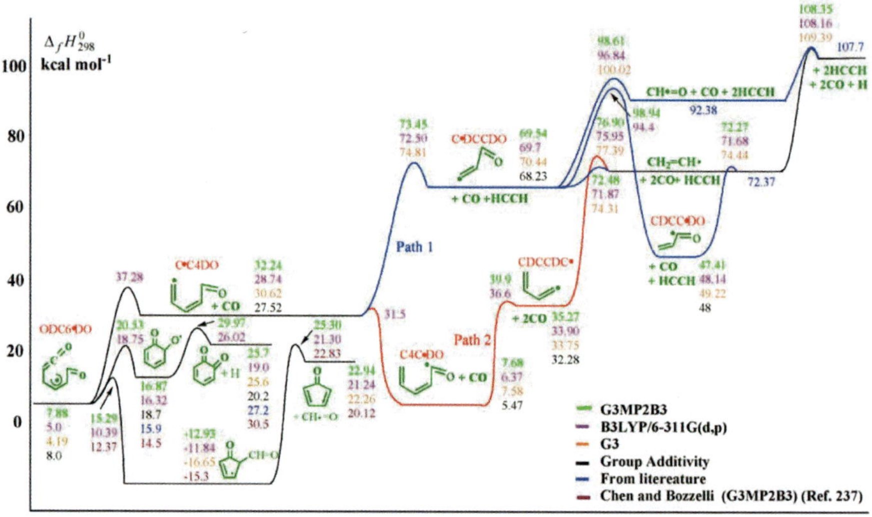

Figure 6.7: Elimination and isomerization reactions of ODC6•DO and resulting intermediates.

The reaction of OCD6•DO to ortho quinone + H product set (Figure 6.7) can proceed via two reaction sequences. The first step requires a 12 kcal mol^{-1} barrier to form a cyclohexadiene-one-alkoxy radical (ODY(C6)O•) and the second step proceeds via a barrier of 12 kcal mol^{-1} to eliminate the H atom and directly form the ortho quinone.

The first CO elimination from ODC•6DO results in the formation of O=CHCH=CHCH=CH• (C•C4DO). The formation of this primary vinylic radical (2,4 pentadien-al -5-yl) is endothermic by 30 kcal mol^{-1}. The C•C4DO can undergo other beta scission reactions via two pathways. The first pathway (Path-1) leads to the elimination of one acetylene to form 2-propenal -1yl (C•DCCDO) by overcoming an energy barrier of 44 kcal mol^{-1}. C•DCCDO is a primary vinylic radical as well. The second path (Path-2) leads to a lower energy isomer through a H-shift reaction resulting in 2,4 pentadiene formyl radical (C4C•DO). This reaction has a low energy barrier of 2.8 kcal mol^{-1} due to exothermicity but it has a tight transition state structure.

C•DCCDO (Path -1) reacts further through three different channels. The most probable path is the barrierless cleavage of CH•=CH—CH=O to form CO + vinyl radical, which dissociates further to acetylene + H. The second reaction could be the direct elimination of the HC•=O radical and the formation of acetylene via a barrier of about 24 kcal mol^{-1}. The third channel is a H atom shift forming the 20 kcal mol^{-1} lower energy isomer CDCC•DO, which can undergo dissociation to CO + vinyl radical. The vinyl radical can dissociate to HCCH + H or react with molecular oxygen.

The pentadiene formyl radical C4C•DO (Path-2) reacts further by beta scission to form CO and butadienyl-1 (CDCCDC•) via a barrier of 39 kcal mol^{-1}. The butadienyl-1 can undergo elimination, which has a 40 kcal mol^{-1} barrier, to form vinyl radical and acetylene or react with molecular oxygen with no barrier.

Alternate loss pathways for the vinyl and acetyl radicals formed in these reactions are the bimolecular association reactions with molecular oxygen. They have similar reaction processes and exothermicity to the energized phenyl-peroxy $[PhOO•]^{\#}$ adduct in this study.

6.2.4.3 Chemical Activation Reaction: Phenyl + O_2

Multi channel, multi-frequency Quantum RRK calculations are performed for *k(E)* with master equation analysis for falloff on the chemical activated phenyl peroxy radical $[PhOO•]^{\#}$ and the intermediates (isomers) in this complex reaction system. This provides an evaluation of the rate constants for the formation of stabilized adducts or reaction products as a function of pressure and temperature. The bi-molecular chemical activated reaction of Phenyl + O_2 system is carried out using the "CHEMASTER" program and incorporates all adducts and product channels illustrated. QRRK with Master equation analysis is used for unimolelcular dissociation of each adduct, but only isomeration to parallel, adjacent products / wells is included in the dissociation of stabilized intermediates. The input file for the phenyl + O_2 reaction system is given in the appendix F.

Note, that 1 atm is an important pressure. Almost all combustion systems run in ambient air at 1 atm. Some scram jet engines run at 0.3 atm, and turbines run at pressure of 10 – 15 atm. It is general that data are illustrated and discussed for 1 atm pressure. Also most experiments run at room temperature and the experiments in super critical water are at several (~ 200) atm and 650 K.

Figures 6.8 a-b-c illustrate the reaction product profiles as a function of temperature at pressure of 0.01, 1 and 10 atm. Figures 6.9 a-b show the product profiles as a function of pressure at 600 K and 1200 K respectively. The figures show that reverse reaction is important at high temperatures because of the very loose transition state structure. We also indicate the importance of the reverse reaction for illustration of mass balance (it is obvious that the reactants can re-react).

The chemical activated reaction of phenyl + O_2 is dominated by the back reaction to Ph• + O_2, followed by the phenoxy + O atom channel. These two channels contribute markedly more than the other channels with increasing temperature and pressure. The contribution to the Ph•OOH channel is the next important channel. We can see from Figure 6.8 three competing channels of significant importance for this Phenyl + O_2 system. Theses channels are ODY(C5) + C•DO, C•C4DO + CO, and Y(C5•) + CO_2 and result from the ring opening of Y(C6•O)DO. The contributions to products Y(C5•O) + CO and *o*-quinone + H are of lower importance in this Phenyl + O_2 system.

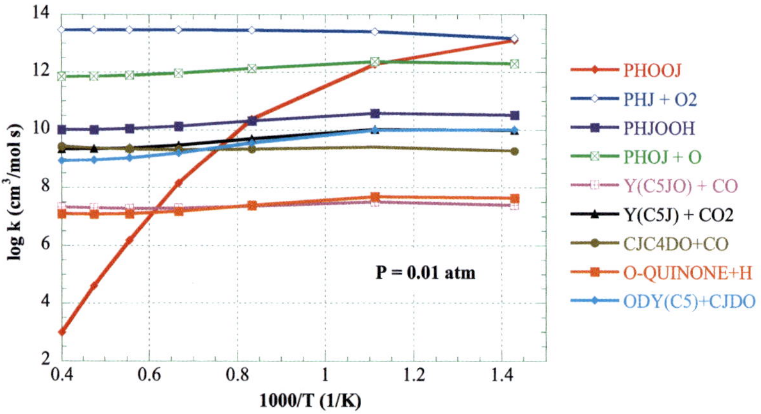

Figure 6.8a: Chemical activation reaction as a function of temperature at 0.01 atm

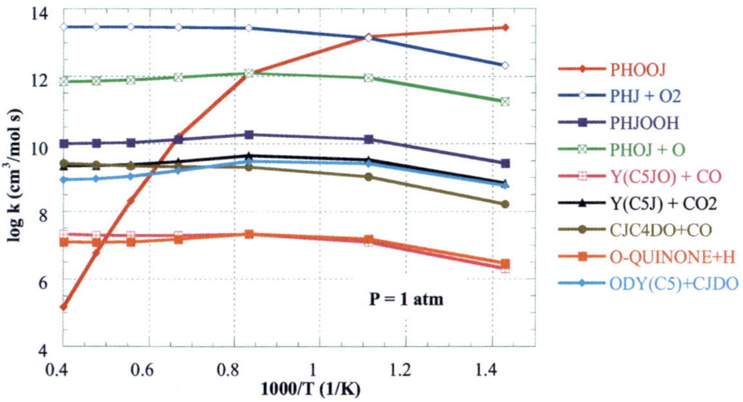

Figure 6.8b: Chemical activation reaction as a function of temperature at 1 atm

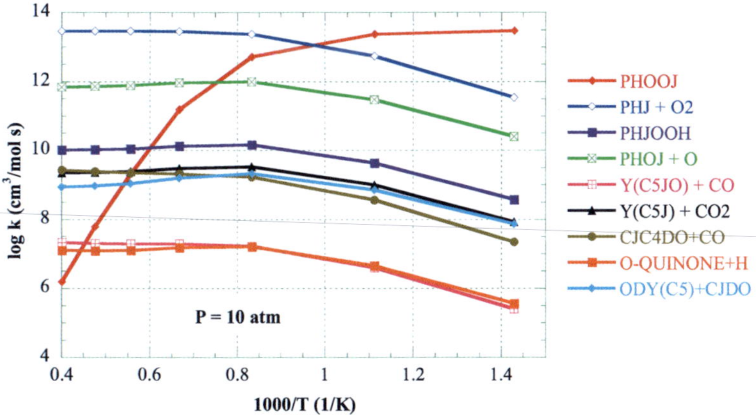

Figure 6.8c: Chemical activation reaction as a function of temperature at 10 atm

At 600 and 1200K, Ph• + O_2 is the overall dominant channel, followed by the phenoxy + O atom channel, both reactions have a noticeably dominance over the other channels. The third important channel is the contribution to Ph•OOH, which is of significant importance at both 600 K and 1200 K. The calculations indicate that the five member ring formation, cyclopentene–yl radical, which dissociates to ODY(C5) + HC•=O, is a more important path than the formation of the linear C•C4DO and CO channel. A competition between two paths, Y(C5•O) + CO, and o-quinone + H, is observable at 1200 K, but at 600 K, the o-quinone + H channel is slightly higher than Y(C5•O) + CO channel. The contributions to products Y(C5•O) + CO and *o*-quinone + H are of lower importance in this Phenyl + O_2 system.

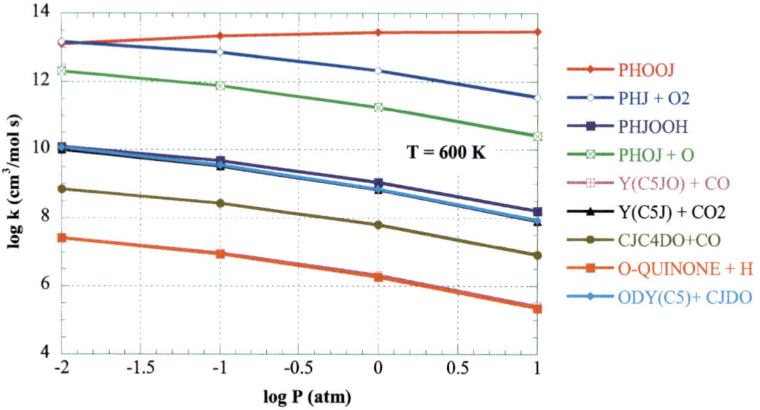

Figure 6.9a: Chemical activation reaction as a function of pressure at 600 K

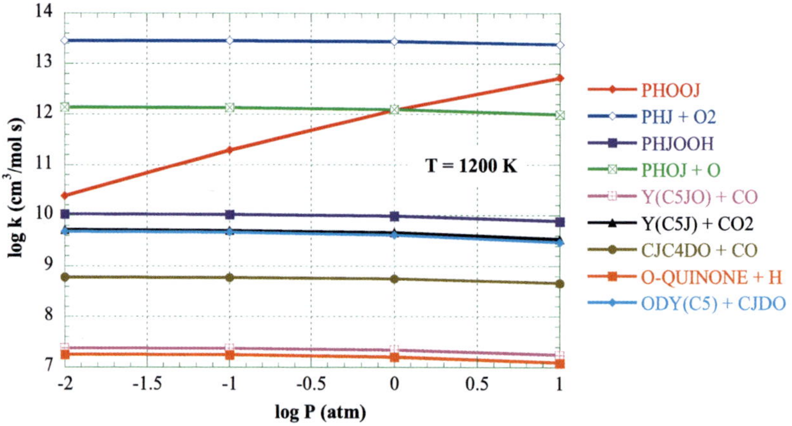

Figure 6.9b: Chemical activation reaction as a function of pressure at 1200 K

6.2.4.4 Unimolecular Dissociation of PhOO•

Figures 6.10, and 6.11 illustrate the dissociation of the stabilized phenyl peroxy adduct to the important product channels as a function of temperature at 1 atm and as a function of pressure at 1200 K. The figures show the competition between the two loose transition state channels: Phenoxy + O atom and Phenyl + O_2; the lower energy channel to the back reaction is dominant in this comparison. A more complete analysis shows that isomer Y(C6•)Y(CO2) is the most important channel.

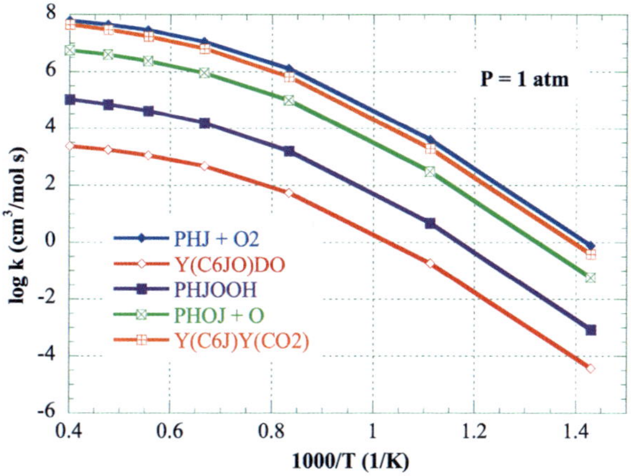

Figure 6.10: Dissociation of PhOO• vs temperature at 1 atm

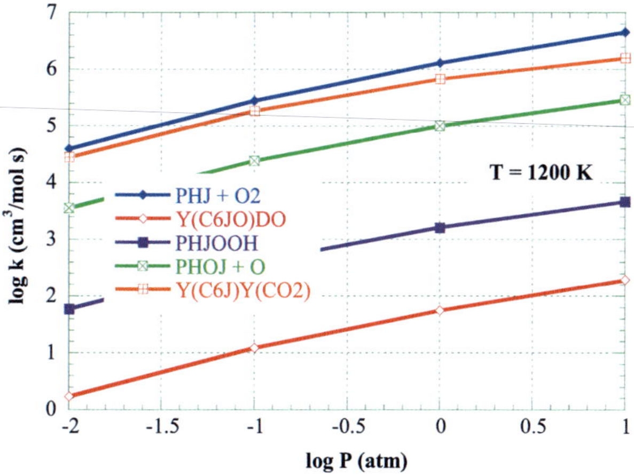

Figure 6.11: Dissociation of phoo• vs pressure at 1200 K

The important reaction of this Y(C6•)Y(CO2) isomer is the reaction to Y(C6•O)DO and subsequently to the two other important (final) dissociation products ODC6•DO and cyclopentadienyl and CO_2.

At high temperature, 1200 K, the three important channels are noticeable too as shown in Figure 6.11, with a small dominance for the back reaction to Ph• + O_2, the contribution of the Y(C6•)Y(CO2) channel is similarly important followed by PhO• + O channel.

The contribution to Ph•OOH channel is higher than that to Y(C6•O)DO, but both are of less importance.

Dissociation reactions (new 5-member ring formation and unzipping) of linear adduct ODC6•DO after ring opening

In the chemical activation reaction the oxy-pinoxy (7–member ring) radical and the ring opening product, ODC6•DO, is formed with some 73 and 58 kcal mol^{-1} of excess energy respectively. The ChemMaster code treats the oxypinoxy (Y(C6•O)DO) and the opened ring (ODC6•DO) as an energized adducts and allow dissociation or further unimolecular reaction.

The opened ring adduct (ODC6•DO) is shown in Figure 6.7 to have two low energy channels: formation of a five member ring, (formyl cyclopentenone-yl radical) and a beta scission to CO and a vinylic O=C—C=C—C=C• radical. This vinylic radical will further beta scission (unzip) to two acetylenes, two CO and one H radical or rapidly react with molecular oxygen.

6.2.4.5 Reduced Mechanism for the Phenyl + O_2 System

The "ThermKin" code described in chapter 2 is used to determine the elementary reaction rate coefficients and express the rate coefficients in several Arrhenius forms. It utilizes canonical transition state theory to determine the rate parameters. Thermodynamic properties of reactants and transition states are required and can be obtained from either literature sources or computational calculations. ThermKin requires the thermodynamic property to be in the NASA polynomial format. ThermKin determines the forward rate constants, *k(T),* based on the canonical transition state theory (CTST).

Table 6.6 lists the high pressure limit kinetic parameters for the elementary reaction steps in this complex phenyl + O_2 reaction system. These parameters are derived from the canonical transition state theory, the statistical mechanics from the DFT and ab initio data and from evaluation of literature data. Rate constants to all channels illustrated are calculated as function of temperature at different pressure. A reduced mechanism is proposed in Appendix F for the Phenyl + O_2 system, for a temperature range of 600K $<T<$ 2500K and at different pressures: 0.01 atm, 0.1atm, 1atm, and 10 atm.

Table 6.6: Calculated Kinetic Parameters (600K <T< 2500K) at 1 atm

Reactions		*A* *($cm^3/mol\ s$)*	*n*	*Ea* *Kcal mol^{-1}*	*ΔH_{RXN}*	*Method/ Ref.*
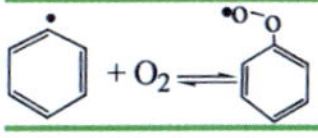	PHJ + O_2 = PHOOJ	6.50E+05	2.708	-		B3LYP
Reverse reaction	PHOOJ = PHJ + O_2	1.07E+11	1.637	47.7		B3LYP
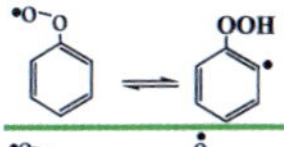	PHOOJ = PHJOOH	4.368E+05	1.941	40.4		B3LYP
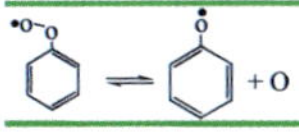	PHOOJ = PHOJ + O	2.024E+08	1.742	41.7		G2M
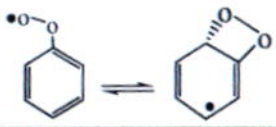	PHOOJ = Y(C6J)Y(C2O2)	1.194E+07	1.458	47.1		B3LYP
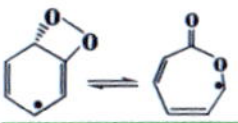	Y(C6J)Y(C2O2) = Y(C6JO)DO	3.10E+08	1.509	-		B3LYP
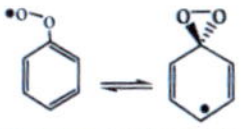	PHOOJ = Y(C6J)Y(CO2)	1.752E+06	1.864	27.98		G2M
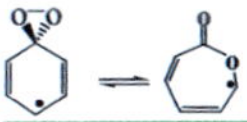	Y(C6J)Y(CO2) = Y(C6JO)DO	6.994E+09	.8563	10.58		G2M
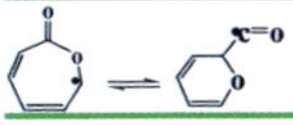	Y(C6JO)DO = Y(C5O)CJDO	7.445E+09	.5233	50.0		B3LYP
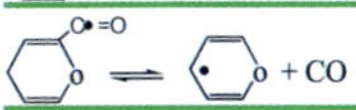	Y(C5O)CJDO = Y(C5JO) + CO	7.34E+11	.3678	8.06		B3LYP
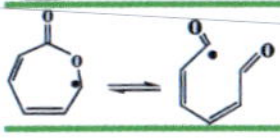	Y(C6JO)DO = ODC6JDO	1.518E+11	.5160	26.20		B3LYP
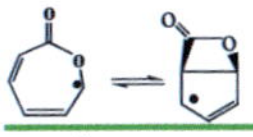	Y(C6JO)DO = Y(C5J)Y(C3O)DO	3.250E+11	.3645	39.37		B3LYP
	Y(C5)CO(OJ) = Y(C5J) + CO_2	2.77E+11	.283	-		B3LYP
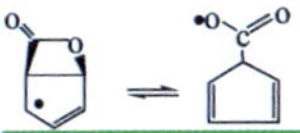	Y(C5J)Y(C3O)DO = Y(C5)CO(OJ)	2.748E+12	.3195	8.76		B3LYP
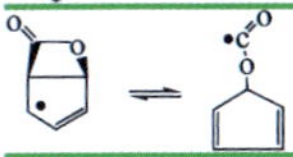	Y(C5J)Y(C3O)DO = Y(C5)OCJDO	2.735E+12	.3651	28.97		B3LYP
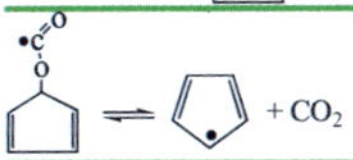	Y(C5)OCJDO = Y(C5J) + CO_2	2.719E+10	0.498	23.13		B3LYP
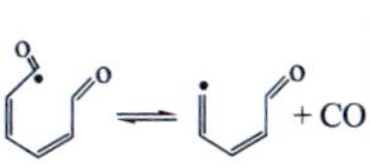	From Af/Ar: CO + C_2H_3 → CH_2=CHCO $1.51x10^{11}\ e^{-4.81\ (kcal/mole)/RT}$ (*Ref. 202*)	3.34E+13	-	29.91 29.45	25.10 24.64	G3MP2B3 B3LYP
	Calculated Stable G3MP2B3 – TS B3LYP Stable G3MP2B3 – TS G3	4.66E+11 2.15E+11	0.892 0.851	31.37 34.97	24.36 26.42	-

Table 6.6 (con't): Calculated Kinetic Parameters (600K <T< 2500K)

Reactions		*A* *(cm^3/mol s)*	*n*	*Ea* *Kcal mol^{-1}*	*ΔH_{RXN}*	*Method/ Ref.*
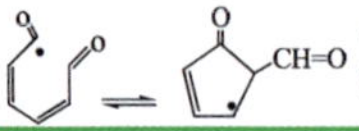	Calculated Stable G3MP2B3 – TS B3LYP	3.79E+09	0.679	5.61	6.8	B3LYP
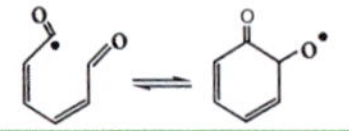	Calculated Stable G3MP2B3 – TS B3LYP	1.14E+09	0.700	19.73	8.5	B3LYP
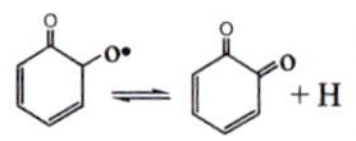	From $CH_3O \rightarrow CH_2{=}O + H$ $4.32x10^{-10}$ (cm^3/molecule s) $e^{-22.06\ (kcal/mole)/RT}$ $2.6x10^{11}$ (l/mol s) $e^{-22.06\ (kcal/mole)/RT}$	2.6E+11	-	10.6	20.3	203
	Calculated	8.45E+08	0.468	9.83	2.7	B3LYP
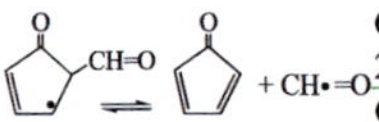	From CH•=CHCH=O → CH•=O + CH≡CH $2.72x10^{13}e^{-32.19(kcal/mole)/RT}$	2.72E+13	.482	30.76	-	B3LYP
	Calculated	1.18E+11	1.20	34.14		B3LYP
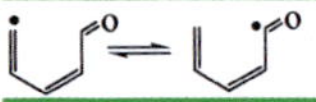	Abstraction; calculated	2.02E+09	0.964	2.82	-22.37	B3LYP
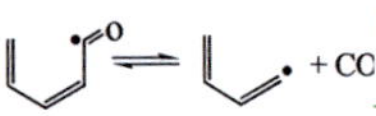	From Af/Ar: CO + $C_2H_3 \rightarrow CH_2{=}CHCO$ $1.51x10^{11}$ $e^{-4.81\ (kcal/mole)/RT}$ (*Ref.* 202)	3.34E+13	-	30.58 30.64 29.22	27.53 27.59 26.17	B3LYP G3MP2B3 G3
	Calculated	6.45E+12 1.35E+13	0.69 0.63	30.74 34.11	27.53 27.59	B3LYP G3MP2B3
	From: HC•=CHCH=CH_2 → CH≡CH + CH_2=CH• $1.0x10^{14}$ (s^{-1}) $e^{-43.92\ (kcal/mole)/RT}$	1.0E+14	-	43.92	-	204
	Calculated	7.82E+12	.869	43.98	38.47	B3LYP
	From: CH_2=CH• → HC≡CH + H $3.16x10^{12}$ (s^{-1}) $e^{-38.35\ (kcal/mole)/RT}$	3.16E+12	-	44.01	-	204
	Calculated	7.61E+12 1.21E+14	.738 -.0618	46.59 44.88	39.82 40.96	G3 B3LYP
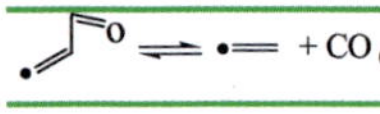	Calculated	1.08E+14 5.41E+14	-.045 -.26	3.38 5.58	2.67	B3LYP G3
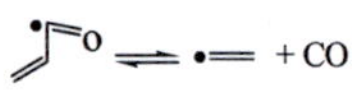	From Af/Ar: CO + $C_2H_3 \rightarrow CH_2{=}CHCO$ $1.51x10^{11}$ $e^{-4.81\ (kcal/mole)/RT}$ (*Ref. 202*)	3.34E+13	-	27.96 29.77 29.04	23.15 24.96 24.23	G3 G3MP2B3 B3LYP
	Calculated	2.88E+12 3.90E+12 2.14E+12	0.568 0.931 0.908	27.72 26.28 26.29	23.15 24.96 24.23	G3 G3MP2B3 B3LYP
	Calculated	2.72E+13	.482	32.19	22.67	B3LYP
	Calculated, Abstraction	2.31E+12	.1353	25.23	-21.56	B3LYP

Table6.6 (con't): Calculated Kinetic Parameters (600K <T< 2500K)

Reactions		*A* *(cm^3/mole s)*	*N*	*Ea* *Kcal mol^{-1}*	*ΔH_{RXN}*	*Method/ Ref.*
•═ ⇌ ≡ + H	6.48×10^{-10} (cm^3/molecule s) $e^{-37.96 (kcal/mole)/RT}$ 3.9×10^{11} (l/mol s) $e^{-37.96 (kcal/mole)/RT}$	3.9E+11	-	37.96	-	204
	1.6×10^{14} (s^{-1}) $e^{-37.96 (kcal/mole)/RT}$	1.6E+14	-	37.96	-	186
	3.94×10^{12} (s^{-1}) $(T/298\ K)^{1.62}$ $e^{-36.96 (kcal/mole)/RT}$	3.94E+12	-	36.96	-	187
	Calculated	3.13E+09 1.36E+10	1.490 1.280	36.86 37.09	35.29	G3 B3LYP
CH•=O ⇌ CO + H	From: CH•=O → CO + H 6.64×10^{-11} (cm^3/molecule s) $e^{-15.54 (kcal/mole)/RT}$ 3.99×10^{10} (l/mole s) $e^{-15.54 (kcal/mole)/RT}$	3.99E+10	-	15.54	-	188
	4.15×10^{-10} (cm^3/molecule s) $e^{-16.81 (kcal/mole)/RT}$ 2.5×10^{11} (l/mol s) $e^{-16.81 (kcal/mole)/RT}$	2.5E+11	-	16.81	-	186
	0.8×10^{-10} (cm^3/molecule s) $e^{-66 (kJ/mole)/RT}$ 4.81×10^{10} (l/mol s) $e^{-15.79 (kcal/mole)/RT}$	4.81E+10	-	15.79	-	205
	Calculated	3.75E+10 1.54E+11	1.044 .978	17.10 15.90	15.29	G3 B3LYP
CH_3O• ⇌ CH_2=O+H	From: CH_3O• → CH_2=O + H 1.66×10^{-10} (cm^3/molecule s) $e^{-25.04 (kcal/mole)/RT}$ 0.999×10^{11} (l/mole s) $e^{-25.04 (kcal/mole)/RT}$	0.99E+11	-	25.04	-	186
	5.8×10^{13} (s^{-1}) $e^{-25.76 (kcal/mole)/RT}$	5.8E+13	-	25.76	-	206
	4.32×10^{-10} (cm^3/molecule s) $e^{-22.06 (kcal/mole)/RT}$ 2.6×10^{11} (l/mol s) $e^{-22.06 (kcal/mole)/RT}$	2.6E+11	-	22.06	-	203
	Calculated	1.83E+11 5.71E+10	0.76 1.033	23.09 22.3	20.17 21.38	G3 B3LYP

6.3 Conclusion

In this study, thermodynamic properties of intermediates, transition states and products important to the formation and destruction of the aromatic ring in the phenyl radical + O_2 reaction system were calculated.

Enthalpy, entropy and heat capacities are determined by using DFT and *ab initio* methods (G3 and G3MP2B3). Our calculated values were compared to the vinyl + O_2 calculation data of Mebel et al., to the phenyl + O_2 calculation data of Hadad et al. and to recent data for phenyl + O_2 by Tokmakov et al. Group additivity calculations were performed as well to test our group additivity values.

We show that the vinyl radical is a good model for phenyl, for which high level calculations on the smaller vinyl system can be used to calibrate *ab initio* and Density Functional Calculations of the phenyl system.

High pressure limit kinetic parameters are obtained from the calculation resulting from the Canonical Transition State Theory. Quantum RRK analysis is utilized to obtain k(E) and master equation analysis is used to evaluate fall-off in this bimolecular, chemically activated reaction system. The Phenyl + O_2 association results in a chemically activated phenyl-peroxy radical with 49 kcal mol^{-1} well depth. Important forward reaction channels of the chemically activated $[PhOO\bullet]^{\#}$ radical are stabilization to phenyl peroxy radical, reaction to phenoxy radical + O, formation of a seven member ring Y(C6•O)DO and intramolecular abstraction to Ph•OOH.

The further reactions of the Y(C6•O)DO radical are more complex and result in several important products through several paths. One of these paths include the ring opening and the unzipping of the non cyclic radical to the formation of two acetylenes, CO and HC•=O. The other paths result in the formation of a cyclopentenyl ketone formaldehyde radical, and to the cyclopentadienyl radical + CO_2 channel. Kinetic parameters for important product channels are reported and a reduced mechanism is proposed.

7. The Dibenzofuran System

The comprehension of the destruction of dibenzofurans is important because these compounds are emitted in the environment and degrade very slowly. For this purpose, this study investigated the dibenzofuranyl + O_2 reaction system to provide an elementary mechanism describing the decomposition of this association. The determination of the thermochemical parameters, the transition state structures and the kinetic parameters on important reaction paths of the dibenzofuranyl + O_2 reaction system are reported in this work.

High Pressure limit kinetic parameters were determined and QRRK analysis was utilized to obtain k(E). The dibenzofuranyl + O_2 reaction system results in a chemically activated dibenzylfuranyl-peroxy radical that is formed with 50 kcal mol^{-1} excess energy available for further reactions. The energized adduct can dissociate to a benzofuran-phenoxy + O atom or undergo unimolecular addition of the peroxy radical to the Π bond system of the ring, or be stabilized. The addition channels further react through several paths to ring opening and ring expansion products. An elementary reaction mechanism, based on the fundamentals of thermochemical kinetics, is developed to describe products formation and reagent loss. The reactions are first analyzed by construction of a potential energy diagram of the system.

7.1 Introduction

It is important to understand the destruction of dibenzofurans and dibenzodioxins as well as their formation paths, because both processes occur under the high temperature combustion conditions. One important class of reactions for aromatic compounds is the loss of hydrogen through abstraction by the radical pool species to form a dibenzofuranyl radical. This radical will rapidly react with the oxygen contained in the combustion environment to form an energized adduct (Figure 7.1, chemical activation). This adduct can undergo further reaction through several complex pathways resulting in a number of intermediates and products. The investigation of this reaction system is important for the evaluation of the kinetics of the aromatics species and for the consideration of the products in incomplete combustion. Several recent publications suggest that new reaction paths are needed to correctly model aromatic oxidation and combustion [207] [208] [209] [210].

The reaction paths describing the formation and subsequent reactions of the active radical (phenyl) via abstraction by radical species (e.g. OH or Cl, H, or O) are important to include in mechanisms because sufficient thermal energy is available to overcome the endothermicity which is around 10 kcal mol^{-1}.

As a consequence of the activity of these radicals, abstraction reactions can occur even at moderate temperatures in several downstream zones of an incinerator.

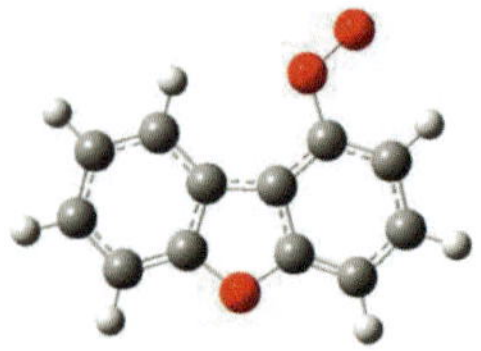

Figure 7.1: Dibenzofuran-peroxy structure.

This study estimates thermodynamic properties of intermediates and products in the decomposition of the dibenzofuranyl + O_2 reaction system. Molecular properties for species are calculated by Density Functional Theory by use of the Gaussian 98 program suites. The reactions are first analyzed by the construction of a potential energy diagram of the system.

Isodesmic reaction analysis is used to calculate the enthalpies of the intermediates and the well depths of adducts. The new Group Additivity parameters developed in a previous chapter, aided to evaluate the enthalpies of formation of large molecules.

We compared and showed that the vinyl radical is a good model for phenyl, which itself is a model for dibenzofuranyl. High-level calculations on the smaller vinyl system or on other small systems can be used to calibrate DFT calculations on the dibenzofuranyl system.

The transition state structures resulting from the decomposition of the dibenzofuran are calculated or derived from our calculated data on the phenyl + O_2 and from the data of Hadad et al. [32]. The results obtained on the vinyl + O_2 reaction system by Mebel et al. [33] are used for comparison as well.

The kinetic parameters of each path are determined as a function of temperature and pressure using the bimolecular chemical activation analysis. High Pressure limit kinetic parameters from the calculation results are obtained with the canonical Transition State Theory. The multifrequency Quantum Rice-Ramsperger-Kassel analysis is utilized to obtain *k(E)* and Master Equation analysis is used for the evaluation of pressure fall-off in this complex bimolecular chemical activation reaction. Results are applicable to elementary experiments at low pressures, ambient combustion studies at one atmosphere, as well as higher-pressure turbine systems.

Nomenclature used in this paper: **DBF** represents dibenzofuran, **BF** represents benzofuran, **Y**(A) represents a cyclic structure (e.g. Y(C5) is cyclopentadiene), and **D** is a double bond (e.g. CDO is CH_2=O). A• represents a radical site on the structure; e.g., Y(C5•) is cyclopentadienyl. In some abbreviated nomenclature the double bond symbol (= or D) is omitted in the name, e.g. O=CHCH=CHCH=CHC•=O is named ODC6•DO.

7.2 Results and Discussion

The DFT method at B3LYP/6-311G(d,p), is used to determine the optimized geometries [211, 212, 213] of the species. Zero-point vibrational energies (ZPVE), vibrational frequencies and thermal correction contributions to enthalpy from harmonic frequencies are scaled in accordance to the scaling factors recommended by Scott and Radom [120]. These geometries, frequencies and moment of inertia are reported in appendix B.

7.2.1 Enthalpy of Formation

7.2.1.1 Enthalpies of Formation of Radicals and Stable Species

Enthalpies of formation were determined using different methods. Intermediates, products and some selected transition states were calculated at B3LYP/6-311G(d,p) level of calculation, and by use of the Group Additivity method. Barrier values from Hadad et al. [32], from Mebel et al. [33] and from other reaction systems are also used. The groups developed in chapter 4 were used to aid in the evaluation of the enthalpies of formation of large size species. The enthalpies of the reference species used in the working reactions are listed in Appendix A.

Table 7.1 gives the calculated enthalpy of formation of benzofuran (BF), dibenzofuran (DBF), benzofuran-phenol (DBFOH), and dibenzofuranyl-hydroperoxide (DBFOOH). The calculated value of BF (3.21 kcal mol^{-1}) shows excellent agreement with the experimental value of Chirico et al. [214]. The enthalpy of DBF calculated in this work and found to be 12.17 kcal mol^{-1} is in very good agreement with the value of Chirico et al. (13.2 kcal mol^{-1}) and Sabbah et al. value (11.3 kcal mol^{-1}) [215].

Table 7.1: Calculated Enthalpy of Formation of Stable Species

Reactions	$\Delta_f H^0_{298}$ (kcal mol^{-1})		
	This work	Literature	GA
Benzofuran + CH_2=CH_2 → C_6H_6 + furan	2.91		
Benzofuran + CH_4 → C_6H_6 + CH_2=CHCH=O	3.52		
Average	**3.21 ± 0.43**	3.25[a]	**3.18**
DBF + CH_2=CH_2 → benzofuran + C_6H_6	12.34		
DBF + 2CH_2=CH_2 → furan + 2C_6H_6	12.00		
Average	**12.17 ± 0.24**	13.2 ± 0.2[a] / 11.3 ± 1.1[b]	**13.27**
DBFOH + CH_2=CH_2 → DBF + CH_2=CHOH	-30.44		
DBFOH + C_6H_6 → DBF + C_6H_5OH	-30.08		
DBFOH + C_6H_5OOH → DBFOOH + C_6H_5OH	-30.13		
Average	**-30.21 ± 0.19**		**-29.33**
DBFOOH + CH_3CH_3 → DBF + CH_3CH_2OOH	-10.83		
DBFOOH + CH_2=CH_2 → DBF + CH_2=CHOOH	-9.61		
DBFOOH + C_6H_6 → DBF + C_6H_5OOH	-10.14		
Average	**-10.19 ± 0.61**		**-9.21**

[a]Chirico et al. [214], [b]Sabbah et al. [215]

Table 7.2: Calculated Enthalpies of Formation of the Radicals

Species	Reactions	$\Delta_f H^0_{298}$ (kcal mol^{-1}) DFT	GA[a]
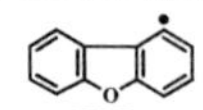	***DBF•*** + CH_3CH_3 → DBF + CH_3CH_2•	74.22	
	DBF• + CH_2=CH_2 → DBF + CH_2=CH•	74.18	
	DBF• + C_6H_6 → DBF + C_6H_5•	73.95	
DBF•	**Average**	**74.12 ± 0.14**	**74.17**
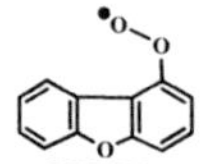	***DBFOO•*** + CH_3CH_2OOH → DBFOOH + CH_3CH_2OO•	23.82	
	DBFOO• + CH_2=CHOOH → DBFOOH + CH_2=CHOO•	24.08	
	DBFOO• + C_6H_5OOH → DBFOOH + C_6H_5OO•	24.20	
DBFOO•	**Average**	**24.04 ± 0.19**	**24.76**
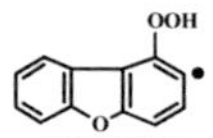	***DBF•OOH*** + CH_2=CH_2 → C_6H_5OOH + CH_2=CH•	53.62	
	DBF•OOH + C_6H_6 → C_6H_5OOH + C_6H_5•	53.38	
DBF•OOH	**Average**	**53.50 ± 0.17**	
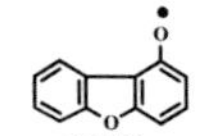	***DBFO•*** + CH_3CH_2OOH → DBFOOH + CH_3CH_2O•	2.33	
	DBFO• + CH_2=CHOOH → DBFOOH + CH_2=CHO•	2.36	
	DBFO• + C_6H_5OOH → DBFOOH + C_6H_5O•	1.59	
DBFO•	**Average**	**2.09 ± 0.43**	**6.11[b] / 1.96[c]**
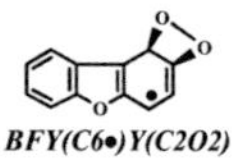	***BFY(C_6H_5•)Y(C2O2)*** + CH_2=CH_2 → Y(C_6H_5•)Y(C2O2) + Benzofuran	87.93	
	BFY(C_6H_5•)Y(C2O2) + CH_2=CH_2 → Benzofuran + Y(C_5H_5•) + CO_2	89.07	
	C_6H_5OO• → ***BFY(C_6H_5•)Y(C2O2)***	83.32	
BFY(C6•)Y(C2O2)	***BFY(C_6H_5•)Y(C2O2)*** + CH_2CH_2 → Benzofuran + CH_2=CHOO•	83.95	
	Average	**86.07 ± 2.86**	**85.41**
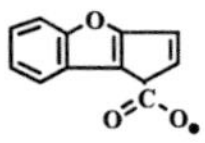	***BFY(C_5H_5)C(O•)=O*** + 2CH_2CH_2 → Furan•1 + C_6H_6 + Y(C_5H_6) + CO_2	-33.11	
	BFY(C_5H_5)C(O•)=O → BFY(C_5H_3•) + CO_2	-32.59	
BFY(C5)C(O•)DO	***BFY(C_5H_5)C(O•)=O*** → Bfuran + CO + CH•=CHCH=CHCH=O	-34.92	
	Average	**-33.54 ± 1.22**	**-25.19**
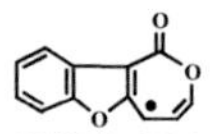	C_6H_5OO• → ***BFY(C6•O)=O***	-31.76	
	BFY(C6•O)=O + H_2 + CH_2=CH_2 → Benzofuran + C_4H_6 + CH•=O + CO	-28.83	
	BFY(C6•O)=O + CH_2CH_2 → Benzofuran + CH_2CHCHCHC•=O + CO	-27.97	
BFY(C6•O)DO	***BFY(C6•O)=O*** + 2CH_4 → Benzofuran + C_3H_6 + CO + CH_3C•=O	-29.27	
	Average	**-29.46 ± 1.63**	**-29.15**
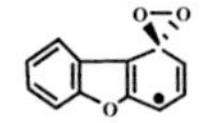	***BFY(C6•)Y(CO_2)*** + CH_4 + CH_2=CH_2 → Benzofuran + Y(C_6H_7•) + Y(CH_2O_2)	45.22	
	BFY(C6•)Y(CO_2) + CH_2=CH_2 → Benzofuran + Y(C6•)Y(CO_2)	46.00	
	BFY(C6•)Y(CO2) → BFY(C_6H_5•O)=O	44.19	
BFY(C6•)Y(CO2)	***BFY(C6•)Y(CO2)*** + CH_2=CH_2 → Benzofuran + Y(C_5H_5•) + CO_2	47.63	
	Average	**45.76 ± 1.45**	**51.75**
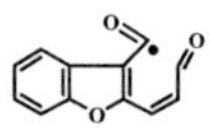	C_6H_5OO• → ***ODC•BFC3DO***	0.37	
	ODC•BFC3DO + H_2 → BFURAN• + CO + CH_2=CHCH=O	0.99	
	ODC•BFC3DO + 2H_2 → BFURAN + CO + CH•=O + CH_2=CH_2	0.80	
ODC•BFC3DO	**Average**	**0.72 ± 0.31**	**1.45**
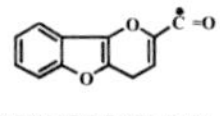	***BFY(C5O)C•=O*** + CH_3CH_3 → Bfuran + CO + CH_2=O + CH•=CHCH=CH_2	15.18	
	BFY(C5O)C•=O + CH_3CH_3 → BFY(C5•O) + CO_2	14.51	
BFY(C5O)C•DO	***BFY(C5O)C•=O*** + CH_2CH_2 → Bfuran + CO + CH•=CHCH=CH_2	12.72	
	Average	**14.14 ± 1.27**	-

[a]Group Additivity; [b]based on structure A; [c]based on structure B, see text and figure 7.2.

The enthalpy of formation of dibenzofuranol and dibenzofuranyl-hydroperoxide are calculated to be -30.21 and -10.19 kcal mol^{-1}, respectively. We have no literature comparison here, but the GA estimation supports the DFT value with -29.33 kcal mol^{-1} for DBFOH and -9.21 kcal mol^{-1} for DBFOOH.

Enthalpies of formation for a series of radicals resulting from the decomposition of the first aromatic ring in were determined using DFT calculations as well and isodesmic reactions whenever possible. Because no literature data is available for these radicals, the Group Additivity method was used to support the DFT values as illustrated in Table 7.2. The enthalpy of formation of the dibenzofuranyl (DBF•) is calculated to be 74.12 kcal mol^{-1}, and is highly supported by GA (74.17 kcal mol^{-1}). The calculated enthalpy of formation of the DBFOO• radical is 24.04 kcal mol^{-1} and is also in very good agreement with the GA value at 24.76 kcal mol^{-1}. The dibenzofuranol radical (DBFO•) shows a difference of ca. 4 kcal mol^{-1} between DFT and GA. This deviation was predictable and can be explained by the final structure of the radical obtained after the optimization (**B**), which is different from the given structure (**A**) before optimization (see Figure 7.2). The high resonance within the dibenzofuran is induced by the phenoxy radical and results in the formation of a double bond in the stable optimized structure. The enthalpy value obtained with GA, 6.11 kcal mol^{-1} is based on structure **A** and a second evaluation of the enthalpy based on structure **B** is found to be 1.96 kcal mol^{-1}. This second value is in good agreement and supports the DFT enthalpy value of the dibenzofuranol radical found to be 2.09 kcal mol^{-1}.

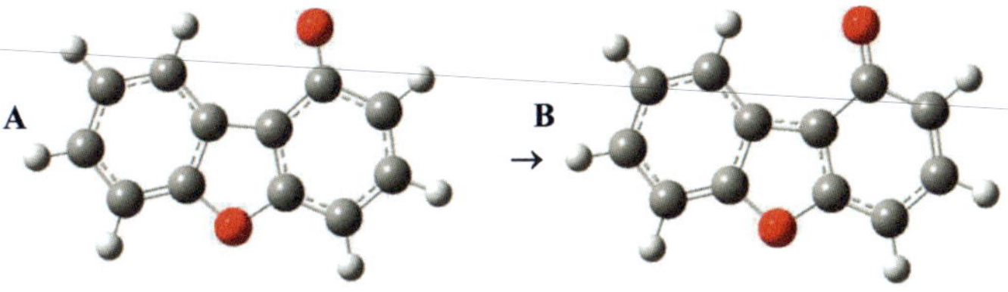

Figure 7.2: Structure of DBFO• after optimization

The DFT values of four radicals deviate reasonably by less than 1 kcal mol^{-1} from the GA values. The remaining twelve eight radicals given in Table 7.2 show higher differences between DFT and GA calculations. In all eight cases we can not explain this deviation at this moment and a review of our GA parameters has to be done.

The subsequent decomposition reactions of ODC•BFC3DO undergoes the formation of a series of radicals through four complex pathways. The enthalpies of formation of these radicals are calculated the same way and the working reactions are reported in Appendix G. The reaction paths identifying these species are given below (section 7.2.5.1).

Table 7.2 (con't): Calculated Enthalpy of Formation of the Species

	Reactions	$\Delta_f H^0_{298}$ (kcal mol^{-1}) DFT	GA[a]
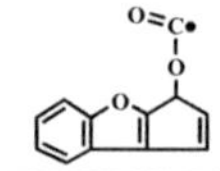 BFY(C5)OC•DO	*BFY(C$_5$)OC•=O* + $2CH_2CH_2$ → Furan•1 + C_6H_6 + Y(C_5H_6) + CO_2	15.60	
	BFY(C$_5$)OC•=O → BFY(C_5H_3•) + CO_2	16.12	
	BFY(C$_5$)OC•=O + CH_2CH_2 → Bfuran + CO + CH•=CHCH=CHCH=O	13.79	
	Average	**15.17 ± 1.22**	-
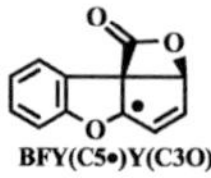 BFY(C5•)Y(C3O)DO	*BFY(C$_5$H$_5$•)Y(C3O)=O* → Benzofuran + Y(C_5H_5•) + CO_2	6.19	
	BFY(C$_5$H$_5$•)Y(C3O)=O + CH_2CH_2 → Benzofuran + CO + CH_2=CHCH=CHC•=O	4.23	
	BFY(C$_5$H$_5$•)Y(C3O)=O → BFY(C_5H_3•) + CO_2	6.55	
	BFY(C$_5$H$_5$•)Y(C3O)=O + CH_3CH_3 → Benzofuran + CH_2=O + CH_2CHCHCHC•=O	6.70	
	Average	**5.92 ± 1.14**	**6.13**
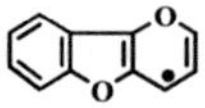 BFY(C5•O)	*BFY(C$_5$H$_5$•O)* + CH_4 → Benzofuran + CO + CH_2=CHCH$_2$•	18.72	
	BFY(C$_5$H$_5$•O) + CH_3CH_3 → Benzofuran + CH•=O + CH_2=CHCH=CH_2	21.28	
	BFY(C$_5$H$_5$•O) + CH_3CH_3 → Benzofuran + O=CHCH=CH• + CH_2=CH_2	19.24	
	BFY(C$_5$H$_5$•O) + CH_3CH_3 → Benzofuran + CH_2=CH• + O=CHCH=CH_2	18.92	
	Average	**19.54 ± 1.18**	**22.31**
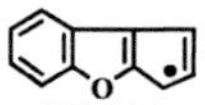 BFY(C5•)	*BFY(C$_5$H$_3$•)* + CH_2=CH_2 → Benzofuran + Y(C_5H_5•)	60.77	
	BFY(C$_5$H$_3$•) + C_6H_6 → Dibenzofuran + Y(C_5H_5•)	60.54	
	BFY(C$_5$H$_3$•) + CH_2=CH_2 → Benzofuran• + CH_2=CHCH$_2$CH=CH_2	61.74	
	BFY(C$_5$H$_3$•) + CH_2=CH_2 → Dibenzofuran + CH_3•	61.43	
	Average	**61.12 ± 0.56**	**57.93**
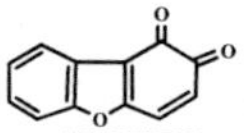 *o*-ODDBFDO	***o*-ODDBFDO** + $2H_2$ + CH_2=CH_2 → Benzofuran + 2CH_2=CHCH=O	-46.86	
	***o*-ODDBFDO** + $2H_2$ + CH_2CH_2 → Benzofuran + CH_2=CHCH=CH_2 + O=CHCH=O	-41.35	
	***o*-ODDBFDO** + $2H_2$ + CH_2=CH_2 → Benzofuran + *o*-quinone	-46.15	
	Average	**-44.79 ± 2.99**	**-38.89**

Table 7.3 gives the calculation of the enthalpies of formation of two other species needed in section 7.2.1.2 as reference species in the working reactions of the enthalpy calculations of the transition state structures. To check their accuracy, these references species have been determined by different methods. For the next calculations, we used the G3 value for •CH_2OCH=O and the B3LYP value for the CH•=C=O.

Table 7.3: Calculated Enthalpy of formation of the Reference Species

	Reactions Series	$\Delta_f H^0_{298}$ (kcal mol^{-1}) B3LYP	G3MP2B3	G3	GA
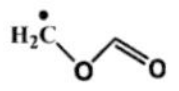 *C•OCDO*	**CH_2•OCH=O** + CH_3CH_3 → CH_3OCH=O + CH_3CH_2•	-36.83	-37.93	-38.42	
	CH_2•OCH=O + CH_3OH → HOCH=O + CH_2•OCH$_3$	-34.18	-34.53	-33.30	
	CH_2•OCH=O + CH_3OCH_3 → CH_3OCH=O + CH_2•OCH$_3$	-36.83	-37.98	-37.14	
	Average	**-35.96 ± 1.53**	**-36.81 ± 1.97**	**-36.29 ± 2.66**	**-35.29**
	CH•=C=O + CH_2=CH_2 → CH=C=O + CH_2=CH•	**33.60**	**34.99**		

7.2.1.2 Enthalpies of Transition State Structures

Enthalpies formation of transition states structures are evaluated via the difference between the calculated value of the TS structure and the calculated value of the stable radical adduct(s) (adjacent product and reactant where both are a single species) at the corresponding levels of calculation. The calculation of the enthalpies of the transition state structures are all performed with DFT, which has been demonstrated in the previous chapters to be an accurate and reliable method. We have calculated one transition state structure with the G3 method because the DFT calculations failed. Zero-point energies (ZPVEs) and thermal corrections to 298.15 K are from DFT calculations. The calculation of $\Delta_f H^0_{298}$ for the transition state structure has been performed via isodesmic reactions as it has been done for the stable species. Table 7.4 gives the enthalpies of ten transition state structures and identifies the reaction path for each transition state structure as well.

Table 7.4: Calculated Enthalpy of Formation of Transition State Structures

	Reactions Series		$\Delta_f H^0_{298}$ (kcal mol^{-1})
			DFT
Identification path: BFC•=O → TSBFC•DO → Benzofuranj1 + CO			
TSBF-CJDO			
	BFC•DO → **TSBFC•DO**		41.30
	TSBFC•DO → Benzofuranj1 + CO		46.90
		Average	**44.10 ± 3.96**
Identification path: BF•CH=O → TSBFC•DO → BFC•=O			
TSBFCJDO			
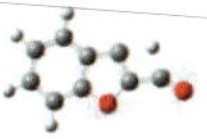	BF•CDO → **TSBFC•DO**		69.95
	TSBFC•DO → BFC•DO		66.90
		Average	**68.43 ± 2.15**
Identification path: CH•=C(CH=C=O)-CH=O → TSCJC2DO-KE[a] → HC≡CCH=O + CH•=C=O			
TSCJC2DO-KE[a]			
	CJC2DO-KE → **TSCJC2DO-KE[a]**		60.97
	TSCJC2DO-KE → HC≡CCH=O + CH•=C=O		59.02
	TSCJC2DO-KE + H_2 → CH_2=C=O + CH•=CHCH=O		59.11
		Average	**59.70 ± 1.1**
Identification path: C•C2DO-R[b] → TSCJC2DO-R[b] → HC≡CCH=O + •OC(=O)CH=CH$_2$			
TSCJC2DO-R[b]			
	C•C2DO-R[*] → **TSCJC2DO-R[b]**		7.53
	TSCJC2DO-R[*] + H_2 → HC≡CCH=O + •OC(=O)CH=CH$_2$		7.35
		Average	**7.44 ± 0.12**
Identification path: C•C4DOR[b] → TSCJC4DO-R[b] → HC•=CHCH=CHCH=O + HOC(=O)CH=CH$_2$			
TSCJC4DO-R[b]			
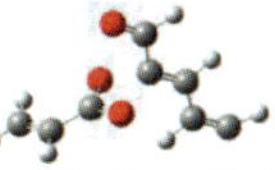	C•C4DOR[b] → **TSCJC4DO-R[b]**		31.83
	TSCJC4DO-R[b] + H_2 → HC•=CHCH=CHCH=O + HOC(=O)CHCH$_2$		31.63
		Average	**31.73 ± 0.14**
Identification path: COC(=O)CH=CH$_2$ → TSC-OC(=O)CH=CH$_2$ → CH$_3$ + OC(=O)CH=CH$_2$			
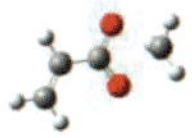	COC(=O)CH=CH$_2$ → **TSC-OC(=O)CH=CH$_2$**		**-26.29[d]**
TSC-OC(=O)CDC			

[a]KE ≡ -C=C=O; [b]R ≡ OC(=O)CH=CH$_2$; [c]G3 method; this value has to be checked

Table 7.4: Calculated $\Delta_f H^0_{298}$ of Transition State Structures

Reactions Series			$\Delta_f H^0_{298}$ (kcal mol^{-1}) B3LYP
Identification path: CH_2•OCH=O → **TSC•O-CDO** → CH_2=O + CH•=O			
	**TSC•O-CDO** → CH_2=O + CH•=O		-3.47
	C•OCDO → **TSC•O-CDO**		-4.77
TSC•O-CDO		**Average**	**-4.12 ± 0.92**[c]
Identification path: •OC(=O)-CDC → **TS•OC(=O)-CDC** → CO_2 + CH_2=CH•			
	•OC(=O)-CDC → **TS•OC(=O)-CDC**		**-21.47**
TS•OC(=O)-CDC			
Identification path: CH_3OC•=O → **TSCO-C•DO** → CO + CH_3O•			
	COC•DO → **TSCO-C•DO**		-24.04
	TSCO-C•DO → CO + CH_3O•		-17.29
TSCO-C•DO		**Average**	**-20.66 ± 4.77**
Identification path: ODC•BFC3DO → TSBFODC6•DO → ODCBFC3•DO			
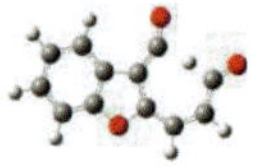	ODC•BFC3DO → **TSBFODC6•DO**		-6.40
	TSBFODC6•DO → ODCBFC3•DO		-7.72
		Average	**-7.06 ± 0.93**
TSBFODC6•DO			

[a]KE ≡ -C=C=O; [b]R ≡ OC(=O)CH=CH_2; [c]G3 method

7.2.2 Entropy and Heat Capacity

Thermodynamic properties calculated for the current study are presented in Table 7.5. Enthalpy of formation and entropy values are reported at 298 K, as most experimental data are referenced or available at 298 K. This facilitates the use of these thermodynamic properties and the use of isodesmic reaction set. Entropies and heat capacities are calculated by statistical mechanics using the harmonic-oscillator approximation for vibrations, based on frequencies and moments of inertia of the optimized B3LYP/6-311G(d,p) structures. Torsional frequencies are not included in the contributions to entropy and heat capacities; instead, they are replaced with values from a separate analysis on each internal rotor analysis (IR).

The contributions from translation, external rotation, and vibrations are summarized as TVR in Table 7.5 along with the IR. The frequencies representing internal rotor torsions are not included in TVR.

Table 7.5 summarizes the entropy and heat capacities of three stable species, benzofuran (BF), dibenzofuran (DBF) and dibenzofuranyl-hydroperoxide (DBFOOH), and of three radicals, dibenzofuranyl (DBF•), benzofuran-phenoxy (DBFO•) and dibenzofuranyl-peroxy (DBFOO•).

Table 7.5: Comparison of Thermochemical Properties from DFT Calculations with GA Values[a]

Species				$\Delta_f H^0_{298}$ [b]	S^0_{298}	Cp (T) cal mol⁻¹ K						
						300 K	400 K	500 K	600 K	800 K	1000 K	1500 K
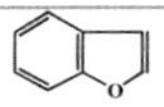	Benzofuran	**DFT**[e]		**3.21**	**77.96**	**27.61**	**37.15**	**45.01**	**51.21**	**60.07**	**66.01**	**74.47**
	(1)[c](1)[d]	**GA**		**3.18**	**76.90**	**26.28**	**35.99**	**44.00**	**50.42**	**59.51**	**65.54**	**72.85**
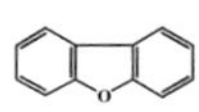	DBF	**DFT**[e]		**12.17**	**90.40**	**39.58**	**53.11**	**64.28**	**73.12**	**85.75**	**94.18**	**106.08**
	(1)[c](1)[d]	**GA**		**13.27**	**90.98**	**38.84**	**52.29**	**63.44**	**72.44**	**85.43**	**94.20**	**106.89**
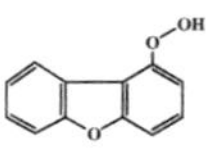	DBFOOH	TVR			96.59	44.52	58.76	70.43	79.62	92.65	101.29	113.46
	(1)[c](1)[d]	IR	DBF—OOH		6.19	3.02	2.72	2.41	2.13	1.74	1.51	1.24
		IR	DBFO—OH		3.31	2.15	1.93	1.78	1.67	1.52	1.41	1.25
		DFT[e]		**-10.19**	**106.09**	**49.69**	**63.41**	**74.62**	**83.42**	**95.91**	**104.21**	**115.95**
		GA		**-9.21**	**106.40**	**47.56**	**61.79**	**73.37**	**82.49**	**95.42**	**103.96**	**115.89**
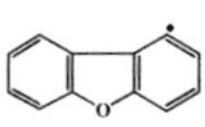	DBF•	**DFT**[e]		**74.12**	**92.12**	**39.23**	**52.15**	**62.78**	**71.23**	**83.54**	**92.02**	**104.50**
	(1)[c](1)[d]	**GA**		**74.17**	**92.46**	**38.43**	**51.11**	**61.51**	**69.75**	**81.68**	**89.72**	**101.65**
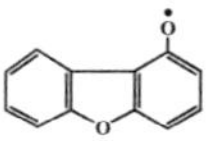	DBFO•	**DFT**[e]		**2.09**	**97.02**	**42.67**	**56.11**	**67.16**	**75.93**	**88.66**	**97.39**	**110.16**
	(1)[c](1)[d]	**GA**		**1.96**	**96.67**	**43.44**	**57.70**	**68.95**	**77.55**	**89.52**	**97.30**	**108.78**
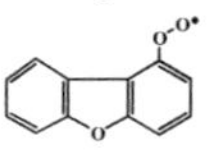	DBFOO•	TVR			98.87	44.23	58.18	69.59	78.55	91.18	99.47	110.93
	(1)[c]	IR	DBF—OO•		4.87	3.10	3.33	3.20	2.92	2.38	1.98	1.48
	(1)[d]	**DFT**[e]		**24.04**	**103.74**	**47.33**	**61.51**	**72.79**	**81.47**	**93.56**	**101.45**	**112.41**
		GA		**24.76**	**108.00**	**45.51**	**58.95**	**69.82**	**78.40**	**90.70**	**98.99**	**110.81**

[a]Thermodynamic properties are referred to a standard state of an ideal gas at 1 atm; [b] ΔH^0_{f298} in kcal mol⁻¹; S^0_{298} in cal mol⁻¹ K⁻¹; [c]optical isomers number; [d]symmetry number; [e]Density Functional Theory at B3LYP/6-311G(d,p).

Entropies and heat capacities are estimated from Density Functional Theory calculations data, and compared with Groups Additivity (GA) estimations. The group needed to estimate the enthalpies of formation, the entropies and the heat capacities of these species were developed in a previous chapter. These groups were estimated from a series of calculated oxygenated hydrocarbons and reported in chapter 4.

We note a general good agreement between DFT and GA. For all but two species, the entropy estimated by GA deviates by less than 0.6 cal mol^{-1} K^{-1}. The entropy difference for DBF between DFT and GA is 0.58 cal mol^{-1} K^{-1}. For the stable hydroperoxide DBFOOH, and the radicals DBF• and DBFOO•, the difference is less than 0.35 cal mol^{-1} K^{-1}. Benzofuran (BF), on the other hand, deviates by 1.06 cal mol^{-1} K^{-1}. The radical DBFOO• shows a serious discrepancy between the two methods which needs to be verified.

Table 7.5 lists the heat capacities, estimated in a range temperature between 300 and 1500 K. With respect to DFT values, we note a good agreement for the heat capacities, varying between 0.02 and 2.85 cal mol^{-1} K^{-1} compared to GA. A deviation of 2.85 cal mol^{-1} K^{-1} can be considered insignificant compared to ca. 100 cal mol^{-1} K^{-1}.

Like for the phenyl system, the entropies and heat capacities of transition state structures are calculated as well. S^0_{298} and $Cp_{298}(T)$ are determined with the SMCPS program (see chapter 2) which doesn't consider the imaginary frequency. These entropies and heat capacities along with the corresponding enthalpies of formation were then converted in NASA polynomial format needed in the "ThermKin" code to determine the kinetic parameters of the elementary reactions. The entropies and heat capacities values of the transition state structures are listed in Table 7.6.

Table 7.6: Thermochemical Properties of Transition State Structures[a]

	Species		B		Cp (T) cal (mol^{-1} K^{-1})						
			$\Delta_f H^0_{298}$	S^0_{298}	**300 K**	**400 K**	**500 K**	**600 K**	**800 K**	**1000 K**	**1500 K**
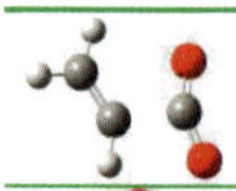	TS•OC(=O)-CDC	DFT		69.20	17.34	21.53	25.46	29.00	34.92	39.54	46.97
		C_d—C(O)O•		5.45	2.05	2.11	2.14	2.11	1.98	1.82	1.49
		Total	**-21.47**	**74.65**	**19.39**	**23.64**	**27.6**	**31.11**	**36.9**	**41.36**	**48.46**
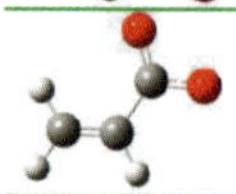	•OC(=O)CDC	DFT		68.05	16.43	20.33	23.59	26.21	30.05	32.70	36.58
		C_d—C(O)O•		5.35	2.05	2.11	2.14	2.11	1.98	1.82	1.49
		Total	**-26.47**	**73.4**	**18.48**	**22.44**	**25.73**	**28.32**	**32.03**	**34.52**	**38.07**
	TSCH•=C(CH=O)-OC(=O)CH=CH_2	DFT		91.27	31.45	37.85	43.07	47.29	53.58	58.01	64.62
		C_d—CH=O[c]		3.32	2.14	2.65	2.98	3.12	3.06	2.80	2.15
		C_d—CH=O[c]		3.32	2.14	2.65	2.98	3.12	3.06	2.80	2.15
		O—CH=O[e]		2.65	1.94	2.32	2.70	3.00	3.27	3.26	2.80
		Total	**8.20**	**100.56**	**37.67**	**45.47**	**51.73**	**56.53**	**62.97**	**66.87**	**71.72**
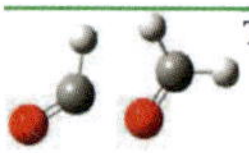	TSC•O-CDO	G3		70.87	17.11	19.25	21.14	22.80	25.52	27.62	30.97
		C•O—CO		2.56	2.14	2.44	2.76	3.01	3.25	3.24	2.81
		Total	**-4.12**	**73.43**	**19.25**	**21.69**	**23.9**	**25.81**	**28.77**	**30.86**	**33.78**
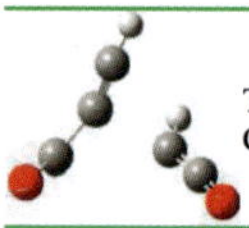	TSC•C2DO-CDCDO	DFT	**59.70**	91.19	27.90	31.90	34.96	37.40	41.12	43.83	47.96

[a]Thermodynamic properties are referred to a standard state of an ideal gas at 1 atm; [b] ΔH^0_{f298} in kcal mol^{-1}; S^0_{f298} in cal mol^{-1} K^{-1}

7.2.3 Internal Rotational Barrier

Potential barriers for internal rotations about the C—OOH, CO—OH and C—OO• bonds in the stable molecules and radical are determined at the B3LYP/6-31G(d,p) calculation level. The potential energy as function of dihedral angle is determined by scanning individual torsion angles from 0° to 360° at 15° intervals and allowing the remaining molecular structural parameters to be optimized (see section 3.4).

The lines in Figures 7.3, 7.4 and 7.5 are the results of the Fourier expansion fit to the data. Values for the coefficients of the Fourier expansion, a_i and b_i in equation 2.19 section (2.1.3), are obtained from the fitting program in Sigma Plot version 2.0, and then used in the "ROTATOR" code (see chapter 2) to calculate the contribution of internal rotors to S^0_{298} and $Cp^0(T)$.

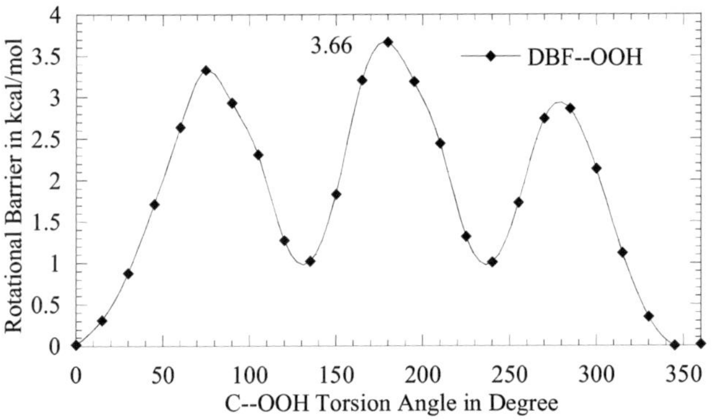

Figure 7.3: Rotational barriers for internal rotations about C_b—OOH

A potential energy diagram for internal rotation versus C_b—OOH torsion angle in DBFOOH is shown in Figure 7.3 ($C_b \equiv C_{benzene\ bond}$). The data show three-fold, potentials with barriers near 3.6 kcal mol^{-1}. This barrier is comparable to those obtained for the vinyl hydroperoxide species reported in chapter 3. The same rotational barrier in the C_b—OO• of the corresponding radical DBF—OO• is however, higher by 1.3 kcal mol^{-1}. The radical DBFOO• shows a two-fold symmetry with a barrier to be found at 4.9 kcal mol^{-1} (Figure 7.4).

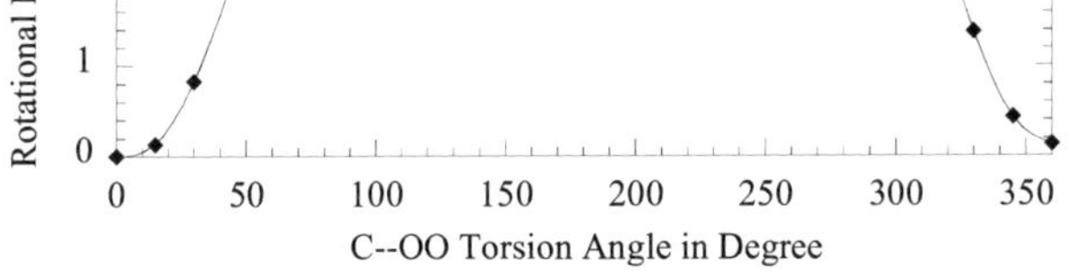

Figure 7.4: Rotational barriers for internal rotations about C_b—OO•

We found interesting to compare these C_b—OO rotational barrier with a similar species and chose the benzofuranol DBFOH for which the C_b—OH rotational energy was investigated (Figure 7.6). This C_b—OH rotation shows a similar two-fold symmetry to the C_b—OO• rotation but its barrier (4 kcal mol^{-1}) is the same as the C_b—OOH barrier demonstrating the influence of the radical on the rotational barriers.

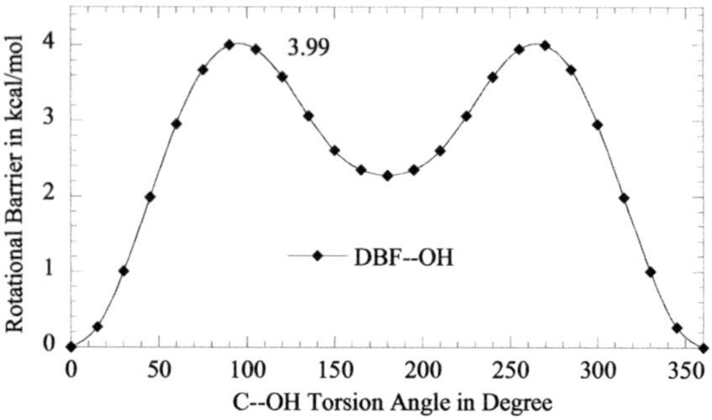

Figure 7.5: Rotational barriers for internal rotations about C_b—OH

7.2.4 Comparison of the Dibenzofuranyl to the Phenyl and Vinyl Systems

One objective of this study is to show and compare the similarities among three systems: dibenzofuranyl + O_2 system, phenyl + O_2 system and vinyl + O_2 system. This comparison suggests that the vinyl + O_2 and the phenyl + O_2 system can be used as model system for larger systems for which it is often difficult or not possible to run higher level methods. The advantage we take from this comparison is that the differences between high and lower level calculation methods may be used to improve calculation values (transition state structures) on the larger dibenzofuranyl + O_2 system.

In this study, the results of the high-level calculations on vinyl + O_2 from Mebel et al. [33] and the higher level calculations carried on the phenyl + O_2 system are used for the determination of more accurate transition states for large reaction systems. We show that the vinyl radical is a good model for phenyl, which is itself a model for dibenzofuranyl. The transition states determined on the phenyl + O_2 system and by Hadad's group [32] are used for the dibenzofuran + O_2 reaction system. For some transition state structures needed for the dibenzofuranyl system and not found in the phenyl or vinyl systems, we chose a small and analogue molecule and calculated the TST structure at a higher level. We show again that the Density Functional Theory calculation method we used, combined with isodesmic reactions, gives very good results.

7.2.4.1 Comparison of the Enthalpies

The calculated enthalpy of formation for DBFOO• is found to be 24.04 ± 0.19 kcal mol^{-1}. In previous chapters enthalpies of formation have been determined for CH_2=CHOO• and C_6H_5OO• using DFT and isodesmic reactions. $\Delta_f H^0_{298}$ of DBFOO• is very close to the calculated enthalpy of CH_2=CHOO• (24.34 ± 0.42 kcal mol^{-1}) and has the same order of magnitude as the calculated enthalpy of C_6H_5OO• (31.28 ± 0.48 kcal mol^{-1}).

7.2.4.2 Comparison of the Bond Strengths

Table 7.7 provides a summary of data to compare the relative bond energies, R—OOH, RO—OH and ROO—H, of vinyl, phenyl and dibenzofuranyl hydroperoxides. The corresponding bonds in vinyl, phenyl and dibenzofuranyl peroxy radicals (RO—O•, and R—OO•) are compared as well. For further comparison, the bond energies R—H in ethylene / ethenol are compared to those in benzene / phenol and dibenzofuran / dibenzofuranol. The data of the DFT calculations demonstrate that the corresponding bond energies in the vinyl, phenyl and dibenzofuranyl radicals and stable species are very similar.

In chapter 6 (section 6.2.3.2) we have pointed out that the C_d—X bond energies (X = OOH or OO•) are lower by ca. 3 kcal mol^{-1} than the C_b—X BE in the phenyl species. This result is confirmed by the calculations done on the dibenzofuranyl system. Table 7.7 show that dibenzofuran hydroperoxide has the same DBF—OOH bond energy (87.56 kcal mol^{-1}) as phenyl-hydroperoxide with C_6H_5—OOH to be 86.93 kcal mol^{-1}. The C_d—OOH bond energy of the vinyl hydroperoxide is lower at 84.5 kcal mol^{-1}. The comparison among the peroxy radicals of the three systems shows the same tend; DBF—OO• (50.08 kcal mol^{-1}) and C_6H_5—OO• (50.22 kcal mol^{-1}) bonds are identical, while CH_2=C—OO• bond is 47.28 kcal mol^{-1}.

Table 7.7: Bond energies derived from DFT calculations at B3LYP/6-311G(d,p) level.

	R—OOH	RO—OH	ROO—H		R—OO•	RO—O•
CH_2=CHOOH	84.5	23.03	86.97	CH_2=CHOO•	47.28	39.65
C_6H_5OOH	87.33	24.32	86.16	C_6H_5OO•	50.22	40.85
DBFOOH	87.56	25.26[a] (21.24[b])	86.33	DBFOO•	50.08	41.62[a] (37.6[b])
	R—H				RO—H	
CH_2=CH_2	111.17			CH_2=CHOH	87.23	
C_6H_6	113.30			C_6H_5OH	87.81	
DBF	114.05			DBFOH	88.42[a] (84.40[b])	

[a]calculated with DBFO• structure A; [b]calculated with DBFO•structure B

The O—O bond energies for DBFO—OH can have two different values depending on the geometry structure used for DBFO• (see section 7.2.1.1). For example O—O is calculated to be 25.26 kcal mol^{-1} with use of structure A and 21.24 kcal mol^{-1} with use of structure B. For all bond energies calculations involving DBFO• we used the enthalpy of formation of structure A.

Table 7.7 shows a consistency in the O—O bond energy values among the three systems. We notice a difference of ca. 1 kcal mol^{-1} between the O—O bond of the smallest system CH_2=CHO—OH and the phenyl hydroperoxide C_6H_5O—OH, which itself differs by only ca. 1 kcal mol^{-1} from the largest system DBFO—OH. The same result is obtained with the corresponding radicals which are found to be 39.65 kcal mol^{-1} for CH_2=CHO—O• 40.85 for C_6H_5O—O• and 41.62 kcal mol^{-1} for DBFO—O•.

In chapter 6 we have seen that the O—H bond energy is independent of the system. This result is verified by the O—H bond in the DBF system. DBF—O calculated to be 86.33 kcal mol^{-1} is the same as the BE in C_6H_5OO—H (86.16 kcal mol^{-1}) and in CH_2=COO—H (86.07 kcal mol^{-1}). The O—H bond energies in dibenzofuranol, phenol and vinyl alcohol confirm the above results with an O—H BE equal to 88.42, 87.81 and 87.23 kcal mol^{-1} respectively. The comparison of dibenzofuranyl, phenyl and vinyl systems goes even further since the C—H bonds in dibenzofuran, benzene and ethylene are calculated to be similar with 114.05, 113.30 and 111.17 kcal mol^{-1} respectively.

The above results mean that the vinyl + O_2 system and the phenyl + O_2 system can be used as model for the dibenzofuranyl + O_2 system. The differences between Density Functional Theory and higher level calculations on smaller molecules can be used to calibrate the DFT calculations as applied to larger dibenzofuranyl + O_2 adduct system.

7.2.4.3 Potential Curves of DBFOO• versus C=COO• and DBFOO• versus PhOO•

A comparison of the calculated potential curves for the dissociation reaction of dibenzofuran-peroxy radical (DBFOO•) to benzofuran-phenoxy + oxygen atom (DBFO• + O) versus O—O bond distance with the corresponding bond cleavage for vinyl-peroxy radical (CH_2=CHOO•) is illustrated in Figure 7.6. At the same time, Figure 7.7, compares the same dissociation of dibenzofuran-peroxy radical (DBFOO• → DBFO• + O) to the dissociation of phenyl-peroxy radical (C_6H_5OO•) to phenoxy + oxygen atom (C_6H_5O• + O) versus O—O bond distance. For technical reasons, we couldn't plot the 3 curves in one figure.

The DBFOO• potential diagram of Figures 7.6 and 7.7 shows that the DFT calculations illustrate the same pattern for the three reaction systems, that there is a very small or no barrier in addition to the reaction of endothermicity, for the dissociation of dibenzofuran-peroxy (DBFOO•) to benzofuran-phenoxy + oxygen atom (DBFO• + O). The Density Functional Theory calculations are also similar to those determined by Mebel et al. [33].

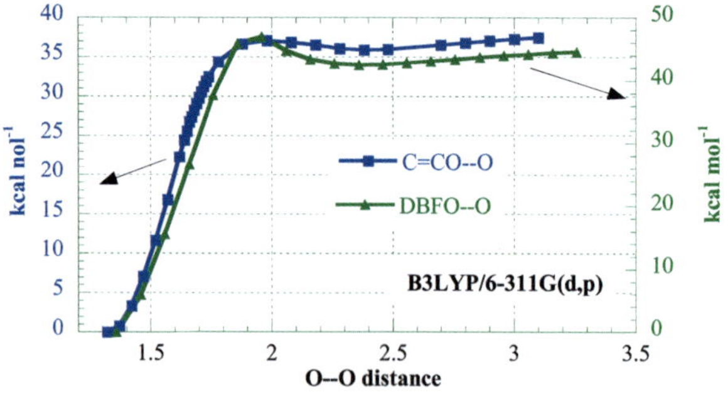

Figure 7.6: DFT calculations at B3LYP/6-311G(d,p) level of theory: Potential energy curve for the dissociation of DBFO—O and C=CO—O versus O—O distance.

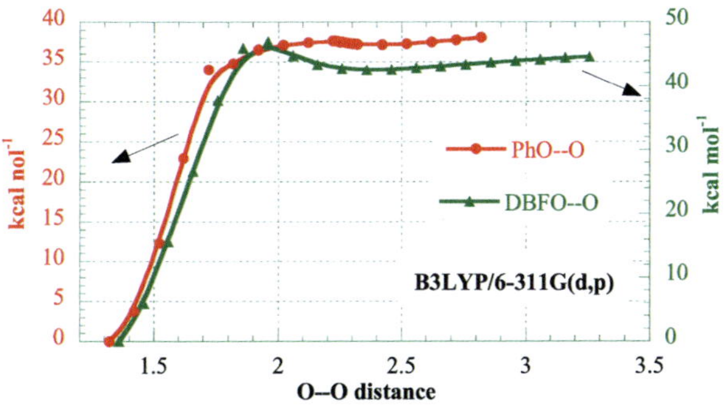

Figure 7.7: DFT calculations at B3LYP/6-311G(d,p) level of theory: Potential energy curve for the dissociation of DBFO—O and PhO—O versus O—O distance.

7.2.5 Potential Energy Surface and Kinetic Parameters

7.2.5.1 Potential Energy Surface

Because of the active phenyl radical contained in the dibenzofuranyl structure, we consider that the dibenzofuranyl + O_2 reaction system behaves the same way as the phenyl + O_2 system. On this basis we have constructed a potential energy surface for the reaction of dibenzofuranyl radical + O_2 as illustrated in Figure 7.8. The major features of this surface are very similar to those calculated for the phenyl + O_2 system. As for the phenyl system, there are five reactions of high importance in the chemical activation (bimolecular reaction) of dibenzofuranyl + O_2:

i. The formation of benzofuranyl peroxy radical DBFOO• (Stabilization)

ii. The Dissociation to benzofuran phenoxy radical plus oxygen atom (DBFO• + O)

iii. The formation of isomer DBF•OOH

iv. The reaction to isomer BFY(C6•)Y(CO2) with the further reaction of this isomer to two product sets through the cyclic oxipinoxy radical BFY(C6•O)DO

v. The reaction back to dibenzofuranyl + O_2.

The dibenzofuranyl + O_2 association results in a chemically activated dibenzofuranyl-peroxy radical with a 50 kcal mol^{-1} well depth at 298K. This chemically activated adduct can dissociate to phenoxy + O, or react back to phenyl + O_2. Both channels have a relatively loose transition states and their barriers are near that of the entrance channel. The dibenzofuranyl + O_2 reaction association can also undergo intramolecular addition of the peroxy radical to several unsaturated carbon sites on the ring via more tight transition states. The reaction to benzofuran-phenoxy + O has a significantly lower barrier of about 10 kcal mol^{-1} below the entrance channel.

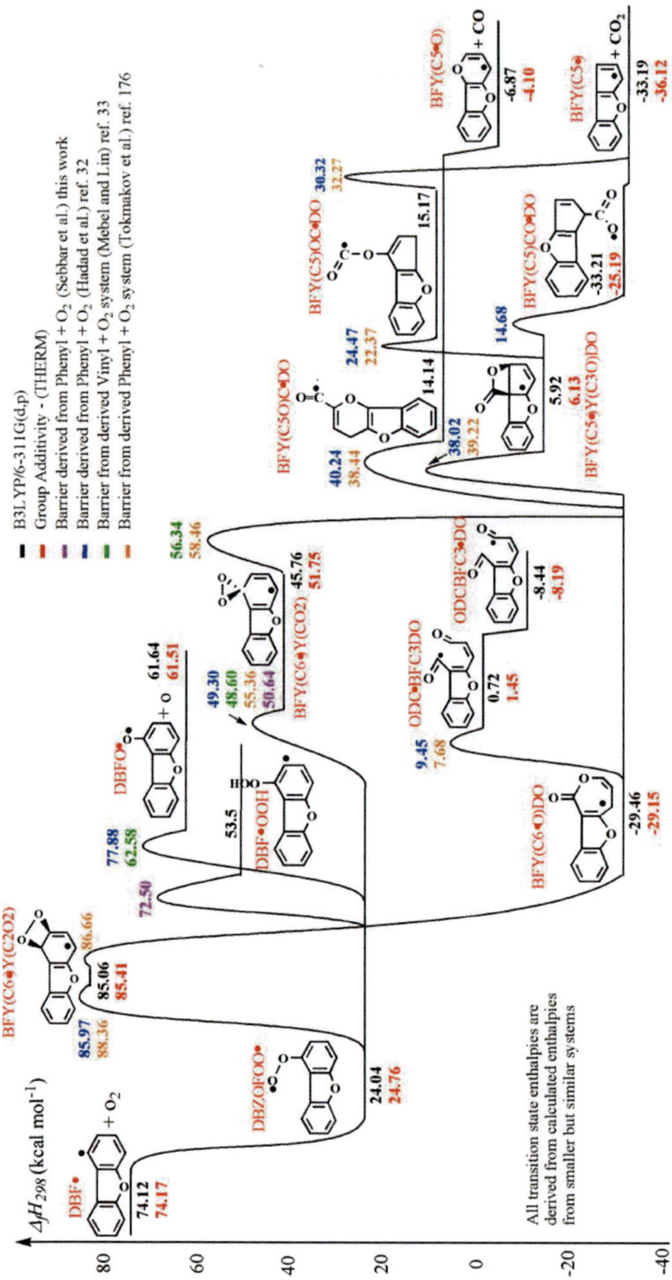

Figure 7.8: Potential Energy Diagram: Dibenzofuranyl + O_2

The abstraction from the ortho position via a five members ring structure is found to have a barrier that is below the entrance channel and thus has some importance. This H-shift reaction or isomerization (DBFOO• to DBF•OOH) results from the intramolecular abstraction of a hydrogen atom from the phenyl ring by the peroxy radical to form a hydroperoxide dibenzofuranyl radical. The H atom is in a bridge structure shifting from the carbon to the radical site. The reaction is calculated to be endothermic by 29 kcal mol^{-1} relatively to the peroxy adduct. The combination of the ring strain and the small reverse barrier brings the forward barrier to 42 kcal mol^{-1} but this is still below the initial energy of $[DBFOO\bullet]^{\#}$ at 50 kcal mol^{-1}. The reverse reaction has a barrier of only 13 kcal mol^{-1}.

The intramolecular addition channels are shown to further react through several paths undergoing the ring opening (unsaturated + carbonyl moieties) as well as the formation of benzofuran-cyclopentadienyl radical + CO_2.

The reaction to the formation of benzofuran-cyclopentadiene + CO_2 has two barriers and is of less importance. It has the same initial barrier as the reaction to the ring opening, but a second isomerization is needed with a tight transition state and barrier. The products resulting from the first ring opening reacts further. Most of the reaction proceeds through the first ring opening channel which forms a chemically activated benzofuran-hexadienyl 2-4, dione 1-6, that will immediately dissociate according to different pathways through elimination reactions or rapidly react with molecular oxygen.

The last important reaction is the dissociation of benzofuran-peroxy adduct back to dibenzofuranyl + O_2. This reaction is possible because of its loose transition state.

The chemically activated ODC•BFC3DO species (Figures 7.8 and 7.9) formed from the ring opening reaction will immediately dissociate according to complex pathways through a series of elimination reactions and through reactions with molecular oxygen. A barrierless abstraction reaction (H-shift), see Table 7.8, will take place immediately to form the isomer ODCBFC3•DO allowing this radical to dissociate according to three complex pathways shown below in Figures 7.10 and 7.11. The Multi channel, multi-frequency Quantum RRK calculations were performed to provides an evaluation of the rate constants of the reactions channels to ODC•BFC3DO and ODCBFC3•DO radicals versus pressure and temperature. Figure 7.10 illustrates the surface plot of the results of the master equation predictions of these reaction channels. The chemical activation channels to ODCBFC3•DO isomer is predicted to dominate markedly at all temperature and pressure which means that most of the ODC•BFC3DO radical will first react through a H-shift reaction, then proceeds to a series of dissociation reactions. The pathway describing the direct dissociation of the ODC•BFC3DO radical is calculated in this work and shown in Figure 7.12 but not considered in our kinetic study.

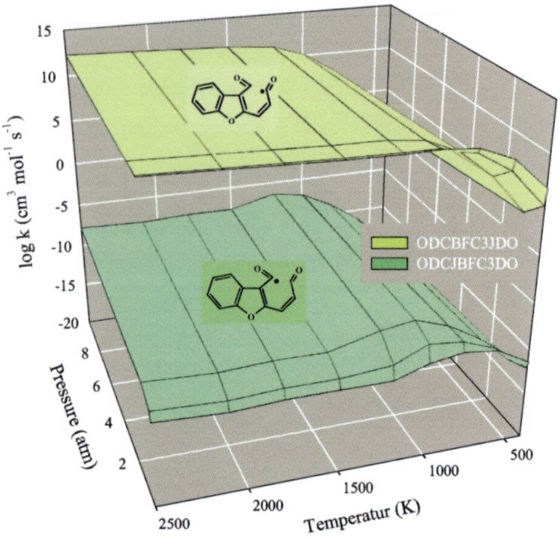

Figure 7.9: Surface plot rate constant for ODC•BFC3DO and ODCBFC3•DO radicals as a function of temperature and pressure.

Pathways 1, 2, and 3 are considered in this chapter and the kinetics parameters are calculated. Enthalpy of formation values of each radical contained in Figures 7.10 and 7.11 are determined using DFT calculation at B3LYP/6-311G(d,p) level combined with isodesmic reactions (whenever possible) as described in section 7.2 and other chapters. These enthalpy values are supported by group additivity analysis. The work reactions used for these species are given in Appendix G. Transition state structures are derived from the vinyl, phenyl or other similar species. Barrier from Hadad et al. work [32] and from Mebel et al. [33] are also used.

Figure 7.12 illustrate the dissociation of ODC•BFC3DO radical (Path-4), the enthalpy values of the radicals are estimated with DFT calculations and group additivity method, but the working reactions are not given in this work. The kinetics of this pathway is not considered in this study, but should be considered in a future work to improve the mechanism of destruction of dibenzofuran.

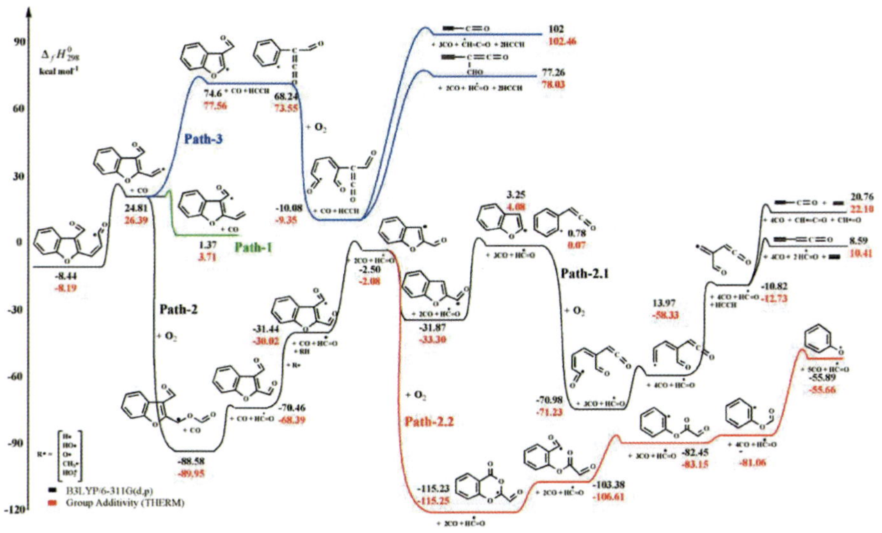

Figure 7.10: Path 2 and Path 3 - Dissociation of ODCBFC3•DO

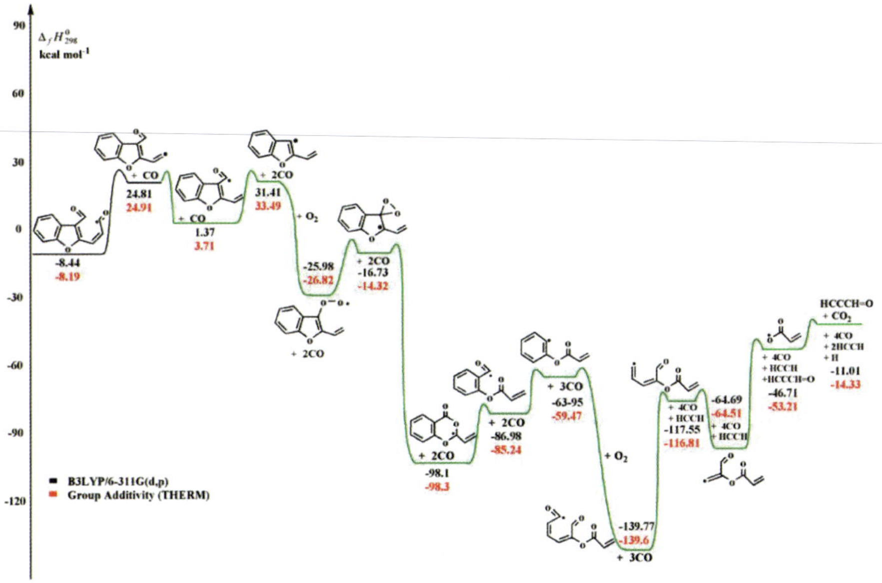

Figure 7.11: Path 1- Dissociation of ODCBFC3•DO

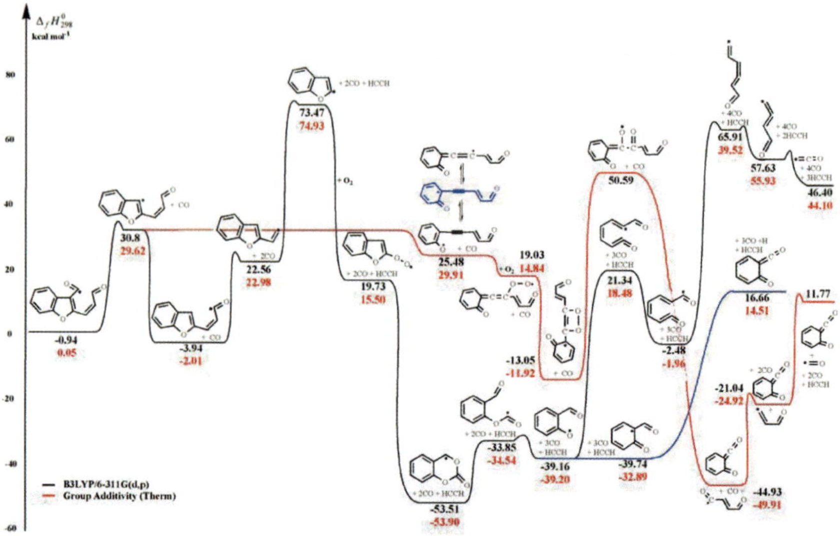

Figure 7.12: Path 4- Dissociation of the ODC•BFC3DO

7.2.5.2 Kinetic Parameters – Mechanism Construction

To apply the results of the chemical activation and the thermal dissociation analysis for comparison to literature or experimental data (when available) it is necessary to construct an elementary chemical reaction mechanism. The mechanism includes all the reactions involved in the chemical activation process, including stabilizations and reactions for thermal dissociation of the stabilized species. The reactions are reversible, so that we implicitly take into account some of the thermal dissociation reactions as the reverse of the forwards (chemically activated) reactions. For example, the dissociation of DBFOO• to DBF• + O_2 is included as being the reverse reaction of DBF• + O_2 ↔ DBFOO•. ThermKin is used to determine the elementary reaction rate coefficients and express the rate coefficients in Arrhenius forms.

The kinetic parameters of the reactions describing the DBF• + O_2 reaction system of the current study are determined with the following methods:

i. From computational results based on thermodynamic properties and frequencies

ii. From the similar system, Vinyl + O_2

iii. From the similar system, Phenyl +O_2

iv. From similar well-known reaction, which reactive part is contained in our unknown reaction.

An example of this method is illustrated below:

The CO elimination reaction of the cyclic radical shown in Figure 6.13 below can be reduced to the simple reaction:

$CH_2{\bullet}OCH{=}O \rightarrow CH_2{=}O + CH{\bullet}{=}O$

Which energy barrier can be easily calculated using high level calculation methods like G3 (see Table 7.3). With the ThermKin code and the G3 calculations, the energy barrier B_1 for the forward reaction $CH_2{\bullet}OCH{=}O \rightarrow CH_2{=}O + CH{\bullet}{=}O$ is calculated to be 31.68 kcal mol^{-1}. The backward reaction B_{-1} can be then estimated as follows:

$$B_{-1} = B_1 - \Delta H^0_{rxn,298} = 31.68 - 18.99 = 12.69 \text{ kcal mol}^{-1} \quad (7.1)$$

The activation energy barrier for the big system is then

$$B_2 = B_{-1} + \Delta H^0_{rxn,298} = 11.09 + 12.69 = 23.78 \text{ kcal mol}^{-1} \quad (7.2)$$

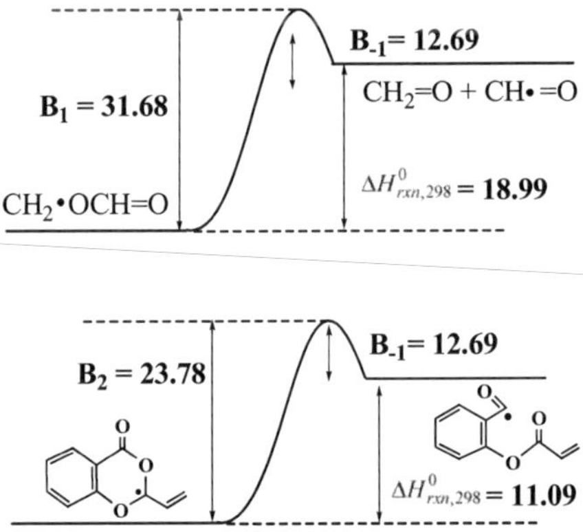

Figure 7.13: Example of energy barrier calculations

The kinetic parameters for the reactions contained in the pathways given above (Path 1, Path 2 and Path 3), in the form of modified Arrhenius rate, are given in Table 7.8. In the preceding text, we reported the approach we used to calculate the kinetic parameters for the dibenzofurany + O_2 system. To test again the appropriateness of the kinetic analysis methods used, we calculated small transition state structures and compared them to the literature data. These calculations allowed us to satisfactory reproduce a range of experimental data on the overall rate.

Table 7.8: Calculated Kinetic Parameters (600K <T< 2500K) for dibenzofuran system + O_2

Reactions		*A* *(cm³/mole s)*	*n*	*Ea* *Kcal mol⁻¹*	*delta*
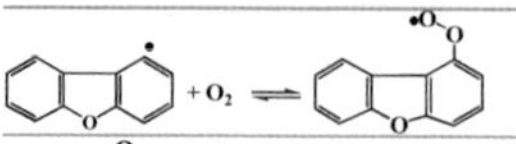	Barrierless	-	-	-	-
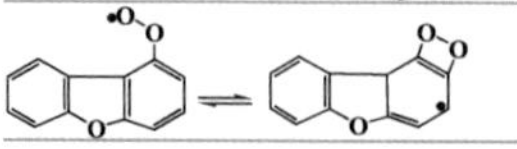	Derived from phenyl system Ea = 47.15	1.19E+07	1.458	61.93	61.02
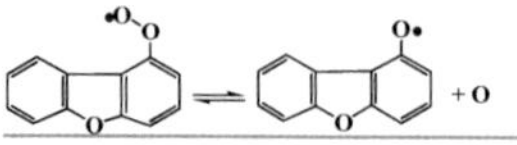	Derived from phenyl system Ea = 41.78	2.02E+08	1.742	38.54	37.61
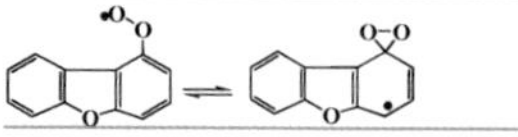	Derived from phenyl system Ea = 27.98	4.00E+11	0.507	30.38	21.72
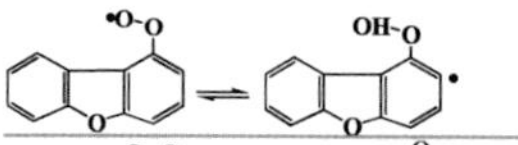	Derived from phenyl system Ea = 40.4	4.33E+05	1.941	41.8	29.46
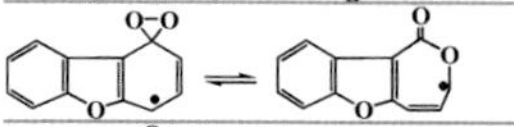	Derived from phenyl system Ea = 10.58	6.99E+09	0.856	10.58	-75.22
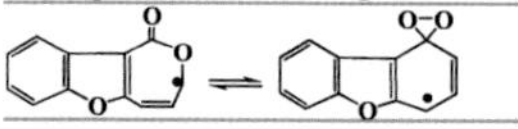	Derived from phenyl system Ea = 79.10	2.43E+08	1.163	84.45	75.22
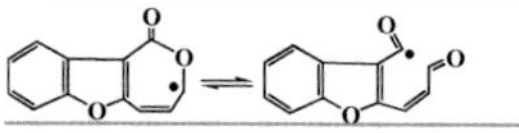	Derived from phenyl system Ea = 26.21	1.52E+11	0.516	38.01	30.18
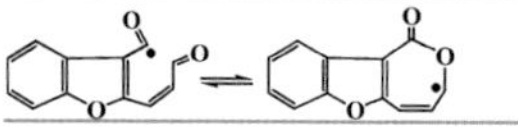	Derived from phenyl system Ea = 4.96	3.46E+10	0.676	4.96	-30.18
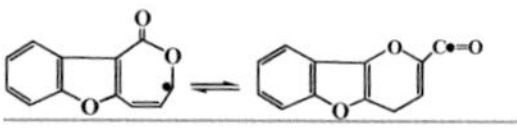	Derived from phenyl system Ea = 50.01	7.44E+09	0.523	69.70	43.60
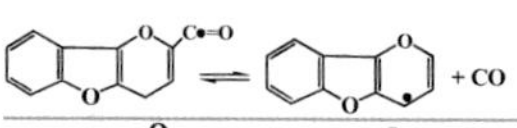	Derived from phenyl system Ea = 8.06	5.68E+11	0.398	8.06	-21.11
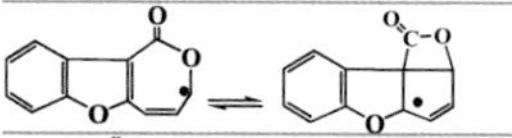	Derived from phenyl system Ea = 39.37	3.25E+11	0.364	63.67	35.38
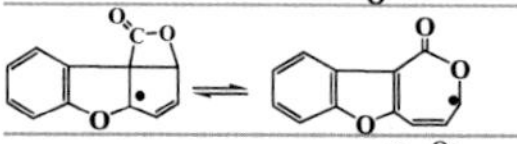	Derived from phenyl system Ea = 29.57	2.2169E+12	0.367	25.58	35.38
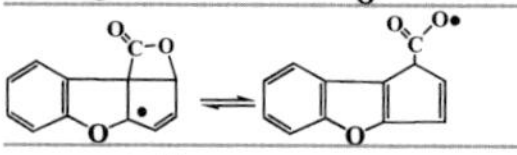	Derived from phenyl system Ea = 8.76	2.74E+12	0.319	8.76	-39.46
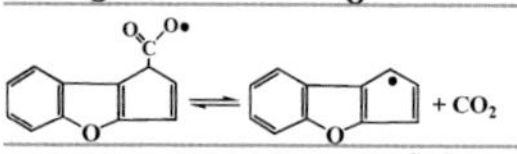	Derived from phenyl system Barrierless	2.77E+11	0.283	Barrierless	0.61
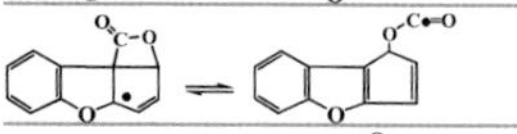	Derived from phenyl system Ea = 28.97	2.73E+12	0.365	21.02	9.25
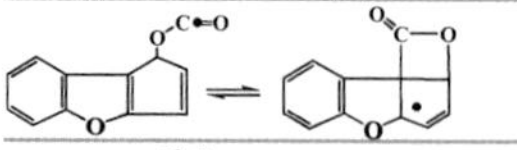	Derived from phenyl system Ea = 17.20	1.1625E+10	0.358	17.20	-9.25
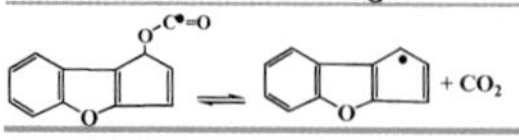	Derived from phenyl system Ea = 23.13	2.72E+10	0.499	23.13	-48.10

Temperature= 600-2500K Pressure = 1 atm

Reactions		A (cm³/mole s)	n	Ea Kcal mol⁻¹	delta
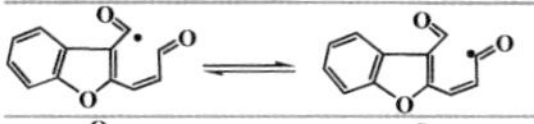	Abstraction reaction – Barrierless	2.63E+12	0.301	-	
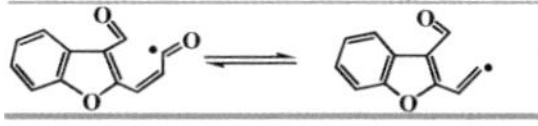	Derived from phenyl system ODC6JDO → CJC4DO + CO Ea = 34.97	2.1524E+11	0.851	42.63	33.25

PATH-1

Reactions		A (cm³/mole s)	n	Ea Kcal mol⁻¹	delta
	Derived from phenyl system: abstraction CJC4DO → C4CJDO Ea = 2.82	2.02E+09	0.964	2.82	-23.44
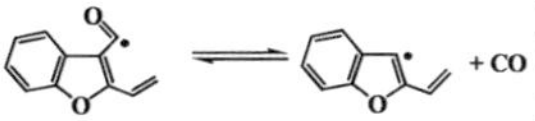	Derived from phenyl system C4CJDO → CDCCDCJ + CO Ea = 30.74 Derived from calculated BFCJDO → BFJ + CO Ea = 36.77	6.45E+12 2.15E+14	0.697 0.667	33.18 31.69	30.04
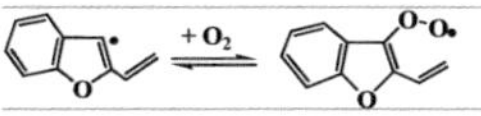	Barrier less				
	Derived from phenyl system PHOOJ → Y(C6J)Y(CO2) Ea = 27.98 based on Lin and Mebel Ea = 29.01 own calculation	1.75E+06 9.68E+06	1.864 0.843	12.13 13.12	9.25
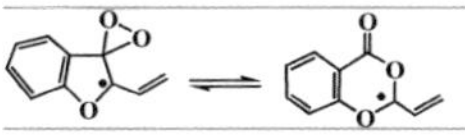	From phenyl system Y(C6J)Y(CO2) → Y(C6JO)DO Ea =10.58	6.99E+09	.856	10.58	-83.97
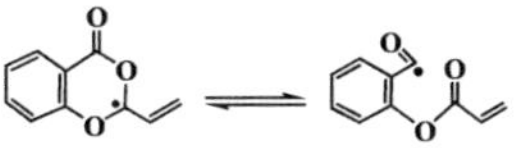	Derived from own calculation CJOCDO → CH•=O + CH₂=O Ea = 31.68	2.29E+08	2.011	23.78	11.09
	Derived from Mebel et al. calculation CJOCDO → CH•=O + CH₂=O Ea = 29.5	-	-	21.60	-
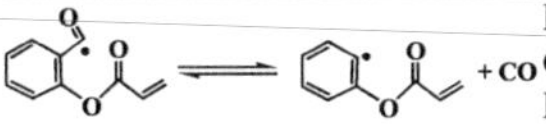	Derived from own calculation CDCCJDO → CDCJ + CO Ea = 26.52 Derived from calculated BFCJDO → BFJ + CO Ea = 36.77	2.88E+12 2.15E+14	0.568 0.667	24.78 24.68	23.03
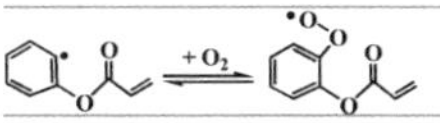	Barrierless				
	Derived from phenyl system PHOOJ → Y(C6J)Y(CO2) Ea = 27.98	1.75E+06	1.864	52.60	43.94
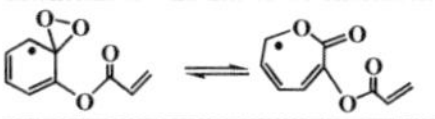	From phenyl system Y(C6J)Y(CO2) → Y(C6JO)DO Ea =10.58	6.99E+09	0.856	10.58	-89.60
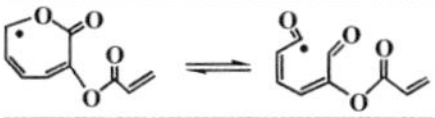	Derived from phenyl system Y(C6JO)DO → ODC6JDO Ea = 26.20	1.52E+11	0.516	21.10	13.24
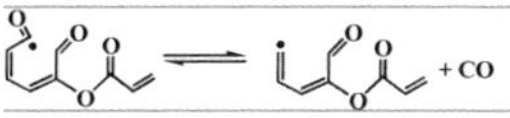	Derived from phenyl system ODC6JDO → CJC4DO + CO Ea = 34.97	2.15E+11	0.851	31.60	22.22
	Derived from reaction CDCCDCJ → CDCJ + CTC Ea = 43.98 Calculated	7.82E+12 8.44E+11	0.869 0.312	58.37 44.10	52.86
	Calculated Ea = 20.70	2.45E+12	0.702	20.70	13.75
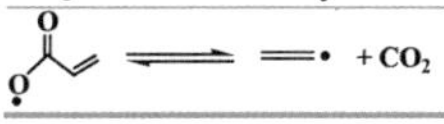	Calculated Ea = 1.52	5.14E-05	5.87	1.52	4.04

PATH-2

Reactions		*A* *(cm³/mole s)*	*n*	*Ea* *Kcal mol⁻¹*	*ref*
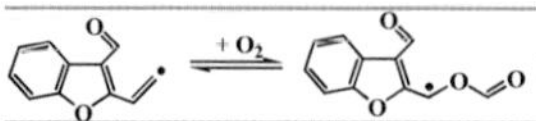	No barrier	-	-	-	-
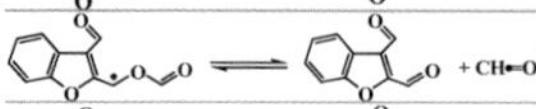	Deriver from calculated CJOCDO → CH•=O + CH_2=O Ea = 31.68	2.29E+08	2.011	31.68	-18.69
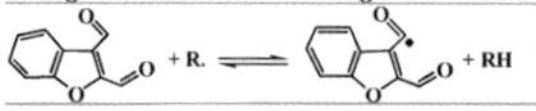	From CH_2=O + H → CH•=O + H_2 $1.0x10^{14}$ (cm^3/mole s) $e^{-10.49\ (kcal/mole)/RT}$	1.0E+14	-	10.49	216
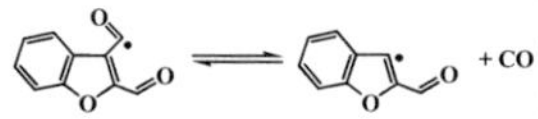	Derived from phenyl calculation: ODC6JDO → CJC4DO + CO Ea = 34.97	2.15E+11	0.851	39.32	29.94
	Derived from calculated BFCJDO → BFJ + CO Ea = 36.77	2.15E+14	0.667	31.59	
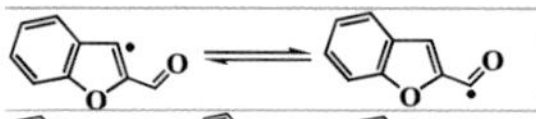	Calculated, Abstraction reaction	6.85E+10	0.872	28.0	-30.37
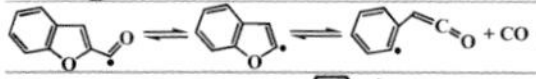	Calculated	2.15E+14	0.667	36.77	35.12
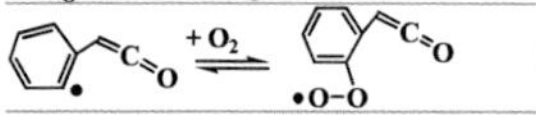	Barrierless	-	-	-	-
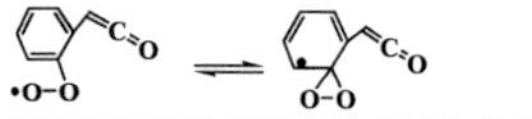	Derived from phenyl system PHOOJ → Y(C6J)Y(CO2) Ea = 27.98	1.75E+06	1.864	41.88	33.22
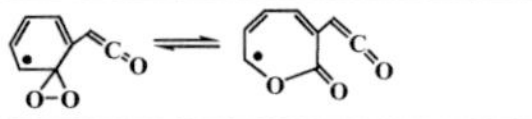	From phenyl system Y(C6J)Y(CO2) → Y(C6JO)DO Ea =10.58	6.99E+09	0.856	10.58	-70.57
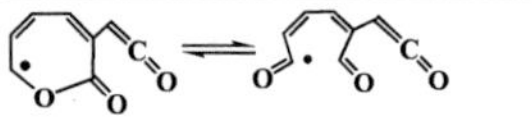	Derived from phenyl system Y(C6JO)DO → ODC6JDO Ea = 26.20	1.5182E+11	0.516	19.22	11.36
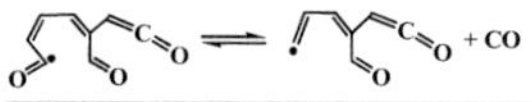	Derived from phenyl system ODC6JDO → CJC4DO + CO Ea = 34.97	2.15E+11	0.851	24.2	14.80
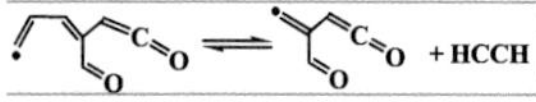	Derived from reaction: CDCCDCJ → CDCJ + CTC Ea = 43.98	7.82E+12	.869	55.87	50.36
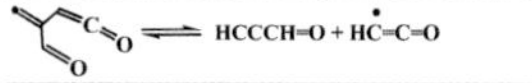	Calculated	6.03E+07	2.911	28.57	
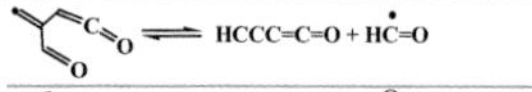	Derived from CJDCCDO → CTC + CJDO Ea = 31.19	2.72E+13	0.482	27.93	19.41
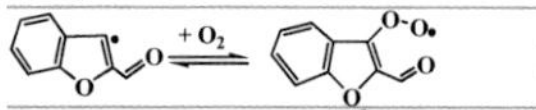	Barrier less	-	-	-	-
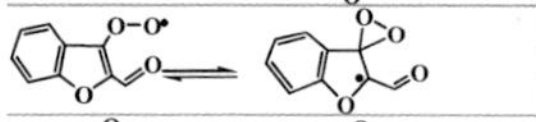	Derived from phenyl system PHOOJ → Y(C6J)Y(CO2) Ea = 27.98	1.75E+06	1.864	23.31	14.65
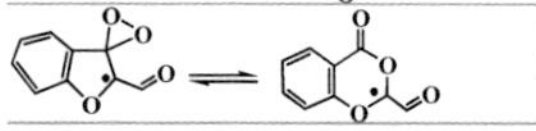	From phenyl system Y(C6J)Y(CO2) → Y(C6JO)DO Ea =10.58	6.99E+09	0.856	10.58	-67.36
	Derived from own calculation CJOCDO → CH•=O + CH_2=O Ea = 31.68	2.29E+08	2.011	24.97	11.95
	From: CJOCDO → CH•=O + CH_2=O Mebel et al. calculations Ea = 29.5	-	-	22.46	
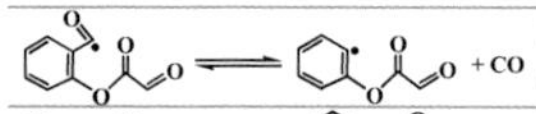	Derived from phenyl system ODC6JDO → CJC4DO + CO Ea = 34.97	2.15E+11	0.851	31.31	21.93
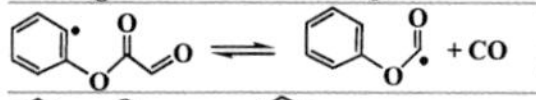	Derived from phenyl system ODC6JDO → CJC4DO + CO Ea = 34.97	2.15E+11	0.851	34.97	-22.56
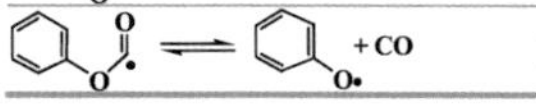	From COCJDO → CH_3O + CO Ea = 21.86	3.50E+12	0.511	6.13	3.96

PATH-3

Reactions		A ($cm^3/mole\ s$)	n	Ea Kcal mol^{-1}	ref
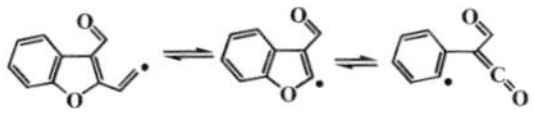	Derived from CDCCDCJ → CTC + CDCJ Ea = 43.98	7.82E+12	0.869	53.34	47.83
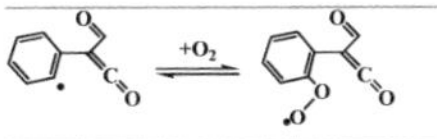	Barrier less				
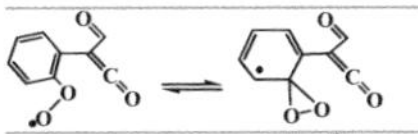	Derived from phenyl system PHOOJ → Y(C6J)Y(CO2) Ea = 27.98	1.75E+06	1.864	42.42	33.76
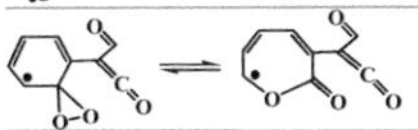	From phenyl system Y(C6J)Y(CO2) → Y(C6JO)DO Ea =10.58	6.99E+09	0.856	10.58	-76.80
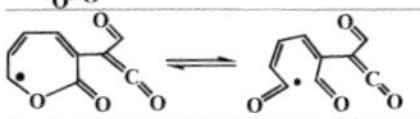	From phenyl system Y(C6JO)DO → ODC6JDO Ea = 26.20	1.5182E+11	.51609	27.74	19.88
	Derived from phenyl calculations ODC6JDO → CJC4DO + CO Ea = 34.97	2.15E+11	0.851	54.14	44.76
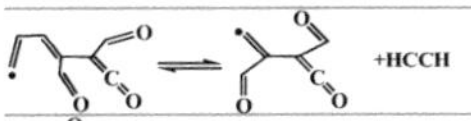	Derived from reaction: CDCCDCJ → CDCJ + CTC Ea = 43.98	7.82E+12	0.869	46.35	40.84
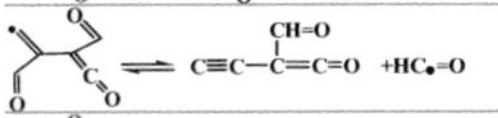	Derived from CJDCCDO → CTC + CJDO Ea = 31.19	2.72E+13	0.482	37.67	28.15
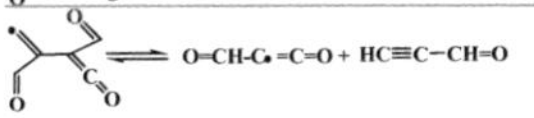	Derived from CJDCCDO → CTC + CJDO Ea = 31.19	2.72E+13	0.482	47.20	37.68
O=CHC•=C=O ⇌ CH•=C=O + CO	Barrierless				

7.2.5.3 Chemical Activation Reaction: Dibenzofuranyl + O_2

Master equation analysis for Fall-off is performed on the DBF• + O_2 reaction system to estimate the rate constants and to determine the important reaction paths as a function of temperature and pressure.

The input file for QRRK analysis with master equation analysis for Fall-off is given in Appendix H and lists the kinetic parameters of the important pathways of the DBF• + O_2 system, calculated with DFT (B3LYP), or derived from the literature by using similar but smaller systems. The method of using and adjusting literature data to our system has one advantage and one goal: it verifies in a first time the accuracy of our calculations and shows in a second time the reliability of using small systems as reference for large systems like dibenzofuranyl + O_2 for which high-level calculations have costly computation time and are difficult if not possible. Because of the active phenyl site contained in dibenzofuranyl, we consider that in a first step the DBF• + O_2 system will react like the Phenyl + O_2 system. Therefore, most of the kinetic parameters determined for the DBF• + O_2 system are derived from the Phenyl + O_2 system, where high-level calculations were still feasible.

The chemical activation of the dibenzofuranyl + O_2 system is carried out using the "CHEMASTER" program. Figures 7.14 and 7.15 illustrate the results for the temperature dependence of the various reaction channels at fixed pressure of 0.01, 1, and 10 atm. Figure 7.14 predicts the master equation results for low temperature (400 K<T<1200 K), while Figure 7.15 illustrates results of high temperature reactions (1200 K<T<2500 K).

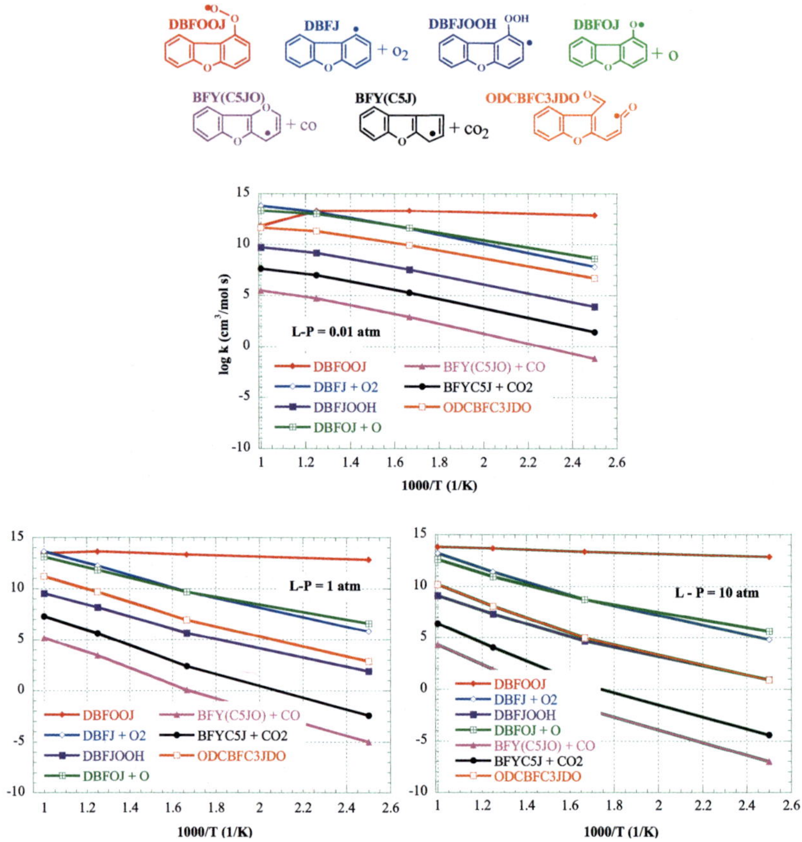

Figure 7.14: Product profiles at a function of temperature (400 K<T<1200 K) by constant pressure.

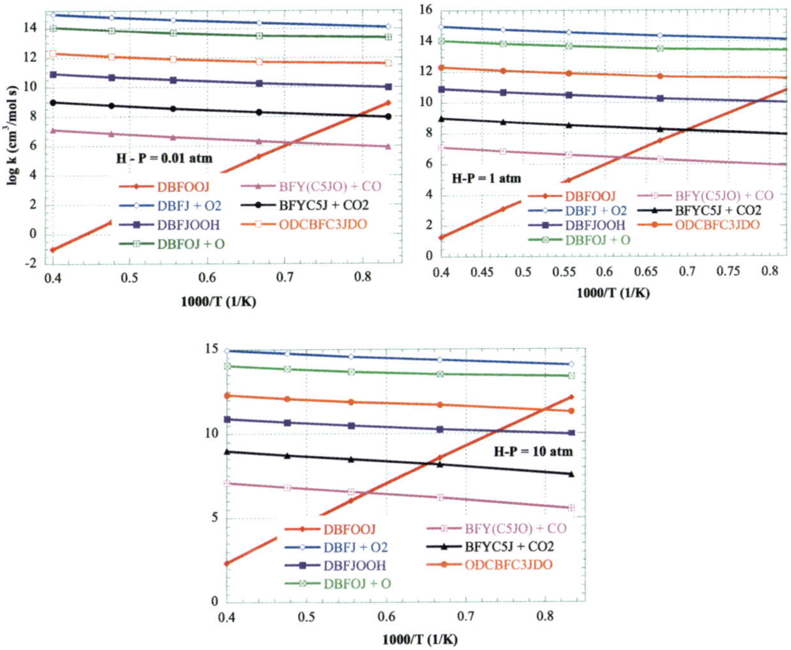

Figure 7.15: Product profiles at a function of temperature (1200 K<T<2500 K) by constant pressure.

At low temperature (below 1200 K), the contribution to the various channels are markedly increased. The reaction to benzofuran-phenoxy + oxygen atom (DBFOJ + O) and to the back reaction, dibenzofuranyl + O_2 (DBFJ + O2) are predicted to dominate and are competing. At high temperature (above 1200 K), these two channel remain the most important but the dissociation back to reactants DFJ + O2 becomes now the dominant pathway. In this chemical activation reaction the ring opening product, ODCBFC3JDO, is formed with about 76 kcal mol^{-1} of excess energy. We treat this as an energized adduct and allow dissociation (beta scission reaction) with CO elimination. This radical will further dissociate according to different possible path or rapidly react with molecular oxygen. The contribution to ODCBFC3JDO is the next important channel and competes with DBFJOOH channel at low temperature at a fixed pressure of 10 atm. Above 1200 K the ring opening channel more important than the DBFJOOH formation.

The BFY(C5J) + CO_2 contribution which is still of importance in this system is relatively low compared to the other channels. BFY(C5JO) + CO formation is less important in this system.

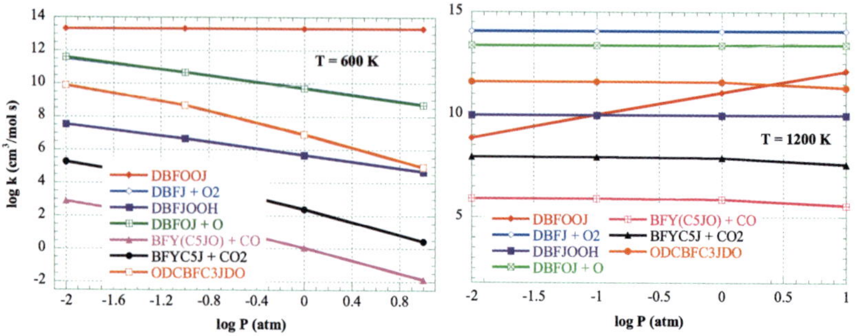

Figure 7.16: Product profiles at a function of pressure by constant temperature.

The reaction product profiles versus pressure at 600 K and 1200 K are illustrated in Figure 7.16. At 600 K the DBFOJ + O channel is in competition with the back reaction to reactants DBFJ + O2, but with increasing temperature, DBFJ + O2 contribution dominates slightly. The contribution to ODCBFC3JDO and DBFJOOH channels are the next important paths and are noticeably higher than the last two channels. The contribution to BFY(C5JO) + CO is noticeably negligible compared to all other channels.

7.2.5.4 Unimolecular Dissociation of DBFOO•

Figure 7.17 illustrates the dissociation reaction of the stabilized dibenzofuran-peroxy adduct to the several important reaction channels as a function of temperature and a pressure of 1 atm. Figure (A) illustrate the predicted results at low temperature range (400 K<T<1200 K), and figure (B) is calculated at temperature range of 1200 K – 2500 K.

Figure 7.18 describes the dissociation of dibenzofuran-peroxy radical (DBFOO•) versus pressure at 600 K and 1200 K.

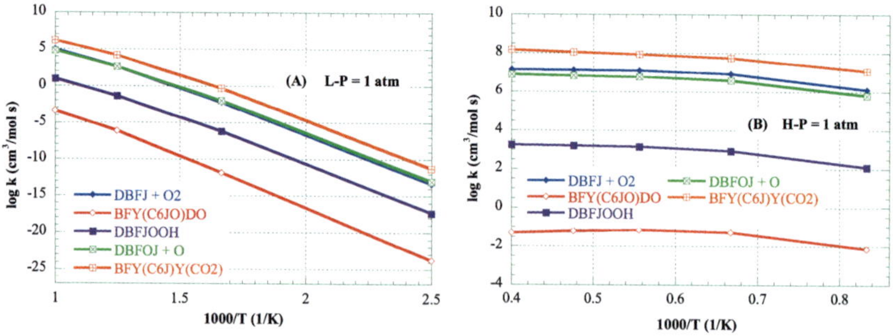

Figure 7.17: Dissociation of dibenzofuran-peroxy radical (DBFOO•) at 1atm versus low range (400 K<T<1200 K) and high range (1200 K<T<2500 K) temperature.

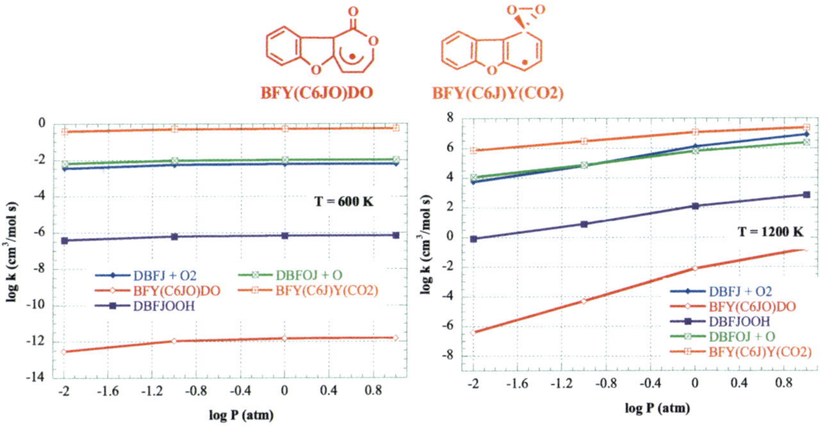

Figure 7.18: Dissociation of dibenzofuran-peroxy radical (DBFOO•) versus pressure at constant temperature.

The overall dominant channel here is the contribution to the BFY(C6J)Y(CO2) channel. The important path of this BFY(C6•)Y(CO2) isomer is the reaction to BFY(C6•O)DO which dissociates to the final products ODCBFC3JDO and BFY(C5J) + CO_2. A more complete analysis shows that the dissociation of this dibenzofuran-peroxy adducts predicts a competition between the two loose transition state channels: benzofuran-phenoxy + O, and the reaction back to DBFJ + O_2. At lower temperature, the lower energy channel to benzofuran-phenoxy + O is slightly dominant, but at higher temperature, sufficient energy is available for the energized radical to overcome the energy associated to the reaction back to reactants, and we see a negligible dominance of this DBFJ + O_2 channel.

7.3 Conclusion

Enthalpies of formation for the dibenzofuranyl + O_2 reaction system are determined with DFT and group additivity methods. In cases for which the DFT method failed, ab initio calculations were performed. Entropies and heat capacities are determined for selected radicals.

We show that the vinyl system is a good model for the phenyl system which it itself used as model for the dibenzofuranyl system. The high-level calculations on the smaller vinyl system can be used to calibrate *ab initio* and Density Functional Theory calculations on the phenyl and the dibenzofuranyl system.

We have determined the PES for the DBF• + O_2 system at the B3LYP level. The dibenzofuranyl + O_2 association results in a chemically activated benzofuran-peroxy radical with some 50 kcal mol^{-1} well depth. This chemically activated adduct can dissociate to benzofuran-phenoxy radical + O, or react back to dibenzofuranyl + O_2. It can also undergo intramolecular addition of peroxy radical to several unsaturated carbon sites on the ring. The intramolecular addition channels are shown to further react through several paths undergoing ring opening as well as formation of benzofuran-cyclopentadienyl radical + CO_2. The radical resulting from the ring opening reacts further through complex reaction paths to a number of unsaturated and carbonyl moieties which have been analysed in this work.

The kinetic analysis approach for the determination of the barriers of the dibenzofuranyl system consisted of reducing a species to ist smallest active unity to allow the calculation of the transition state structures. Further improvements were achieved by using literature data, calculated data of Mebel et al. on the vinyl + O_2 system and those of Hadad et al. on the phenyl + O_2 system. This approche confirmed the satisfactory accuracy of DFT with isodesmic reactions method.

An elementary reaction mechanism for the dibenzofuranyl + O_2 reaction system is constructed to describe the formation of products and reagent loss.

8. Conclusion

The emission of dibenzofurans and dioxins from industrial processes is a major environmental concern. One option to reduce emission levels is to vary the operation conditions in such way that dibenzofurans and dioxins are largely destroyed prior to their release into the environment. Focussing on dibenzofuran, the main objective of this work is to improve our understanding of the general oxidation chemistry and to provide a mechanism suitable for future modelling studies.

To achieve this goal a major obstacle related to the size of the aromatic species involved in the dibenzofuranyl + O_2 system had to be overcome. For large size species, only rather low level quantum mechanic calculations are affordable, which are not able to provide accurate thermochemical data. To circumvent this problem, the investigation of the dibenzofuranyl + O_2 system has been approached as follows:

- We used the DFT method which is fast, can handle large compounds and yield reasonably accurate energies. When combined with isodesmic work reactions we showed that the accuracy improves significantly and the results become suitable for thermochemical evaluations.
- We reduced the system dibenzofuranyl + O_2 to a smaller but similar system, phenyl + O_2. Later reaction can further be reduced to the smallest possible unity, vinyl + O_2.

For that purpose, the behaviour of three systems, *Vinyl* + O_2, *Phenyl* + O_2 and *Dibenzofuranyl* + O_2 and their products were investigated.

The investigation of the vinyl system consisted in the calculation of thermochemical properties of a series of peroxides and peroxy species using DFT combined with isodesmic reactions. We showed that the vinyl radical for which high-level calculations can be performed is a good model for the phenyl. The accuracy of the DFT method, which is the primary method for this work, was checked through a number of comparisons with high-level calculations. At this point, it is worthwhile mentioning the importance of the vinyl + O_2 reaction system:

(a) Vinyl is interesting in its own right, (b) calculations can be performed at different levels leading to the confirmation of the methodology and allowing comparison with literature.

The investigation of the phenyl radical with O_2 reaction system was the second step in our systematic approach. This part is of value, because the electronic structure of the phenyl radical is analogue to that of the dibenzofuranyl radical. The phenyl system can therefore serve as a model from which the oxidation of other heteroatomic aromatic radicals can be considered. Thermodynamic properties of intermediates, transition states and products, important to formation and destruction of the aromatic ring were calculated. The vinyl system was used to calibrate *ab initio* and DFT calculations for the phenyl system. Partially based on the results obtained with the vinyl system, high-pressure-limit kinetic parameters are obtained using cannonical Transition State Theory. An elementary reaction mechanism was constructed to model experimental data obtained in a combustor at 1 atm, and in high-pressure turbine systems (5-20 atm), as well as in supercritical water.

In the final phase, we evaluated enthalpies of formation of a series of stables species and radicals related to the dibenzofuranyl + O_2 reaction system. Calculations were performed on the decomposition of the rings with DFT and group additivity methods. After showing that the phenyl system is a good model for the dibenzofuranyl system, transition state species related to the dibenzofuranyl reaction system were derived from those calculated for the phenyl system.

The addition of O_2 to dibenzofuranyl results in a chemically activated benzofuranyl-peroxy radical with some 50 kcal mol^{-1} well depth. This chemically activated adduct can dissociate to the benzofuran phenoxy radical + O, or react back to dibenzofuranyl + O_2, both with relatively loose transition states and barriers near that of the entrance channel. It can also undergo intramolecular addition of peroxy radical to several unsaturated carbon sites on the ring via somewhat more tight transition states; but one path has a significantly lower barrier, 10 kcal mol^{-1} below the entrance channel. The intramolecular addition channels are shown to further react through several paths undergoing ring opening (unsaturated + carbonyl moieties) as well as benzofuran cyclopentadienyl radical + CO_2. Finally, a simplified mechanism for the association of dibenzofuranyl with O_2 has been developed in this work.

In addition to the systems described above, related species have been investigated as well. The reason is that very little thermochemical data on peroxy and peroxides species are available in the literature and these data are essential for this study. Therefore, we have developed thermochemical properties of a series of species (stable molecules and radicals). Group Additivity was developed as well because it is an excellent tool for the thermochemical estimation of large molecules. Taking into account the importance of accuracy of the groups developed with DFT, is was necessary to validate the GA results with another computational method. We choose G3MP2B3 method which is a modified version of the well regarded G3 method. The overall results showed good agreement between the G3MP2B3 and DFT values. Data reported in this study suggest that the DFT analysis in connection with isodesmic working reactions to reduce errors provide a reasonable method for the determination of enthalpies of formation.

With the completion of this work, an elementary submechanism is now available that can be used in the future to model the decomposition of dibenzofuran in the exhaust. One straightforward application would be to implement it into larger mechanism and compare its prediction to experimental compounds.

In conclusion, this work shows that ab intio and DFT methods are excellent tools for the determination of unknown thermochemical properties. The proven accuracy of quantum mechanics calculations make them cost-effective alternatives to time-consuming, difficult experiments and we expect their role to steadily increase in the future.

9. Appendices

Appendix A: Literature Reference Data

We have summarized here, the literature references of the species needed and used in the previous chapters. Table A.1 gives the enthalpy of formation of the stable hydrocarcons and Table A.2 summarizes the enthalpy of formation of the radicals. Table A.3 summarizes the enthalpy of formations of the oxygen containing radicals available in the literature. Table A.4 lists the literature enthalpy of formation data of oxygen containing species used in the above chapters.

Table A.1: Enthalpies of Formation for Standard Species in Work Reactions

Species	ΔH^0_{f298} (kcal mol^{-1})	Source
CH_4	-17.89 ± 0.07	ref. 217
CH_3CH_3	-20.24 ± 0.12[a] / -20.04 ± 0.07[b]	ref. 217[a] / ref. 218[b]
$CH_3CH_2CH_3$	-24.82 ± 0.14 / -25.02 ± 0.12	ref. 217 / ref. 218[b]
CH_2=CH_2	12.55 ± 0.1	ref. 219
CH_3CH=CH_2	4.83 ± 0.09[a] / 4.88[b]	ref. 220[a] / ref. 221[b]
CH_3CH=$CHCH_3$	-2.58 ± 0.24 / -1.83 ± 0.30	ref. 222
CH_2=$CHCH_2CH_2CH_3$	-5.33 ± 0.31[a] / -5.03[b]	ref. 223[a] / Therm[b]
CH_2=$CHCH_2CH$=CH_2	25.41 ± 0.31	ref. 224
CH_2=$CHCH_2CH_3$	-0.15 ± 0.19	ref. 222
CH_2=$CHCH$=CH_2	26.0 ± 0.19	ref. 222
CH_2=$CHCH$=$CHCH$=CH_2	40.1 ± 0.6	ref. 225
CH_3CH=$CHCH$=$CHCH_3$	11.5 ± 0.4	ref. 225
$CH_3(CH_3)C$=CH_2	-4.29 ± 0.26	ref. 222
HC≡CH	54.19 ± 0.19	ref. 226
CH_3C≡CH	44.32 ± 0.21	ref. 226
CH≡CCH_2CH_3	39.48 ± 0.21	ref. 222
C_6H_6	19.81 ± 0.12	ref. 227
$C_6H_5CH_3$	11.95 ± 0.15	ref. 227
$C_6H_5CH_2CH_3$	7.12 ± 0.20	ref. 227
C_6H_5CH=CH_2	35.11 ± 0.24	ref. 228
Y(C_5H_6) (YC5)	31.89 ± not given[a] / 31.26[b]	ref. 229[a] / Therm[b]
C_6H_8 (YC6)	26.1 ± not given[a] / 26.05[b]	ref. 230[a] / Therm[b]

Table A.2: Enthalpy of Formation for Radicals in Work Reactions

Species	ΔH^0_{f298} (kcal mol^{-1})	Source
H•	52.10 ± 0.001	ref. 231
CH_3•	34.82 ± 0.2	ref. 232
CH_3CH_2•	28.4 ± 0.5	ref. 233
$CH_3CH_2CH_2$•	23.9 ± 0.5	ref. 233
CH_2=CH•	71.62 ± 0.05[a] / 71.0 ± 1.0[b] / 71.72 ± 0.4[c]	ref. 219[a] / ref. 233[b] / ref. 26[c]
CH_2=CHCH_2•	40.9 ± 0.7[a]	ref. 233
CH_2=C•CH_3	58.89 ± 0.12	ref. 26
HC≡C•	140.75 ± 0.26[a] / 133.0 ± 2.0[b]	ref. 26[a] / ref. 233[b]
C_6H_5•	81.4 ± 0.16[a] / 81.0 ± 2.0[b]	ref. 26[a] / ref. 233[b]
Y(C_5H_5•)	62	ref. 26

Table A.3: Enthalpy of Formation for Radicals used in Work Reactions

Species	$\Delta_f H^0_{298}$ (kcal mol^{-1})	Source
CH•=O	10.40	ref. 219
CH_3C•=O	-2.9 ± 0.7	ref. 233
CH_3O•	4.1 ± 1.0	ref. 233
CH_2=CHO•	4.44 ± 0.31	ref. 26
CH_2•OH	-2.0 ± 1.0	ref. 233
O•CH=O	-31 ± 3 / -29.61	ref. 234 / ref. 29
CH_2=CHC(O•)=O	-22.47	ref. 26
CH_3CH_2OO•	-6.5 ± 2.36	ref. 235
CH_3CH_2O•	-2.03 ± 0.39 / -3.90 ± 1.27	ref. 26 / ref. 236
CH_2=CHOO•	24.45 ± 1.15 / 24.34 ± 0.42	ref. 237 / ref. 26
C_6H_5O•	12.94 ± 0.56 / 13.0 ± 1.0	ref. 26 / ref. 233
O•	59.55 ± 0.024	ref. 238
HO•	8.96 ± 0.09	ref. 239
HOO•	3.25 ± 0.5	ref. 240
CH_2=CHCH_2O•	23.48 ± 0.31	ref. 26
CH_3OO•	2.15 ± 1.22	ref. 235
CH_2=CHCH_2OO•	20.61 ± 1.08	ref. 26
C_6H_5OO•	35.38 /31.28 ± 0.48	ref. 241 / ref. 26

Table A.4: Enthalpies of Formation for Reference Species used in Work Reactions

Species	$\Delta_f H^0_{298}$ (kcal mol^{-1})	Source
CO	-26.41 ± 0.04	ref. 238
CO_2	-94.05 ± 0.03	ref. 238
CH_2=O	-27.70[a] / -27.77 ± 0.61[b]	ref. 219[a] / ref. 242[b]
CH_3CH=O	-40.80 ± 0.35[a] / -39.70 ± 0.12[b]	ref. 243[a] / ref. 244
CH_3CH_2CH=O	-45.09 ± 0.18	ref. 243
CH_3OH	-48.08 ± 0.05	ref. 245
CH_3CH_2OH	-56.24 ± 0.07	ref. 223
CH_2=$CHCH_2OH$	-29.55 ± 0.35	ref. 246
CH_2=C(CH_3)OH	-42.06[a] / -42.1 ± not given[b]	ref. 26[a] / 247[b]
Syn-CH_2=CHOH	-30.59 ± 0.55	ref. 247
cis-CH_3CH=CHOH	-38.84[a] / -41.6 ± not given[b]	ref. 26[a] / ref. 248[b]
trans-CH_3CH=CHOH	-38.85[a] / -40.4 ± not given[b]	ref. 26[a] / ref. 248[b]
CH_3CH=C(CH_3)OH	-47.65[a] /-50.7 ± not given[b]	ref. 26[a] / ref. 249[b]
CH_2=CHC(=O)OH	-80.51 ± 0.55	ref. 250
CH_3CH_2C(=O)OH	-108.4 ± 0.5	ref. 222
$CH_2(OH)_2$	-91.03	Therm
OHCH=O	-90.57 ± 0.14[a] / -90.49	ref. 223[a] / 251[b]
CH_3C(=O)OH	-103.26 ± 0.12	ref. 223
CH_3OCH_3	-43.99 ± 0.12	ref. 252
$CH_3CH_2OCH_3$	-51.73 ± 0.16	ref. 252
CH_3OCH=O	-84.97[a] /-83.39 ± 1.19[b]	ref. 253[a] / ref. 30[b]
CH_2=$CHOCH_3$	-24.54[a] / -23.48[b]	Therm[a] / ref. 254[b]
CH_2=$CHCH_2OH$	-29.55 ± 0.35	ref. 255
CH_2=C(CH_3)OH	-42.1	ref. 256
CH_2=C(OH)CH=CH_2	-18.0 ± 1.0	ref. 257
CH_2=CHOH	-29.8 ± 2.0	ref. 258
C_6H_5OH	-23.03 ± 0.14	ref. 259
$C_6H_5CH_2OH$	-22.6 ± 0.72	ref. 260
$C_6H_5OCH_3$	-17.27 ± 0.93	ref. 223
C_6H_5OCH=O	-51.55 ± 0.74[a] / -51.49 ± 0.26[b]	ref. 261[a]/ ref. 30[b]
Furan	-8.29	ref. 262
Furan-CH=O	-36.1	ref. 263
Y(C_5H_6O)	-1.37 ± 0.23	ref. 30
CH_3CH_2OOH	-39.5 ± 0.53	ref. 86
CH_3OOH	-31.8 ± 0.94	ref. 264
O=CHCH=O	-50.66[a] ± 0.19 / -50.6[b]	ref. 265[a] / Therm[b]
CH_2O_2	9.42	Therm
CH_2=CHOOH	-9.65	ref. 266
C_6H_5OOH	-2.68 ± 0.49	ref. 26
CH_2=C=O	-20.85	ref. 267

Appendix B: Geometries Parameters

B.1 Species used for the Hydroperoxide Study

Table B.1: Geometry Parameters for vinyl, allyl, ethynyl and phenyl hydroperoxides[a]

Structure		Bond length (angstrom)		Bond angle (degree)		Dihedral angle (degree)
CH_2=CHOOH						
C						
C	1	r21=1.327				
O	2	r32=1.368	1	a321=120.64		
O	3	r43=1.460	2	a432=107.62	1	d4321=151.2
H	1	r51=1.082	2	a512=121.37	3	d5123=-4.9
H	1	r61=1.081	2	a612=119.77	3	d6123=172.6
H	2	r72=1.084	1	a721=125.09	5	d7215=178.5
H	4	r84=0.967	3	a843=99.22	2	d8432=133.6
***trans*-CH_3CH=CHOOH**						
C						
C	1	r21=1.330				
C	2	r32=1.499	1	a321=123.89		
O	1	r41=1.369	2	a412=120.23	3	d4123=-171.2
O	4	r54=1.471	1	a541=107.22	2	d5412=-139.4
H	5	r65=0.967	4	a654=99.023	1	d6541=-133.5
H	1	r71=1.086	2	a712=124.79	3	d7123=4.5
H	2	r82=1.086	3	a823=118.28	1	d8231=-177.6
H	3	r93=1.092	2	a932=111.64	1	d9321=-4.8
H	3	r103=1.095	2	a1032=110.87	1	d10321=-125.4
H	3	r113=1.095	2	a1132=111.44	1	d11321=115.7
***cis*-CH_3CH=CHOOH**						
C						
C	1	r21=1.331				
C	2	r32=1.499	1	a321=125.65		
H	1	r41=1.084	2	a412=124.41	3	d4123=-178.5
H	2	r52=1.085	3	a523=118.06	1	d5231=177.7
H	3	r63=1.090	2	a632=111.16	1	d6321=11.6
H	3	r73=1.096	2	a732=111.37	1	d7321=-108.7
H	3	r83=1.094	2	a832=110.65	1	d8321=132.5
O	1	r91=1.371	2	a912=120.97	3	d9123=5.2
O	9	r109=1.467	1	a1091=107.34	2	d10912=142.2
H	10	r1110=0.967	9	a11109=99.06	1	d111091=133.2
CH_2=CHCH_2OOH						
C						
C	1	r21=1.328				
C	1	r31=1.498	2	a312=124.03		
O	3	r43=1.430	1	a431=106.67	2	d4312=-123.7
O	4	r54=1.454	3	a543=106.67	1	d5431=-169.4
H	1	r61=1.086	3	a613=115.46	4	d6134=57.1
H	2	r72=1.084	1	a721=121.58	3	d7213=-177.8
H	2	r82=1.085	1	a821=121.65	3	d8213=1.6
H	3	r93=1.096	1	a931=110.71	2	d9312=118.3
H	3	r103=1.097	1	a1031=111.54	2	d10312=-2.9
H	5	r115=0.966	4	a1154=100.12	3	d11543=-116.2
CH_3(CH_3)C=CHOOH						
C						
C	1	r21=1.336				
O	1	r31=1.369	2	a312=121.19		
O	3	r43=1.477	1	a431=107.23	2	d4312=-133.9
C	2	r52=1.504	1	a521=120.05	3	d5213=-171.8
C	2	r62=1.503	5	a625=117.27	1	d6251=-177.9
H	1	r71=1.086	2	a712=123.85	5	d7125=4.2
H	4	r84=0.967	3	a843=98.91	1	d8431=-132.4
H	5	r95=1.096	2	a952=111.13	6	d9526=-60.9
H	5	r105=1.092	2	a1052=112.04	6	d10526=178.2
H	5	r115=1.096	2	a1152=110.52	6	d11526=57.3
H	6	r126=1.089	2	a1262=111.46	5	d12625=-172.2
H	6	r136=1.096	2	a1362=110.98	5	d13625=67.1
H	6	r146=1.095	2	a1462=110.28	5	d14625=-51.0

[a] Distances in angstroms and angles in degrees. Geometry parameters optimized at the B3LYP/6-311G(d,p) level of theory.

Table B.1 (cont): Geometry Parameters for vinyl, allyl, ethynyl and phenyl hydroperoxides[a]

Structure		Bond length (angstrom)		Bond angle (degree)		Dihedral angle (degree)
CH_2=C(CH_3)OOH						
C						
C	1	r21=1.332				
C	2	r32=1.498	1	a321=126.48		
O	2	r42=1.381	3	a423=117.08	1	d4231=174.3
O	4	r54=1.470	2	a542=109.34	3	d5423=48.2
H	5	r65=0.967	4	a654=99.27	2	d6542=-127.3
H	1	r71=1.081	2	a712=120.71	3	d7123=4.7
H	1	r81=1.082	2	a812=120.49	3	d8123=-178.2
H	3	r93=1.091	2	a932=110.75	1	d9321=118.8
H	3	r103=1.090	2	a1032=110.29	1	d10321=-1.6
H	3	r113=1.095	2	a1132=110.48	1	d11321=-122.0
CH_3CH=C(CH_3)OOH						
C						
C	1	r21=1.337				
C	1	r31=1.498	2	a312=125.97		
C	2	r42=1.498	1	a421=126.07	3	d4213=178.3
O	2	r52=1.382	4	a524=116.33	1	d5241=-174.6
O	5	r65=1.479	2	a652=108.93	4	d6524=-55.6
H	1	r71=1.086	3	a713=117.31	2	d7132=-177.7
H	3	r83=1.096	1	a831=111.25	2	d8312=110.9
H	3	r93=1.089	1	a931=111.24	2	d9312=-9.6
H	3	r103=1.095	1	a1031=110.52	2	d10312=-130.7
H	4	r114=1.090	2	a1142=110.66	1	d11421=0.5
H	4	r124=1.092	2	a1242=110.60	1	d12421=-119.8
H	4	r134=1.095	2	a1342=110.54	1	d13421=121.1
H	6	r146=0.967	5	a1465=99.14	2	d14652=126.8
CH≡COOH						
H						
O	1	r21=0.969				
O	2	r32=1.502	1	a321=98.20		
C	3	r43=1.290	2	a 432=109.21	1	d4321=122.4
C	4	r54=1.202	2	a542=145.52	1	d5421=114.4
H	5	r65=1.061	2	a652=157.62	1	d6521=109.6
CH_3CH≡COOH						
C						
H	1	r21=1.093				
H	1	r31=1.094	2	a312=108.35		
H	1	r41=1.094	2	a412=107.67	3	d4123=115.7
C	1	r51=1.456	2	a512=112.00	3	d5123=-123.2
C	5	r65=1.204	4	a654=150.82	3	d6543=145.8
O	6	r76=1.292	5	a765=176.24	4	d7654=-164.6
O	7	r87=1.513	1	a871=111.84	4	d8714=8.5
H	8	r98=0.969	7	a987=97.92	6	d9876=124.8
C_6H_5OOH						
C						
C	1	r21=1.389				
C	1	r31=1.395	2	a312=120.59		
C	3	r43=1.391	1	a431=119.31	2	d4312=0.02
C	4	r54=1.395	3	a543=121.02	1	d5431=-0.15
C	5	r65=1.390	4	a654=118.75	3	d6543=-0.12
H	3	r73=1.083	1	a731=120.29	2	d7312=179.8
H	2	r82=1.083	1	a821=121.52	3	d8213=-179.7
H	1	r91=1.084	2	a912=119.22	6	d9126=-179.8
O	6	r106=1.381	5	a1065=124.68	4	d10654=-179.4
O	10	r1110=1.444	6	a11106=111.34	5	d111065=-4.3
H	11	r1211=0.967	10	a121110=99.82	6	d1211106=117.6
H	5	r135=1.0794	4	a1354=121.41	3	d13543=179.5
H	4	r144=1.0843	3	a1443=120.03	1	d14431=179.6

[a] Distances in angstroms and angles in degrees. Geometry parameters optimized at the B3LYP/6-311G(d,p) level of theory.

B.2 Species used for the Peroxide Study

Table B.2: Geometry Parameters for vinyl, allyl, ethynyl, and phenyl peroxides[a]

Structure		Bond length (angstrom)		Bond angle (degree)		Dihedral angle (degree)
CH_2=CHOOCH$_3$						
C						
C	1	r21=1.327				
O	2	r32=1.366	1	a321=129.04		
O	3	r43=1.453	2	a432=110.53	1	d4321=1.4
C	4	r54=1.420	3	a543=105.73	2	d5432=-126.6
H	1	r61=1.079	2	a612=122.13	3	d6123=-0.5
H	1	r71=1.081	2	a712=118.79	3	d7123=-179.8
H	2	r82=1.085	1	a821=123.92	6	d8216=178.92
H	5	r95=1.094	4	a954=111.52	3	d9543=64.481
H	5	r105=1.093	4	a1054=111.12	3	d10543=-59.0
H	5	r115=1.092	4	a1154=104.10	3	d11543=-177.0
***trans*-CH_3CH=CHOOCH$_3$**						
C						
C	1	r21=1.330				
O	2	r32=1.367	1	a321=120.40		
C	1	r41=1.499	2	a412=123.88	3	d4123=-171.2
O	3	r53=1.477	2	a532=106.99	1	d5321=-140.1
C	5	r65=1.415	3	a653=105.20	2	d6532=-145.4
H	1	r71=1.086	4	a714=118.23	2	d7142=-177.6
H	2	r82=1.086	1	a821=124.56	4	d8214=4.4
H	4	r94=1.095	1	a941=111.48	2	d9412=115.5
H	4	r104=1.092	1	a1041=111.62	2	d10412=-5.0
H	4	r114=1.095	1	a1141=110.91	2	d11412=-125.6
H	6	r126=1.094	5	a1265=111.67	3	d12653=63.7
H	6	r136=1.094	5	a1365=111.21	3	d13653=-59.5
H	6	r146=1.092	5	a1465=104.31	3	d14653=-177.5
***cis*-CH_3CH=CHOOCH$_3$**						
C						
C	1	r21=1.332				
C	1	r31=1.499	2	a312=125.61		
O	2	r42=1.369	1	a421=121.16	3	d4213=-5.0
O	4	r54=1.474	2	a542=107.15	1	d5421=143.5
H	1	r61=1.085	3	a613=118.03	2	d6132=-177.6
H	2	r72=1.084	1	a721=124.14	3	d7213=178.7
H	3	r83=1.096	1	a831=111.39	2	d8312=108.4
H	3	r93=1.090	1	a931=111.14	2	d9312=-11.8
H	3	r103=1.094	1	a1031=110.70	2	d10312=-132.7
C	5	r115=1.416	4	a1154=105.30	2	d11542=143.5
H	11	r1211=1.094	5	a12115=111.22	4	d121154=59.6
H	11	r1311=1.094	5	a13115=111.64	4	d131154=-63.6
H	11	r1411=1.092	5	a14115=104.24	4	d141154=177.6
CH_3(CH_3)C=CHOOCH$_3$						
C						
C	1	r21=1.336				
O	2	r32=1.367	1	a321=121.33		
O	3	r43=1.485	2	a432=106.98	1	d4321=-134.2
C	4	r54=1.414	3	a543=105.16	2	d5432=-144.7
C	1	r61=1.503	2	a612=122.62	3	d6123=5.8
C	1	r71=1.504	6	a716=117.24	2	d7162=177.7
H	2	r82=1.086	1	a821=123.63	6	d8216=-178.2
H	5	r95=1.095	4	a954=111.75	3	d9543=63.6
H	5	r105=1.094	4	a1054=111.33	3	d10543=-59.6
H	5	r115=1.092	4	a1154=104.33	3	d11543=-177.6
H	6	r126=1.089	1	a1261=111.43	7	d12617=-172.4
H	6	r136=1.096	1	a1361=111.02	7	d13617=66.9
H	6	r146=1.095	1	a1461=110.31	7	d14617=-51.2
H	7	r157=1.096	1	a1571=111.16	6	d15716=-61.1
H	7	r167=1.092	1	a1671=112.02	6	d16716=178.0
H	7	r177=1.096	1	a1771=110.54	6	d17716=57.2

[a]Distances in angstroms and angles in degrees. Geometry parameters optimized at the B3LYP/6-311G(d.p) level of theory.

Table B.2 (cont): Geometry Parameters for vinyl, allyl, acetyl, and phenyl peroxides[a]

Structure angle		Bond length (angstrom)		Bond angle (degree)		Dihedral (degree)
$CH_3CH{=}C(CH_3)OOCH_3$						
C						
C	1	r21=1.337				
O	2	r32=1.379	1	a321=117.54		
O	3	r43=1.488	2	a432=108.63	1	d4321=129.6
C	4	r54=1.413	3	a543=105.12	2	d5432=144.9
C	2	r62=1.499	1	a621=125.73	3	d6213=-174.6
H	1	r71=1.086	2	a712=116.71	6	d7126=-4.0
H	5	r85=1.094	4	a854=111.28	3	d8543=58.5
H	5	r95=1.095	4	a954=111.81	3	d9543=-64.7
H	5	r105=1.092	4	a1054=104.37	3	d10543=176.5
H	6	r116=1.090	2	a1162=110.53	1	d11621=0.831
H	6	r126=1.091	2	a1262=110.73	1	d12621=-119.5
H	6	r136=1.094	2	a1362=110.50	1	d13621=121.3
C	1	r141=1.498	2	a1412=125.93	6	d14126=178.3
H	14	r1514=1.096	1	a15141=111.2	2	d151412=110.7
H	14	r1614=1.089	1	a16141=111.18	2	d161412=-9.8
H	14	r1714=1.095	1	a17141=110.56	2	d171412=-130.8
$CH_2{=}CHCH_2OOCH_3$						
C						
C	1	r21=1.328				
C	2	r32=1.498	1	a321=124.01		
O	3	r43=1.425	2	a432=106.83	1	d4321=-123.8
O	4	r54=1.464	3	a543=105.42	2	d5432=-168.9
C	5	r65=1.413	4	a654=105.28	3	d6543=-150.4
H	1	r71=1.084	2	a712=121.58	3	d7123=-177.8
H	1	r81=1.085	2	a812=121.63	3	d8123=1.6
H	2	r92=1.086	3	a9230115.51	4	d9234=57.0
H	3	r103=1.097	2	a1032=110.46	1	d10321=118.2
H	3	r113=1.096	2	a1132=111.64	1	d11321=-2.7
H	6	r126=1.095	5	a1265=111.49	4	d12654=62.9
H	6	r136=1.095	5	a1365=111.24	4	d13654=-59.9
H	6	r146=1.092	5	a1465=104.63	4	d14654=-178.1
$CH_2{=}C(CH_3)OOCH_3$						
C						
C	1	r21=1.333				
O	2	r32=1.377	1	a321=116.37		
O	3	r43=1.478	2	a432=109.06	1	d4321=138.5
C	4	r54=1.414	3	a543=105.01	2	d5432=148.1
C	2	r62=1.499	1	a621=126.15	3	d6213=-174.6
H	1	r71=1.081	2	a712=120.57	6	d7126=178.1
H	1	r81=1.081	2	a812=120.65	6	d8126=-5.0
H	5	r95=1.094	4	a954=111.18	3	d9543=58.6
H	5	r105=1.094	4	a1054=111.71	3	d10543=-64.5
H	5	r115=1.092	4	a1154=104.32	3	d11543=176.7
H	6	r126=1.090	2	a1262=110.11	1	d12621=3.0
H	6	r136=1.091	2	a1362=110.90	1	d13621=-117.4
H	6	r146=1.094	2	a1462=110.46	1	d14621=123.4
$CH{\equiv}COOCH_3$						
O						
C	1	r21=1.283				
C	2	r32=1.203	1	a321=176.64		
H	3	r43=1.060	2	a432=179.47	1	d4321=0.6
O	1	r51=1.533	2	a512=108.62	3	d5123=-179.6
C	5	r65=1.412	1	a651=104.16	2	d6512=-179.9
H	6	r76=1.093	5	a765=103.31	1	d7651=179.9
H	6	r86=1.094	5	a865=111.68	1	d8651=62.2
H	6	r96=1.094	5	a965=111.68	1	d9651=-62.3

[a]Distances in angstroms and angles in degrees. Geometry parameters optimized at the B3LYP/6-311G(d.p) level of theory.

Table B.2 (cont): Geometry Parameters for vinyl, allyl, acetyl, and phenyl peroxides[a]

Structure		Bond length (angstrom)		Bond angle (degree)		Dihedral angle (degree)
$CH_3CH{\equiv}COOCH_3$						
O						
C	1	r21=1.285				
C	2	r32=1.206	1	a321=176.91		
O	1	r41=1.544	2	a412=108.70	3	d4123=-179.8
C	4	r54=1.410	1	a541=104.03	2	d5412=-179.8
H	5	r65=1.094	4	a654=103.62	1	d6541=-179.9
H	5	r75=1.094	4	a754=111.76	1	d7541=62.2
H	5	r85=1.094	4	a854=111.76	1	d8541=-62.1
C	3	r93=1.456	2	a932=179.75	1	d9321=-17.7
H	9	r109=1.094	3	a1093=111.96	2	d10932=83.4
H	9	r119=1.094	3	a1193=111.64	2	d11932=-38.2
H	9	r129=1.094	3	a1293=109.81	2	d12932=-157.1
$C_6H_5OOCH_3$						
C						
C	1	r21=1.389				
C	1	r31=1.395	2	a312=120.60		
C	3	r43=1.391	1	a431=119.26	2	d4312=0.1
C	4	r54=1.395	3	a543=121.08	1	d5431=0.2
C	5	r65=1.392	4	a654=118.79	3	d6543=-0.04
H	1	r71=1.084	2	a712=119.21	6	d7126=179.8
H	2	r82=1.083	1	a821=121.47	3	d8213=179.5
H	3	r93=1.083	1	a931=120.32	2	d9312=-179.6
H	4	r104=1.084	3	a1043=120.00	1	d10431=-179.7
H	5	r115=1.079	4	a1154=121.29	3	d11543=-179.5
O	6	r126=1.377	5	a1265=124.70	4	d12654=179.2
O	12	r1312=1.452	6	a13126=110.92	5	d131265=7.04
C	13	r1413=1.419	12	a141312=127.04	6	d1413126=-116.6
H	14	r1514=1.092	13	a151413=141.77	12	d15141312=68.6
H	14	r1614=1.093	13	a161413=92.50	12	d16141312=-59.1
H	14	r1714=1.093	13	a171413=89.39	12	d17141312=-169.3

B.3 Species used for the Hydrogenated Unsaturated Study

Table B.3: Geometry Parameters for unsaturated oxygenated species[a]

Structure		Bond length (angstrom)		Bond angle (degree)		Dihedral angle (degree)
$CH(OH)_2CH{=}O$						
C						
C	1	r21=1.531				
O	2	r32=1.202	1	a321=121.38		
O	1	r41=1.387	2	a412=111.22	3	d4123=4.9
O	1	r51=1.415	2	a512=103.55	3	d5123=-118.0
H	1	r61=1.099	2	a612=110.03	3	d6123=123.6
H	2	r72=1.105	1	a721=116.02	4	d7214=-177.4
H	4	r84=0.969	1	a841=106.87	2	d8412=-24.5
H	5	r95=0.963	1	a951=107.39	2	d9512=179.2
$CH_2{=}CHCH(CH{=}O)CH{=}O$						
C						
C	1	r21=1.534				
C	2	r32=1.498	1	a321=113.57		
C	3	r43=1.327	2	a432=123.84	1	d4321=118.07
C	2	r52=1.534	1	a521=107.90	3	d5213=-126.8
O	5	r65=1.199	2	a652=124.36	1	d6521=122.89
H	1	r71=1.112	2	a712=114.52	3	d7123=-176.3
H	2	r82=1.101	1	a821=104.95	7	d8217=-54.66
H	3	r93=1.084	2	a932=115.82	1	d9321=-61.92
H	4	r104=1.083	3	a1043=121.39	2	d10432=179.9
H	4	r114=1.086	3	a1143=121.94	2	d11432=0.004
H	5	r125=1.112	2	a1252=114.52	1	d12521=-56.8
O	1	r131=1.199	2	a1312=124.36	3	d13123=3.935

[a]Distances in angstroms and angles in degrees. Geometry parameters optimized at the B3LYP/6-311G(d,p) level of theory.

Table B.3 (con't): Geometry Parameters for unsaturated oxygenated species[a]

Structure		Bond length (angstrom)		Bond angle (degree)		Dihedral angle (degree)
$CH_3C(OH)_2CH=O$						
C						
C	1	r21=1.524				
C	2	r32=1.539	1	a321=112.32		
O	3	r43=1.202	2	a432=121.61	1	d4321=-125.63
O	2	r52=1.393	1	a521=108.77	3	d5213=-121.9
O	2	r62=1.424	1	a621=111.43	3	d6213=113.6
H	1	r71=1.091	2	a712=110.61	3	d7123=-60.2
H	1	r81=1.093	2	a812=108.91	3	d8123=-179.5
H	1	r91=1.091	2	a912=110.78	3	d9123=61.7
H	3	r103=1.10	2	a1032=115.96	1	d10321=56.3
H	5	r115=0.96	2	a1152=106.79	1	d11521=145.
H	6	r126=0.96	2	a1262=106.83	1	d12621=69.9
$CH_2=CHCH(CH=O)CH=CH_2$						
C						
C	1	r21=1.329				
C	2	r32=1.512	1	a321=124.53		
C	3	r43=1.512	2	a432=111.68	1	d4321=-127.2
C	4	r54=1.329	3	a543=124.52	2	d5432=127.2
C	3	r63=1.533	2	a632=108.96	1	d6321=112.2
O	6	r76=1.201	3	a763=124.16	2	d7632=-118.8
H	1	r81=1.085	2	a812=121.73	3	d8123=0.6
H	1	r91=1.083	2	a912=121.37	3	d9123=-179.7
H	2	r102=1.088	3	a1023=115.94	4	d10234=53.1
H	3	r113=1.092	2	a1132=110.28	1	d11321=-4.2
H	4	r124=1.088	3	a1243=115.95	2	d12432=-53.1
H	5	r135=1.083	4	a1354=121.38	3	d13543=179.7
H	5	r145=1.085	4	a1454=121.73	3	d14543=-0.6
H	6	r156=1.113	3	a1563=114.65	2	d15632=61.1
$CH_2=CHCH_2CH=O$						
C						
C	1	r21=1.328				
C	2	r32=1.505	1	a321=124.58		
C	3	r43=1.519	2	a432=112.30	1	d4321=-118.3
O	4	r54=1.202	3	a543=124.39	2	d5432=132.3
H	1	r61=1.085	2	a612=121.69	3	d6123=-0.6
H	1	r71=1.083	2	a712=121.44	3	d7123=179.8
H	2	r82=1.088	3	a823=116.17	4	d8234=61.07
H	3	r93=1.100	2	a932=109.36	1	d9321=123.8
H	3	r103=1.091	2	a1032=112.12	1	d10321=3.9
H	4	r114=1.113	3	a1143=114.78	2	d11432=-47.9
$CH_2=CHCH(CH_3)CH=CH_2$						
C						
C	1	r21=1.329				
C	2	r32=1.511	1	a321=125.29		
C	3	r43=1.511	2	a432=110.44	1	d4321=-119.71
C	4	r54=1.329	3	a543=125.29	2	d5432=119.702
H	1	r61=1.084	2	a612=121.66	3	d6123=-179.4
H	1	r71=1.085	2	a712=121.60	3	d7123=0.6
H	2	r82=1.089	3	a823=115.46	4	d8234=60.7
H	3	r93=1.095	2	a932=108.35	1	d9321=-1.1
H	4	r104=1.089	3	a1043=115.46	2	d10432=-60.7
H	5	r115=1.084	4	a1154=121.66	3	d11543=179.5
H	5	r125=1.085	4	a1254=121.60	3	d12543=-0.6
C	3	r133=1.544	2	a1332=110.50	1	d13321=117.7
H	13	r1413=1.09	3	a14133=110.89	2	d141332=61.2
H	13	r1513=1.09	3	a15133=110.66	2	d151332=-58.4
H	13	r1613=1.09	3	a16133=110.66	2	d161332=-179.0′
$CH_2=CHCH=O$						
C						
C	1	r21=1.333				
C	2	r32=1.475	1	a321=121.25		
O	3	r43=1.208	2	a432=124.31	1	d4321=-179.9
H	1	r51=1.086	2	a512=120.95	3	d5123=0.001
H	1	r61=1.083	2	a612=122.29	3	d6123=179.9
H	2	r72=1.085	1	a721=122.41	5	d7215=-179.9
H	3	r83=1.112	2	a832=114.49	1	d8321=-0.0

[a]Distances in angstroms and angles in degrees. Geometry parameters optimized at the B3LYP/6-311G(d,p) level of theory.

Table B.3 (con't): Geometry Parameters for unsaturated oxygenated species[a]

Structure		Bond length (angstrom)		Bond angle (degree)		Dihedral angle (degree)
$CH_2=C(CH=O)CH=CH_2$						
C						
C	1	r21=1.336				
C	2	r32=1.467	1	a321=127.73		
C	3	r43=1.343	2	a432=121.71	1	d4321=176.3
C	3	r53=1.499	2	a532=120.67	1	d5321=-4.09
O	5	r65=1.208	3	a653=124.13	2	d6532=177.8
H	1	r71=1.082	2	a712=123.50	3	d7123=-0.03
H	1	r81=1.083	2	a812=120.41	3	d8123=-179.8
H	2	r92=1.087	1	a921=117.94	7	d9217=179.55
H	4	r104=1.083	3	a1043=119.98	2	d10432=179.9
H	4	r114=1.084	3	a1143=121.73	2	d11432=-0.02
H	5	r125=1.106	3	a1253=115.94	2	d12532=-2.67
$CH_2=CHC(OH)_2CH=CH_2$						
C						
C	1	r21=1.328				
C	2	r32=1.514	1	a321=125.17		
C	3	r43=1.514	2	a432=110.61	1	d4321=114.2
C	4	r54=1.328	3	a543=125.17	2	d5432=114.1
O	3	r63=1.418	2	a632=107.96	1	d6321=-6.07
O	3	r73=1.418	2	a732=109.89	1	d7321=-126.
H	1	r81=1.082	2	a812=121.05	3	d8123=-0.78
H	1	r91=1.084	2	a912=121.12	3	d9123=179.1
H	2	r102=1.085	3	a1023=113.39	4	d10234=-66.1
H	4	r114=1.085	3	a1143=113.40	2	d11432=-66.1
H	5	r125=1.084	4	a1254=121.12	3	d12543=179.1
H	5	r135=1.082	4	a1354=121.06	3	d13543=-0.76
H	6	r146=0.964	3	a1463=106.51	2	d14632=176.0
H	7	r157=0.964	3	a1573=106.50	2	d15732=55.2
$CH_2=C(CH=O)C\equiv CH$						
C						
C	1	r21=1.326				
C	2	r32=1.513	1	a321=125.62		
C	3	r43=1.546	2	a432=111.88	1	d4321=-135.4
O	4	r54=1.198	3	a543=123.89	2	d5432=-18.08
H	1	r61=1.084	2	a612=121.99	3	d6123=0.5
H	1	r71=1.083	2	a712=120.70	3	d7123=-179.7
H	2	r82=1.084	3	a823=113.65	4	d8234=44.0624
H	3	r93=1.105	2	a932=107.84	1	d9321=112.126
H	4	r104=1.109	3	a1043=114.05	2	d10432=163.8
C	3	r113=1.459	2	a1132=115.69	1	d11321=-8.1
C	11	r1211=1.201	3	a12113=178.02	2	d121132=-132.8
H	12	r1312=1.062	11	a131211=179.60	3	d1312113=-174.0
$CH_2=CHCH(OH)CH=CH_2$						
C						
C	1	r21=1.345				
C	2	r32=1.467	1	a321=121.49		
C	3	r43=1.336	2	a432=127.09	1	d4321=179.9
C	2	r52=1.491	1	a521=115.58	3	d5213=179.9
O	5	r65=1.207	2	a652=126.54	1	d6521=179.9
H	1	r71=1.083	2	a712=121.95	3	d7123=-0.0
H	1	r81=1.084	2	a812=121.17	3	d8123=179.9
H	3	r93=1.087	2	a932=114.18	1	d9321=-0.0
H	4	r104=1.083	3	a1043=120.00	2	d10432=179.9
H	4	r114=1.080	3	a1143=121.98	2	d11432=0.0
H	5	r125=1.111	2	a1252=113.04	1	d12521=0.05
$CH_2=CHOCH=CH_2$						
C						
C	1	r21=1.327				
O	2	r32=1.367	1	a321=121.87		
C	3	r43=1.367	2	a432=118.44	1	d4321=179.9
C	4	r54=1.327	3	a543=121.87	2	d5432=180.
H	1	r61=1.082	2	a612=121.33	3	d6123=-0.0
H	1	r71=1.080	2	a712=119.77	3	d7123=179.9
H	2	r82=1.087	1	a821=122.99	6	d8216=-179.9
H	4	r94=1.087	5	a945=122.99	3	d9453=180.
H	5	r105=1.080	4	a1054=119.77	3	d10543=180.
H	5	r115=1.082	4	a1154=121.33	3	d11543=0.

[a]Distances in angstroms and angles in degrees. Geometry parameters optimized at the B3LYP/6-311G(d,p) level of theory.

Table B.3 (con't): Geometry Parameters for unsaturated oxygenated species[a]

Structure		Bond length (angstrom)		Bond angle (degree)		Dihedral angle (degree)
CH_2=CHCH(CH=O)C≡CH						
C						
C	1	r21=1.326				
C	2	r32=1.513	1	a321=125.62		
C	3	r43=1.546	2	a432=111.88	1	d4321=-135.45
O	4	r54=1.198	3	a543=123.89	2	d5432=-18.08
H	1	r61=1.084	2	a612=121.99	3	d6123=0.49
H	1	r71=1.083	2	a712=120.70	3	d7123=-179.74
H	2	r82=1.084	3	a823=113.65	4	d8234=44.06
H	3	r93=1.105	2	a932=107.84	1	d9321=112.12
H	4	r104=1.109	3	a1043=114.05	2	d10432=163.80
C	3	r113=1.459	2	a1132=115.69	1	d11321=-8.10
C	11	r1211=1.201	3	a12113=178.02	2	d121132=-132.80
H	12	r1312=1.062	11	a131211=179.6	3	d1312113=-174.00
CH_2=C(OH)$_2$						
C						
C	1	r21=1.327				
O	2	r32=1.367	1	a321=122.83		
O	2	r42=1.367	1	a421=122.83	3	d4213=180.
H	1	r51=1.079	2	a512=119.94	3	d5123=-7.4
H	1	r61=1.079	2	a612=119.94	3	d6123=172.5
H	3	r73=0.965	2	a732=108.19	1	d7321=136.6
H	4	r84=0.965	2	a842=108.19	1	d8421=136.6
$C_6H_5C(OH)_2CH_3$						
C						
C	1	r21=1.393				
C	2	r32=1.393	1	a321=120.33		
C	3	r43=1.396	2	a432=120.28	1	d4321=0.1
C	4	r54=1.398	3	a543=119.15	2	d5432=0.02
C	5	r65=1.392	4	a654=120.48	3	d6543=-0.086
C	4	r74=1.527	3	a743=121.38	2	d7432=-178.4
C	7	r87=1.529	4	a874=112.11	3	d8743=97.6
O	7	r97=1.418	8	a978=110.94	4	d9784=119.4
O	7	r107=1.420	8	a1078=105.51	4	d10784=-121.4
H	1	r111=1.084	2	a1112=120.21	3	d11123=-179.9
H	2	r122=1.084	1	a1221=120.02	6	d12216=179.6
H	3	r133=1.081	2	a1332=120.52	1	d13321=179.3
H	5	r145=1.083	4	a1454=119.23	7	d14547=-2.1
H	6	r156=1.084	5	a1565=119.75	4	d15654=179.8
H	8	r168=1.090	7	a1687=110.73	4	d16874=58.8
H	8	r178=1.090	7	a1787=109.94	4	d17874=-61.7
H	8	r188=1.094	7	a1887=109.46	4	d18874=178.0
H	9	r199=0.963	7	a1997=106.78	8	d19978=54.520
H	10	r2010=0.965	7	a20107=106.87	8	d201078=178.2
$C_6H_5C(=CH_2)CH=CH_2$						
C						
C	1	r21=1.393				
C	2	r32=1.390	1	a321=120.26		
C	3	r43=1.402	2	a432=120.91	1	d4321=0.3
C	4	r54=1.402	3	a543=118.14	2	d5432=-0.9
C	5	r65=1.391	4	a654=120.94	3	d6543=1.05
C	4	r74=1.488	3	a743=121.43	2	d7432=178.9
C	7	r87=1.480	4	a874=116.96	3	d8743=144.3
C	8	r98=1.333	7	a987=125.31	4	d9874=139.9
C	7	r107=1.342	4	a1074=121.82	3	d10743=-35.1
H	1	r111=1.084	2	a1112=120.23	3	d11123=179.7
H	2	r122=1.084	1	a1221=120.04	6	d12216=-179.
H	3	r133=1.083	2	a1332=119.74	1	d13321=-178.
H	5	r145=1.084	4	a1454=119.33	3	d14543=179.8
H	6	r156=1.084	5	a1565=119.70	4	d15654=179.3
H	8	r168=1.088	7	a1687=115.74	4	d16874=-38.9
H	9	r179=1.084	8	a1798=121.27	7	d17987=178.2
H	9	r189=1.084	8	a1898=121.62	7	d18987=-2.4
H	10	r1910=1.083	7	a19107=121.13	4	d191074=176.0
H	10	r2010=1.083	7	a20107=121.65	4	d201074=-4.07

[a]Distances in angstroms and angles in degrees. Geometry parameters optimized at the B3LYP/6-311G(d,p) level of theory.

Table B.3 (con't): Geometry Parameters for unsaturated oxygenated species[a]

Structure angle		Bond length (angstrom)		Bond angle (degree)		Dihedral (degree)
$C_6H_5C(CH_3)_2CH{=}O$						
C						
C	1	r21=1.393				
C	2	r32=1.390	1	a321120.14		
C	3	r43=1.403	2	a432121.19	1	d4321=0.01
C	4	r54=1.398	3	a543117.97	2	d5432=0.47
C	5	r65=1.394	4	a654120.93	3	d6543=-0.7
C	4	r74=1.537	3	a743118.94	2	d7432=-179.2
C	7	r87=1.542	4	a874113.00	3	d8743=-171.6
C	7	r97=1.538	8	a978109.93	4	d9784=124.42
C	7	r107=1.533	8	a1078=106.16	9	d10789=119.3
O	10	r1110=1.20	7	a11107=125.31	8	d111078=-121.6
H	1	r121=1.084	2	a1212=120.32	3	d12123=179.9
H	2	r132=1.084	1	a1321=120.16	6	d13216=179.6
H	3	r143=1.084	2	a1432=119.21	1	d14321=179.8
H	5	r155=1.082	4	a1554=120.35	7	d15547=-0.7
H	6	r166=1.084	5	a1665=119.51	4	d16654=-179.6
H	8	r178=1.093	7	a1787=112.09	9	d17879=-178.1
H	8	r188=1.092	7	a1887=110.79	9	d18879=60.8
H	8	r198=1.092	7	a1987=109.92	9	d19879=-57.9
H	9	r209=1.092	7	a2097=110.06	8	d20978=-63.4
H	9	r219=1.090	7	a2197=111.53	8	d21978=176.2
H	9	r229=1.092	7	a2297=110.50	8	d22978=56.7
H	10	r2310=1.112	7	a23107=113.75	8	d231078=58.2
$C_6H_5CH{=}O$						
C						
C	1	r21=1.393				
C	2	r32=1.391	1	a321=119.70		
C	3	r43=1.398	2	a432=120.18	1	d4321=-0.01
C	4	r54=1.400	3	a543=119.84	2	d5432=0.025
C	5	r65=1.388	4	a654=119.93	3	d6543=-0.01
C	4	r74=1.481	3	a743=119.83	2	d7432=-179.
O	7	r87=1.209	4	a874=124.88	3	d8743=179.9
H	1	r91=1.084	2	a912=119.80	3	d9123=-179.9
H	2	r102=1.083	1	a1021=120.10	6	d10216=179.9
H	3	r113=1.085	2	a1132=120.32	1	d11321=-179.9
H	5	r125=1.083	4	a1254=118.46	3	d12543=179.9
H	6	r136=1.084	5	a1365=120.08	4	d13654=179.9
H	7	r147=1.112	4	a1474=114.39	3	d14743=-0.08
$C_6H_5CH{=}CH_2$						
C						
C	1	r21=1.3917				
C	2	r32=1.3921	1	a321=120.03		
C	3	r43=1.4028	2	a432=121.26	1	d4321=-0.00
C	4	r54=1.404	3	a543=117.92	2	d5432=0.01
C	5	r65=1.3884	4	a654=120.89	3	d6543=-0.01
C	4	r74=1.4712	3	a743=118.91	2	d7432=-179.9
H	1	r81=1.0849	2	a812=120.32	3	d8123=-179.9
H	2	r92=1.0847	1	a921=120.15	6	d9216=179.9
H	3	r103=1.085	2	a1032=119.67	1	d10321=179.9
H	5	r115=1.083	4	a1154=119.87	3	d11543=179.9
H	6	r126=1.084	5	a1265=119.65	4	d12654=-179.9
C	7	r137=1.335	4	a1374=127.64	3	d13743=179.9
H	13	r1413=1.083	7	a14137=120.79	4	d141374=-179.9
H	13	r1513=1.084	7	a15137=122.81	4	d151374=-0.028
H	7	r167=1.087	4	a1674=114.41	3	d16743=-0.08

[a]Distances in angstroms and angles in degrees. Geometry parameters optimized at the B3LYP/6-311G(d,p) level of theory.

Table B.3 (con't): Geometry Parameters for unsaturated oxygenated species[a]

Structure		Bond length (angstrom)		Bond angle (degree)		Dihedral angle (degree)
$C_6H_5OCH=O$						
C						
C	1	r21=1.394				
C	2	r32=1.392	1	a321=120.35		
C	3	r43=1.390	2	a432=119.18	1	d4321=0.3
C	4	r54=1.392	3	a543=121.18	2	d5432=0.42
C	5	r65=1.394	4	a654=119.09	3	d6543=-0.9
O	4	r74=1.393	3	a743=117.75	2	d7432=177.8
C	7	r87=1.365	4	a874=118.39	3	d8743=123.9
O	8	r98=1.189	7	a987=122.03	4	d9874=175.1
H	1	r101=1.083	2	a1012=120.09	3	d10123=179.9
H	2	r112=1.083	1	a1121=120.13	6	d11216=179.7
H	3	r123=1.083	2	a1232=121.66	1	d12321=-179.9
H	5	r135=1.083	4	a1354=120.07	3	d13543=177.5
H	6	r146=1.083	5	a1465=119.48	4	d14654=179.9
H	8	r158=1.103	7	a1587=112.53	4	d15874=-5.3
$CH_3C_6H_4CH=O$						
C						
C	1	r21=1.405				
C	2	r32=1.397	1	a321=118.43		
C	3	r43=1.391	2	a432=120.76	1	d4321=0.
C	4	r54=1.396	3	a543=120.39	2	d5432=0.
C	5	r65=1.401	4	a654=119.24	3	d6543=0.
C	5	r75=1.478	4	a754=120.14	3	d7543=180.
O	7	r87=1.209	5	a875=125.00	4	d8754=180.
C	2	r92=1.507	1	a921=120.35	6	d9216=180.
H	1	r101=1.085	2	a1012=119.29	9	d10129=0.
H	3	r113=1.084	2	a1132=119.45	9	d11329=0.
H	4	r124=1.085	3	a1243=120.06	2	d12432=180.
H	6	r136=1.083	5	a1365=118.46	4	d13654=180.
H	7	r147=1.112	5	a1475=114.33	4	d14754=0.
H	9	r159=1.094	2	a1592=111.02	1	d15921=59.5
H	9	r169=1.094	2	a1692=111.02	1	d16921=-59.5
H	9	r179=1.091	2	a1792=111.54	1	d17921=179.9
$C_6H_5C(=CH_2)CH=O$						
C						
C	1	r21=1.392				
C	2	r32=1.392	1	a321=120.21		
C	3	r43=1.402	2	a432=120.77	1	d4321=-0.3
C	4	r54=1.402	3	a543=118.37	2	d5432=0.7
C	5	r65=1.391	4	a654=120.82	3	d6543=-0.7
C	4	r74=1.482	3	a743=120.70	2	d7432=-179.1
H	1	r81=1.083	2	a812=120.17	3	d8123=-179.7
H	2	r92=1.084	1	a921=120.09	6	d9216=-179.7
H	3	r103=1.084	2	a1032=119.33	1	d10321=-178.5
H	5	r115=1.084	4	a1154=119.26	3	d11543=178.0
H	6	r126=1.084	5	a1265=119.72	4	d12654=179.8
C	7	r137=1.339	4	a1374=123.90	3	d13743=136.9
C	7	r147=1.501	4	a1474=118.15	3	d14743=-44.0
O	14	r1514=1.207	7	a15147=124.52	4	d151474=177.7
H	13	r1613=1.083	7	a16137=119.48	4	d161374=178.7
H	13	r1713=1.083	7	a17137=122.17	4	d171374=-2.0
H	14	r1814=1.109	7	a18147=114.85	4	d181474=-3.3
$C_6H_5C(=CH_2)OH$						
C						
C	1	r21=1.392				
C	2	r32=1.392	1	a321=120.19		
C	3	r43=1.403	2	a432=120.73	1	d4321=0.2
C	4	r54=1.402	3	a543=118.49	2	d5432=-0.6
C	5	r65=1.390	4	a654=120.66	3	d6543=0.7
C	4	r74=1.482	3	a743=120.23	2	d7432=179.8
C	7	r87=1.336	4	a874=124.75	3	d8743=-145.1
O	7	r97=1.376	4	a974=116.10	3	d9743=31.610
H	1	r101=1.084	2	a1012=120.22	3	d10123=179.8
H	2	r112=1.084	1	a1121=120.12	6	d11216=179.5
H	3	r123=1.084	2	a1232=119.65	1	d12321=178.0
H	5	r135=1.083	4	a1354=119.25	3	d13543=-177.8
H	6	r146=1.084	5	a1465=119.66	4	d14654=-179.7
H	8	r158=1.081	7	a1587=120.29	4	d15874=-179.1
H	8	r168=1.080	7	a1687=120.91	4	d16874=3.5
H	9	r179=0.963	7	a1797=108.64	4	d17974=25.9

[a]Distances in angstroms and angles in degrees. Geometry parameters optimized at the B3LYP/6-311G(d,p) level of theory

Table B.3 (con't): Geometry Parameters for unsaturated oxygenated species[a]

Structure		Bond length (Angstrom)		Bond angle (degree)		Dihedral angle (degree)
$C_6H_5C(=CH_2)CH_3$						
C						
C	1	r21=1.390				
C	2	r32=1.393	1	a321=120.33		
C	3	r43=1.404	2	a432=121.51	1	d4321=-0.0
C	4	r54=1.406	3	a543=117.12	2	d5432=0.0
C	5	r65=1.387	4	a654=121.46	3	d6543=-0.0
C	4	r74=1.490	3	a743=120.73	2	d7432=-179.9
C	7	r87=1.339	4	a874=123.05	3	d8743=179.9
C	7	r97=1.512	4	a974=117.73	3	d9743=-0.03
H	1	r101=1.084	2	a1012=120.49	3	d10123=-179.9
H	2	r112=1.084	1	a1121=120.14	6	d11216=179.9
H	3	r123=1.082	2	a1232=118.52	1	d12321=-179.9
H	5	r135=1.082	4	a1354=119.99	3	d13543=179.9
H	6	r146=1.084	5	a1465=119.51	4	d14654=179.9
H	8	r158=1.083	7	a1587=120.73	9	d15879=-0.0
H	8	r168=1.082	7	a1687=123.22	9	d16879=179.9
H	9	r179=1.094	7	a1797=111.53	4	d17974=59.8
H	9	r189=1.094	7	a1897=111.53	4	d18974=-59.8
H	9	r199=1.090	7	a1997=110.53	4	d19974=-179.9
$CH_3OCH=O$						
C						
O	1	r21=1.435				
C	2	r32=1.348	1	a321=116.52		
O	3	r43=1.192	2	a432=122.80	1	d4321=-179.9
H	1	r51=1.087	2	a512=105.90	3	d5123=-179.9
H	1	r61=1.093	2	a612=111.23	3	d6123=61.5
H	1	r71=1.093	2	a712=111.23	3	d7123=-61.3
H	3	r83=1.107	2	a832=113.10	1	d8321=0.00
$Y(C_6H_6)=O$						
C						
C	1	r21=1.469				
C	2	r32=1.349	1	a321=121.38		
C	3	r43=1.456	2	a432=122.61	1	d4321=-0.007
C	4	r54=1.340	3	a543=120.73	2	d5432=-0.002
C	5	r65=1.497	4	a654=121.66	3	d6543=0.005
O	1	r71=1.217	6	a716=120.68	5	d7165=179.9
H	2	r82=1.083	1	a821=116.40	6	d8216=-179.9
H	3	r93=1.085	2	a932=119.54	1	d9321=179.9
H	4	r104=1.083	3	a1043=118.47	2	d10432=-179.9
H	5	r115=1.085	6	a1156=117.69	1	d11561=179.9
H	6	r126=1.097	5	a1265=110.24	4	d12654=122.69
H	6	r136=1.097	5	a1365=110.23	4	d13654=-122.6
$Y(C_5H_5)C(OH)=O$						
C						
C	1	r21=1.343				
C	2	r32=1.470	1	a321=109.52		
C	3	r43=1.343	2	a432=109.57	1	d4321=0.2
C	4	r54=1.515	3	a543=108.60	2	d5432=1.2
C	5	r65=1.520	4	a654=114.65	3	d6543=-128.2
O	6	r76=1.201	5	a765=126.35	4	d7654=126.8
O	6	r86=1.356	5	a865=110.89	4	d8654=-54.5
H	1	r91=1.079	5	a915=123.41	4	d9154=-177.5
H	2	r102=1.082	1	a1021=126.04	5	d10215=178.9
H	3	r113=1.082	2	a1132=124.51	1	d11321=-179.8
H	4	r124=1.079	5	a1245=123.28	1	d12451=178.9
H	5	r135=1.102	4	a1354=108.22	3	d13543=114.1
H	8	r148=0.968	6	a1486=106.42	5	d14865=-179.4

[a]Distances in angstroms and angles in degrees. Geometry parameters optimized at the B3LYP/6-311G(d,p) level of theory.

Table B.3 (con't): Geometry Parameters for unsaturated oxygenated species[a]

Structure		Bond length (angstrom)		Bond angle (degree)		Dihedral angle (degree)
Y(C_5H_5O)CH=O						
C						
O	1	r21=1.442				
C	2	r32=1.360	1	a321=115.94		
C	3	r43=1.341	2	a432=123.09	1	d4321=-23.5
C	4	r54=1.457	3	a543=118.18	2	d5432=-4.7
C	1	r61=1.507	2	a612=112.50	3	d6123=40.1
C	1	r71=1.523	6	a716=113.03	5	d7165=-151.1
O	7	r87=1.201	1	a871=123.06	6	d8716=-46.9
H	1	r91=1.106	7	a917=106.76	8	d9178=74.64
H	3	r103=1.082	4	a1034=124.95	5	d10345=170.57
H	4	r114=1.080	3	a1143=119.92	2	d11432=-178.4
H	5	r125=1.084	4	a1254=119.31	3	d12543=-162.4
H	6	r136=1.082	1	a1361=117.60	7	d13617=30.052
H	7	r147=1.109	1	a1471=114.02	6	d14716=133.52
Y(C_5H_5)OCH=O						
O						
C	1	r21=1.441				
C	2	r32=1.513	1	a321=116.66		
C	3	r43=1.342	2	a432=108.54	1	d4321=-135.5
C	4	r54=1.476	3	a543=109.56	2	d5432=3.9
C	2	r62=1.513	3	a623=103.38	4	d6234=-6.0
C	1	r71=1.354	2	a712=116.08	3	d7123=61.4
O	7	r87=1.192	1	a871=122.08	2	d8712=179.9
H	2	r92=1.097	3	a923=108.25	4	d9234=108.6
H	3	r103=1.080	2	a1032=123.27	6	d10326=178.2
H	4	r114=1.082	3	a1143=126.23	2	d11432=-175.8
H	5	r125=1.082	4	a1254=124.20	3	d12543=-179.8
H	6	r136=1.080	2	a1362=123.27	3	d13623=-178.2
H	7	r147=1.103	1	a1471=113.20	2	d14712=0.1
CH_3OC•=O						
C						
O	1	r21=1.445				
C	2	r32=1.334	1	a321=116.00		
O	3	r43=1.181	2	a432=126.92	1	d4321=-179.9
H	1	r51=1.087	2	a512=105.36	3	d5123=179.9
H	1	r61=1.090	2	a612=110.44	3	d6123=61.08
H	1	r71=1.090	2	a712=110.44	3	d7123=-61.1
CH_2=C•CH=O						
C						
C	1	r21=1.314				
C	2	r32=1.462	1	a321=132.53		
O	3	r43=1.204	2	a432=126.97	1	d4321=179.9
H	1	r51=1.093	2	a512=121.39	3	d5123=-0.03
H	1	r61=1.086	2	a612=122.47	3	d6123=-179.9
H	3	r73=1.118	2	a732=111.92	1	d7321=0.0
CH_2=CHC•=O						
C	1	r21=1.333				
C	2	r32=1.476	1	a321=120.40		
O	3	r43=1.185	2	a432=128.33	1	d4321=180.
H	1	r51=1.084	2	a512=120.57	3	d5123=0.
H	1	r61=1.083	2	a612=121.77	3	d6123=180.
H	2	r72=1.089	1	a721=122.38	5	d7215=180.
C•$(OH)_2$CH=O						
C						
C	1	r21=1.409				
O	2	r32=1.255	1	a321=117.84		
O	1	r41=1.330	2	a412=118.71	3	d4123=-0.01
O	1	r51=1.331	2	a512=124.44	3	d5123=-179.9
H	2	r62=1.096	1	a621=118.49	4	d6214=-179.9
H	4	r74=0.982	1	a741=103.24	2	d7412=0.02
H	5	r85=0.966	1	a851=108.88	2	d8512=179.9

[a]Distances in angstroms and angles in degrees. Geometry parameters optimized at the B3LYP/6-311G(d,p) level of theory.

Table B.3 (con't): Geometry Parameters for unsaturated oxygenated radicals[a]

Structure		Bond length (angstrom)		Bond angle (degree)		Dihedral angle (degree)
CH_2=CHC•(CH=O)C≡CH						
C						
C	1	r21=1.355				
C	2	r32=1.426	1	a321=125.53		
C	3	r43=1.474	2	a432=118.77	1	d4321=179.99
O	4	r54=1.215	3	a543=123.66	2	d5432=0.0330
H	1	r61=1.084	2	a612=121.65	3	d6123=0.0293
H	1	r71=1.082	2	a712=120.87	3	d7123=-179.9
H	2	r82=1.085	3	a823=114.33	4	d8234=0.0217
H	4	r94=1.106	3	a943=114.95	2	d9432=-179.9
C	3	r103=1.401	2	a1032=123.31	1	d10321=-0.01
C	10	r1110=1.211	3	a11103=178.24	2	d111032=175.5
H	11	r1211=1.062	10	a121110=179.87	3	d1211103=172.7
CH_2=CHC•(CH=O)CH=O						
C						
C	1	r21=1.470				
C	2	r32=1.413	1	a321=118.23		
C	3	r43=1.365	2	a432=126.37	1	d4321=-179.9
C	2	r52=1.466	1	a521=116.30	3	d5213=-179.9
O	5	r65=1.216	2	a652=126.17	1	d6521=179.96
H	1	r71=1.108	2	a712=115.16	3	d7123=179.99
H	3	r83=1.086	2	a832=114.39	1	d8321=-0.009
H	4	r94=1.083	3	a943=120.00	2	d9432=179.97
H	4	r104=1.080	3	a1043=121.38	2	d10432=-0.005
H	5	r115=1.109	2	a1152=113.94	1	d11521=-0.029
O	1	r121=1.216	2	a1212=124.44	3	d12123=-0.006
Furanj1						
C						
C	1	r21=1.356				
C	2	r32=1.445	1	a321=107.19		
C	3	r43=1.354	2	a432=103.40	1	d4321=-0.03
O	4	r54=1.328	3	a543=114.12	2	d5432=0.01
H	1	r61=1.075	2	a612=134.65	3	d6123=-179.9
H	2	r72=1.079	1	a721=126.33	5	d7215=-179.9
H	3	r83=1.076	2	a832=128.45	1	d8321=179.9
Furanj2						
C						
C	1	r21=1.362				
C	2	r32=1.421	1	a321=103.75		
C	3	r43=1.348	2	a432=109.81	1	d4321=-0.01
O	4	r54=1.375	3	a543=107.86	2	d5432=-0.007
H	1	r61=1.077	2	a612=132.98	3	d6123=-179.9
H	2	r72=1.077	1	a721=127.27	5	d7215=-179.9
H	4	r84=1.074	3	a843=135.61	2	d8432=-179.9
Y(C_6H_5•)=O						
C						
C	1	r21=1.452				
C	2	r32=1.374	1	a321=120.94		
C	3	r43=1.407	2	a432=120.25	1	d4321=0.02
C	4	r54=1.407	3	a543=120.68	2	d5432=-0.03
C	5	r65=1.374	4	a654=120.25	3	d6543=0.02
O	1	r71=1.251	6	a716=121.53	5	d7165=179.9
H	2	r82=1.083	1	a821=116.86	6	d8216=179.9
H	3	r93=1.084	2	a932=120.30	1	d9321=-179.9
H	4	r104=1.083	3	a1043=119.65	2	d10432=179.9
H	5	r115=1.084	6	a1156=120.30	1	d11561=-179.9
H	6	r126=1.083	5	a1265=122.17	4	d12654=179.9

[a]Distances in angstroms and angles in degrees. Geometry parameters optimized at the B3LYP/6-311G(d,p) level of theory.

Table 11.3 (con't): Geometry Parameters for unsaturated oxygenated radicals[a]

Structure		Bond length (angstrom)		Bond angle (degree)		Dihedral angle (degree)
$C_6H_5C\bullet{=}O$						
C						
C	1	r21=1.399				
C	2	r32=1.391	1	a321=119.80		
C	3	r43=1.397	2	a432=119.87	1	d4321=0.003
C	4	r54=1.403	3	a543=120.22	2	d5432=-0.02
C	5	r65=1.388	4	a654=119.68	3	d6543=0.029
C	4	r74=1.481	3	a743=119.31	2	d7432=179.9
O	7	r87=1.186	4	a874=128.52	3	d8743=179.8
H	1	r91=1.084	2	a912=119.78	3	d9123=179.9
H	2	r102=1.083	1	a1021=120.09	6	d10216=-179.9
H	3	r113=1.083	2	a1132=121.13	1	d11321=179.9
H	5	r125=1.083	4	a1254=119.03	3	d12543=-179.9
H	6	r136=1.084	5	a1365=120.05	4	d13654=179.9
$Y(C_5H_5)C(O\bullet){=}O$						
C						
C	1	r21=1.372				
C	2	r32=1.433	1	a321=109.46		
C	3	r43=1.371	2	a432=109.05	1	d4321=0.7
C	4	r54=1.505	3	a543=108.53	2	d5432=-2.7
C	5	r65=1.601	4	a654=104.32	3	d6543=-109.3
O	6	r76=1.205	5	a765=122.80	4	d7654=-176.1
O	6	r86=1.281	5	a865=106.24	4	d8654=3.7
H	1	r91=1.080	5	a915=123.68	4	d9154=-178.7
H	2	r102=1.08	1	a1021=125.59	5	d10215=-177.4
H	3	r113=1.08	2	a1132=125.08	1	d11321=-178.3
H	4	r124=1.07	5	a1245=124.40	1	d12451=-175.1
H	5	r135=1.09	4	a135=117.53	3	d13543=133.02
$Y(C_5H_5O)C\bullet{=}O$						
C						
O	1	r21=1.425				
C	2	r32=1.356	1	a321=116.38		
C	3	r43=1.343	2	a432=123.17	1	d4321=21.68
C	4	r54=1.452	3	a543=118.18	2	d5432=4.559
C	1	r61=1.492	2	a612=113.41	3	d6123=-38.2
C	1	r71=1.595	6	a716=109.38	5	d7165=-91.4
O	7	r87=1.176	1	a871=123.74	6	d8716=-66.7
H	1	r91=1.091	6	a916=113.42	5	d9165=151.9
H	3	r103=1.082	4	a1034=124.95	5	d10345=-171.0
H	4	r114=1.080	3	a1143=119.94	2	d11432=178.69
H	5	r125=1.083	4	a1254=119.49	3	d12543=164.28
H	6	r136=1.082	1	a1361=118.28	2	d13612=-153.1
$Y(C_5H_5)OC\bullet{=}O$						
C						
C	1	r21=1.510				
C	2	r32=1.340	1	a321=108.49		
C	3	r43=1.480	2	a432=109.49	1	d4321=3.5
C	1	r51=1.509	2	a512=103.70	3	d5123=-5.3
O	1	r61=1.452	2	a612=112.79	3	d6123=-127.7
C	6	r76=1.334	1	a761=116.68	2	d7612=-121.4
O	7	r87=1.181	6	a876=127.24	1	d8761=-179.9
H	1	r91=1.098	2	a912=109.82	3	d9123=111.9
H	2	r102=1.080	1	a1021=123.27	5	d10215=177.4
H	3	r113=1.081	2	a1132=126.27	1	d11321=-176.5
H	4	r124=1.081	3	a1243=124.24	2	d12432=179.9
H	5	r135=1.080	1	a1351=123.27	2	d13512=-177.4

[a]Distances in angstroms and angles in degrees. Geometry parameters optimized at the B3LYP/6-311G(d,p) level of theory.

B.4 Species used for the Phenyl + O_2 Study

Table B.4: Geometry Parameters for unsaturated oxygenated radicals[a]

Structure		Bond length (angstrom)		Bond angle (degree)		Dihedral angle (degree)
Y(C5O)C•DO						
C						
C	1	r21=1.510				
C	2	r32=1.340	1	a321=108.49		
C	3	r43=1.480	2	a432=109.49	1	d4321=3.5
C	1	r51=1.509	2	a512=103.70	3	d5123=-5.3
O	1	r61=1.452	2	a612=112.79	3	d6123=-127.7
C	6	r76=1.334	1	a761=116.68	2	d7612=-121.4
O	7	r87=1.181	6	a876=127.24	1	d8761=-179.9
H	1	r91=1.098	2	a912=109.82	3	d9123=111.9
H	2	r102=1.080	1	a102=123.271	5	d10215=177.4
H	3	r113=1.081	2	a113=126.272	1	d11321=-176.
H	4	r124=1.081	3	a124=124.243	2	d12432=179.9
H	5	r135=1.080	1	a135=123.271	2	d13512=-177.4
C•DCCDO						
C						
C	1	r21=1.311				
C	2	r32=1.495	1	a321=122.53		
O	3	r43=1.205	2	a432=123.19	1	d4321=180.
H	1	r51=1.080	2	a512=137.88	3	d5123=0.
H	2	r62=1.088	3	a623=114.98	4	d6234=0.
H	3	r73=1.108	2	a732=115.01	1	d7321=0.
Y(C5)OC•DO						
C						
C	1	r21=1.510				
C	2	r32=1.340	1	a321=108.49		
C	3	r43=1.480	2	a432=109.49	1	d4321=3.5
C	1	r51=1.509	2	a512=103.70	3	d5123=-5.31
O	1	r61=1.452	2	a612=112.79	3	d6123=-127.7
C	6	r76=1.334	1	a761=116.68	2	d7612=-121.4
O	7	r87=1.181	6	a876=127.24	1	d8761=-179.9
H	1	r91=1.098	2	a912=109.82	3	d9123=111.9
H	2	r102=1.080	1	a1021=123.270	5	d10215=177.4
H	3	r113=1.082	2	a1132=126.275	1	d11321=-176.5
H	4	r124=1.082	3	a1243=124.243	2	d12432=179.94
H	5	r135=1.080	1	a1351=123.276	2	d13512=-177.4
Y(C5•)Y(C3O)DO						
C						
C	1	r21=1.499				
C	2	r32=1.548	1	a321=104.13		
C	3	r43=1.490	2	a432=105.11	1	d4321=0.7
C	4	r54=1.390	3	a543=109.35	2	d5432=-1.3
O	3	r63=1.498	2	a632= 89.55	1	d6321=-113.6
C	2	r72=1.545	3	a723= 83.74	4	d7234=115.3
O	7	r87=1.187	2	a872=137.19	3	d8723=178.9
H	1	r91=1.080	2	a912=124.06	3	d9123=177.7
H	2	r102=1.091	3	a1023=119.63	4	d10234=-130.6
H	3	r113=1.088	2	a1132=119.72	1	d11321=135.1
H	4	r124=1.080	3	a1243=124.27	2	d12432=-178.7
H	5	r135=1.083	4	a1354=124.30	3	d13543=-177.8
Y(C5)CO•DO						
C						
C	1	r21=1.506				
C	2	r32=1.371	1	a321=108.53		
C	3	r43=1.433	2	a432=109.04	1	d4321=-2.7
C	1	r51=1.471	2	a512=103.51	3	d5123=3.6
C	1	r61=1.603	2	a612=104.41	3	d6123=-109.2
O	6	r76=1.205	1	a761=122.67	2	d7612=-176.1
O	6	r86=1.280	1	a861=106.20	2	d8612=3.7
H	1	r91=1.091	2	a912=117.55	3	d9123=133.1
H	2	r102=1.078	1	a1021=124.39	5	d10215=-175.3
H	3	r113=1.081	2	a1132=125.86	1	d11321=176.2
H	4	r124=1.081	3	a1243=124.93	2	d12432=179.9
H	5	r135=1.080	1	a1351=123.70	2	d13512=-178.7

[a]Distances in angstroms and angles in degrees. Geometry parameters optimized at the B3LYP/6-311G(d,p) level of theory.

Table B.4 (con't): Geometry Parameters for unsaturated oxygenated radicals[a]

Structure		Bond length (angstrom)		Bond angle (degree)		Dihedral angle (degree)
Y(C6•)Y(C2O2)						
C						
C	1	r2=1.481				
C	2	r3=1.370	1	a321=116.67		
C	3	r4=1.426	2	a432=123.54	1	d4321=9.7
C	4	r5=1 416	3	a543=120.64	2	d5432=8.4
C	1	r6=1.466	2	a612=112.96	3	d6123=-22.1
O	1	r7=1.518	2	a712=125.34	3	d7123=-122.8
O	6	r8=1.365	1	a861=95.212	2	d8612=-140.8
H	1	r9=1.100	2	a912=112.10	3	d9123=110.3
H	2	r102=1.082	1	a1021=120.63	6	d10216=167.6
H	3	r113=1.084	2	a1132=118.76	1	d11321=-170.8
H	4	r124=1.082	3	a1243=119.17	2	d12432=-179.2
H	5	r135=1.082	4	a1354=122.20	3	d13543=161.1
Y(C5•)						
C						
C	1	r21=1.435				
C	2	r32=1.368	1	a321=107.24		
C	3	r43=1.481	2	a432=108.23	1	d4321=0.007
C	1	r51=1.435	2	a512=109.02	3	d5123=-0.01
H	1	r61=1.082	2	a612=125.48	3	d6123=179.9
H	2	r72=1.079	1	a721=125.66	5	d7215=-179.
H	3	r83=1.081	2	a832=126.93	1	d8321=179.9
H	4	r94=1.081	3	a943=124.82	2	d9432=179.9
H	5	r1051.079	1	a1051=125.66	2	d10512=179.9
Y(C6•O)DO						
C						
C	1	r21=1.457				
C	2	r32=1.376	1	a321=131.35		
C	3	r43=1.399	2	a432=129.34	1	d4321=-0.02
C	4	r54=1.413	3	a543=125.97	2	d5432=-0.1
C	5	r65=1.360	4	a654=128.90	3	d6543=-0.08
O	6	r76=1.351	5	a765=131.44	4	d7654=0.037
O	1	r81=1.198	2	a812=123.06	3	d8123=-179.7
H	2	r92=1.082	1	a921=110.38	7	d9217=-179.6
H	3	r103=1.086	2	a1032=115.00	1	d10321=179.9
H	4	r114=1.083	3	a1143=117.14	2	d11432=179.9
H	5	r125=1.084	6	a1256=113.88	7	d12567=179.9
H	6	r136=1.083	5	a1365=119.81	4	d13654=-179.9
Y(C6•)Y(CO2)						
C						
C	1	r21=1.415				
C	2	r32=1.365	1	a321=120.85		
C	3	r43=1.468	2	a432=120.35	1	d4321=-0.0
C	4	r54=1.468	3	a543=117.19	2	d5432=-0.0
C	5	r65=1.365	4	a654=120.35	3	d6543=0.00
O	4	r74=1.424	3	a743=116.42	2	d7432=144.4
O	4	r84=1.424	3	a843=116.40	2	d8432=-144.4
H	1	r91=1.083	2	a912=119.80	3	d9123=-179.9
H	2	r102=1.084	1	a1021=119.31	6	d10216=179.9
H	3	r113=1.082	4	a1134=116.69	5	d11345=-179.9
H	5	r125=1.082	4	a1254=116.69	3	d12543=179.9
H	6	r136=1.084	5	a1365=119.82	4	d13654=179.9
ODC6•DO						
C						
C	1	r21=1.471				
C	2	r32=1.346	1	a321=121.55		
C	3	r43=1.444	2	a432=123.96	1	d4321=180.
C	4	r54=1.346	3	a543=124.05	2	d5432=-179.9
C	5	r65=1.468	4	a654=120.41	3	d6543=179.99
O	6	r76=1.186	5	a765=128.36	4	d7654=-179.9
O	1	r81=1.209	2	a812=123.86	3	d8123=-179.9
H	1	r91=1.111	2	a912=114.85	3	d9123=-0.003
H	2	r102=1.086	1	a1021=116.51	8	d10218=0.003
H	3	r113=1.088	2	a1132=118.90	1	d11321=0.000
H	4	r124=1.086	3	a1243=117.65	2	d12432=0.000
H	5	r135=1.090	4	a1354=121.93	3	d13543=-0.00

[a]Distances in angstroms and angles in degrees. Geometry parameters optimized at the B3LYP/6-311G(d,p) level of theory.

Table B.4 (con't): Geometry Parameters for unsaturated oxygenated radicals[a]

Structure		Bond length (angstrom)		Bond angle (degree)		Dihedral angle (degree)
Y(C5•O)						
C						
C	1	r21=1.356				
O	2	r32=1.380	1	a321=122.65		
C	3	r43=1.380	2	a432=116.96	1	d4321=0.02
C	4	r54=1.356	3	a543=122.65	2	d5432=-0.02
C	5	r65=1.416	4	a654=120.69	3	d6543=0.02
H	1	r71=1.083	6	a716=121.27	5	d7165=-179.9
H	2	r82=1.079	1	a821=125.97	6	d8216=179.9
H	4	r94=1.079	5	a945=125.97	6	d9456=-179.9
H	5	r105=1.083	6	a1056=121.29	1	d10561=179.9
H	6	r116=1.079	5	a1165=121.83	4	d11654=-179.9
C•C4DO						
C						
C	1	r21=1.318				
C	2	r32=1.458	1	a321=124.25		
C	3	r43=1.345	2	a432=124.47	1	d4321=180.
C	4	r54=1.466	3	a543=121.28	2	d5432=180.
O	5	r65=1.210	4	a654=124.39	3	d6543=180.
H	1	r71=1.080	2	a712=138.46	3	d7123=0.
H	2	r82=1.090	3	a823=116.46	4	d8234=0.
H	3	r93=1.088	2	a932=116.73	1	d9321=0.
H	4	r104=1.086	3	a1043=122.01	2	d10432=0.
H	5	r115=1.113	4	a1154=114.59	3	d11543=0.
C4C•DO						
C						
C	1	r21=1.338				
C	2	r32=1.448	1	a321=123.17		
C	3	r43=1.346	2	a432=124.79	1	d4321=179.9
C	4	r54=1.463	3	a543=120.44	2	d5432=-179.9
O	5	r65=1.187	4	a654=128.86	3	d6543=179.9
H	1	r71=1.085	2	a712=121.37	3	d7123=0.0
H	1	r81=1.083	2	a812=121.65	3	d8123=-179.9
H	2	r92=1.086	1	a921=119.77	7	d9217=-179.9
H	3	r103=1.087	2	a1032=117.30	1	d10321=0.0
H	4	r114=1.090	3	a1143=121.86	2	d11432=0.0
CDCCDC•						
C						
C	1	r21=1.336				
C	2	r32=1.467	1	a321=123.77		
C	3	r43=1.316	2	a432=125.17	1	d4321=180.
H	1	r51=1.085	2	a512=121.54	3	d5123=0.
H	1	r61=1.083	2	a612=121.49	3	d6123=180.
H	2	r72=1.086	1	a721=119.76	5	d7215=180.
H	3	r83=1.091	2	a832=115.85	1	d8321=0.
H	4	r94=1.080	3	a943=138.36	2	d9432=0.

[a]Distances in angstroms and angles in degrees. Geometry parameters optimized at the B3LYP/6-311G(d,p) level of theory.

B.5 Species used for the Dibenzofuranyl + O_2 Study

Table B.5: Geometry Parameters for unsaturated oxygenated species[a]

Structure		Bond length (angstrom)		Bond angle (degree)		Dihedral angle (degree)
Furan						
C						
C	1	r21=1.357				
C	2	r32=1.434	1	a321=106.10		
C	3	r43=1.35	2	a432=106.10	1	d4321=0.003
O	4	r54=1.362	3	a543=110.49	2	d5432=-0.03
H	1	r61=1.076	2	a612=133.70	3	d6123=179.9
H	2	r72=1.078	1	a721=126.50	5	d7215=-179.
H	3	r83=1.078	2	a832=127.39	1	d8321=179.9
H	4	r94=1.076	3	a943=133.70	2	d9432=179.9
BFURAN						
C						
C	1	r21=1.405				
C	1	r31=1.388	2	a312=121.34		
C	3	r43=1.401	1	a431=118.42	2	d4312=-0.00
C	4	r54=1.405	3	a543=118.84	1	d5431=0.000
C	5	r65=1.387	4	a654=123.59	3	d6543=0.000
O	5	r75=1.367	4	a754=110.25	3	d7543=179.9
C	7	r87=1.372	5	a875=106.05	4	d8754=0.005
C	8	r98=1.352	7	a987=112.29	5	d9875=-0.005
H	1	r101=1.084	2	a1012=119.11	6	d10126=-179.9
H	2	r112=1.083	1	a1121=119.40	3	d11213=-179.9
H	3	r123=1.084	1	a1231=120.73	2	d12312=-179.9
H	6	r136=1.082	5	a1365=121.27	4	d13654=179.9
H	8	r148=1.077	9	a1489=132.66	4	d14894=179.9
H	9	r159=1.078	8	a1598=126.20	7	d15987=-179.9
DBF						
C						
C	1	r21=1.402				
C	1	r31=1.391	2	a312=121.03		
C	3	r43=1.397	1	a431=118.64	2	d4312=-0.009
C	4	r54=1.405	3	a543=119.00	1	d5431=0.003
C	5	r65=1.385	4	a654=123.26	3	d6543=0.006
H	1	r71=1.083	2	a712=119.33	6	d7126=-179.9
H	2	r82=1.083	1	a821=119.48	3	d8213=-179.9
H	3	r93=1.084	1	a931=120.54	2	d9312=179.9
H	6	r106=1.082	5	a1065=121.15	4	d10654=179.9
C	4	r114=1.451	3	a1143=135.60	1	d11431=-179.9
O	5	r125=1.375	4	a1254=111.57	3	d12543=179.9
C	11	r1311=1.405	4	a13114=105.39	3	d131143=179.9
C	13	r1413=1.385	11	a141311=123.27	4	d1413114=179.9
C	14	r1514=1.393	13	a151413=116.73	11	d15141311=0.0005
C	15	r1615=1.402	14	a161514=121.31	13	d16151413=-0.001
C	16	r1716=1.391	15	a171615=121.03	14	d17161514=-0.0004
H	15	r1815=1.083	14	a181514=119.20	13	d18151413=179.9
H	16	r1916=1.083	15	a191615=119.33	14	d19161514=-179.9
H	17	r2017=1.083	16	a201716=120.54	15	d20171615=-179.9
H	14	r2114=1.082	13	a211413=121.15	11	d21141311=-179.9

[a] Distances in angstroms and angles in degrees. Geometry parameters optimized at the B3LYP/6-311G(d,p) level of theory.

Table B.5 (con't): Geometry Parameters for unsaturated oxygenated species[a]

Structure		Bond length (angstrom)		Bond angle (degree)		Dihedral angle (degree)
DBFOH						
C						
C	1	r21=1.402				
C	1	r31=1.391	2	a312=121.21		
C	3	r43=1.397	1	a431=118.36	2	d4312=-0.005
C	4	r54=1.405	3	a543=119.19	1	d5431=0.007
C	5	r65=1.384	4	a654=123.26	3	d6543=-0.003
H	1	r71=1.083	2	a712=119.26	6	d7126=179.9
H	2	r82=1.083	1	a821=119.48	3	d8213=-179.9
H	3	r93=1.082	1	a931=121.34	2	d9312=179.9
H	6	r106=1.082	5	a1065=121.16	4	d10654=179.9
C	4	r114=1.447	3	a1143=135.72	1	d11431=179.9
O	5	r125=1.377	4	a1254=111.47	3	d12543=179.9
C	11	r1311=1.403	4	a13114=106.07	3	d131143=179.9
C	13	r1413=1.387	11	a141311=123.80	4	d1413114=-179.9
C	14	r1514=1.392	13	a151413=116.14	11	d15141311=0.03
C	15	r1615=1.401	14	a161514=121.86	13	d16151413=-0.02
C	16	r1716=1.394	15	a171615=120.76	14	d17161514=0.011
H	14	r1814=1.081	13	a181413=121.45	11	d18141311=-179.9
H	16	r1916=1.085	15	a191615=119.68	14	d19161514=-179.9
H	15	r2015=1.083	14	a201514=119.36	13	d20151413=179.9
O	17	r2117=1.363	16	a211716=123.60	15	d21171615=179.9
H	21	r2221=0.963	17	a222117=109.25	16	d22211716=0.09
DBFOOH						
C						
C	1	r21=1.402				
C	1	r31=1.391	2	a312=121.21		
C	3	r43=1.398	1	a431=118.39	2	d4312=0.04
C	4	r54=1.406	3	a543=119.13	1	d5431=-0.13
C	5	r65=1.384	4	a654=123.27	3	d6543=0.14
H	1	r71=1.083	2	a712=119.26	6	d7126=-179.85
H	2	r82=1.083	1	a821=119.48	3	d8213=179.98
H	3	r93=1.082	1	a931=121.15	2	d9312=-179.78
H	6	r106=1.082	5	a1065=121.12	4	d10654=-179.98
C	4	r114=1.447	3	a1143=135.84	1	d11431=-179.80
O	5	r125=1.376	4	a1254=111.51	3	d12543=-179.78
C	11	r1311=1.402	4	a13114=106.09	3	d131143=179.69
C	13	r1413=1.387	11	a141311=123.77	4	d1413114=-179.79
C	14	r1514=1.391	13	a151413=116.20	11	d15141311=-0.06
C	15	r1615=1.402	14	a161514=122.28	13	d16151413=-0.13
C	16	r1716=1.388	15	a171615=119.78	14	d17161514=-0.13
H	15	r1815=1.083	14	a181514=119.27	13	d18151413=179.69
H	14	r1914=1.081	13	a191413=121.37	11	d19141311=179.78
H	16	r2016=1.079	15	a201615=120.74	14	d20161514=179.53
O	17	r2117=1.377	16	a211716=125.51	15	d21171615=-179.40
O	21	r2221=1.443	17	a222117=111.29	16	d22211716=-3.89
H	22	r2322=0.968	21	a232221=99.89	17	d23222117=116.23

Table B.5 (con't): Geometry Parameters for unsaturated oxygenated species[a]

Structure		Bond length (angstrom)		Bond angle (degree)		Dihedral angle (degree)
DBF•						
C						
C	1	r21=1.402				
C	1	r31=1.391	2	a312=121.05		
C	3	r43=1.397	1	a431=118.63	2	d4312=-0.006
C	4	r54=1.405	3	a543=118.96	1	d5431=0.0016
C	5	r65=1.384	4	a654=123.34	3	d6543=0.0053
H	1	r71=1.083	2	a712=119.33	6	d7126=-179.9
H	2	r82=1.083	1	a821=119.50	3	d8213=-179.9
H	3	r93=1.083	1	a931=120.53	2	d9312=179.99
H	6	r106=1.082	5	a1065=121.17	4	d10654=179.9
C	4	r114=1.451	3	a1143=135.53	1	d11431=-179.9
O	5	r125=1.377	4	a1254=111.53	3	d12543=179.99
C	11	r1311=1.408	4	a13114=105.23	3	d131143=179.9
C	13	r1413=1.393	11	a141311=123.23	4	d1413114=179.9
C	14	r1514=1.374	13	a151413=113.38	11	d15141311=0.002
C	15	r1615=1.381	14	a161514=127.37	13	d16151413=-0.002
C	16	r1716=1.399	15	a171615=117.44	14	d17161514=-0.001
H	16	r1816=1.083	15	a181615=121.78	14	d18161514=-179.9
H	17	r1917=1.084	16	a191716=120.14	15	d19171615=-179.9
H	14	r2014=1.082	13	a201413=122.26	11	d20141311=-179.9
DBFO•						
C						
C	1	r21=1.407				
C	1	r31=1.386	2	a312=121.31		
C	3	r43=1.402	1	a431=118.02	2	d4312=0.0004
C	4	r54=1.411	3	a543=119.43	1	d5431=-0.0007
C	5	r65=1.385	4	a654=123.16	3	d6543=0.001
H	1	r71=1.083	2	a712=119.05	6	d7126=179.9
H	2	r82=1.083	1	a821=119.22	3	d8213=-179.9
H	3	r93=1.082	1	a931=122.09	2	d9312=179.9
H	6	r106=1.082	5	a1065=121.37	4	d10654=179.9
C	4	r114=1.431	3	a1143=135.29	1	d11431=-179.9
O	5	r125=1.373	4	a1254=111.26	3	d12543=-179.9
C	11	r1311=1.394	4	a13114=106.22	3	d131143=179.9
C	13	r1413=1.389	11	a141311=123.78	4	d1413114=179.9
C	14	r1514=1.415	13	a151413=117.07	11	d15141311=0.012
C	15	r1615=1.375	14	a161514=121.80	13	d16151413=-0.008
C	16	r1716=1.461	15	a171615=122.21	14	d17161514=-0.01
H	15	r1815=1.083	14	a181514=118.50	13	d18151413=179.9
H	14	r1914=1.082	13	a191413=121.13	11	d19141311=-179.9
H	16	r2016=1.083	15	a201615=121.48	14	d20161514=179.9
O	17	r2117=1.244	16	a211716=122.39	15	d21171615=179.9
DBFOO•						
C						
C	1	r21=1.403				
C	1	r31=1.390	2	a312=121.20		
C	3	r43=1.398	1	a431=118.28	2	d4312=0.0005
C	4	r54=1.406	3	a543=119.27	1	d5431=0.0008
C	5	r65=1.384	4	a654=123.21	3	d6543=-0.001
H	1	r71=1.083	2	a712=119.26	6	d7126=179.9
H	2	r82=1.083	1	a821=119.40	3	d8213=179.9
H	3	r93=1.082	1	a931=121.20	2	d9312=179.9
H	6	r106=1.082	5	a1065=121.17	4	d10654=-179.9
C	4	r114=1.446	3	a1143=135.73	1	d11431=179.99
O	5	r125=1.376	4	a1254=111.62	3	d12543=179.9
C	11	r1311=1.405	4	a13114=106.03	3	d131143=-179.9
C	13	r1413=1.385	11	a141311=123.52	4	d1413114=-179.9
C	14	r1514=1.395	13	a151413=116.83	11	d15141311=-0.00
C	15	r1615=1.398	14	a161514=121.89	13	d16151413=0.000
C	16	r1716=1.388	15	a171615=119.34	14	d17161514=0.002
H	15	r1815=1.083	14	a181514=119.26	13	d18151413=-179.9
H	14	r1914=1.081	13	a191413=121.10	11	d19141311=179.99
H	16	r2016=1.080	15	a201615=121.66	14	d20161514=179.99
O	17	r2117=1.394	16	a211716=123.69	15	d21171615=-179.9
O	21	r2221=1.322	17	a222117=115.58	16	d22211716=-0.085

Table B.6: Vibrational Frequencies[a] (cm^{-1}) for CH_2=CHOOH, *trans*-CH_3CH=CHOOH, *cis*-CH_3CH=CHOOH, $CH_3(CH_3)$C=CHOOH CH_2=C(CH_3)OOH and CH_3CH=C(CH_3)OOH

Molecules	Frequencies ν^a
CH_2=CHOOH	109.2, 216.2, 344.0, 526.5, 709.3, 862.7, 871.7, 962.9, 990.3, 1176.7, 1312.0, 1397.1, 1417.0, 1690.1, 3158.8, 3179.0, 3255.2, 3778.4
Trans-CH_3CH=CHOOH	107.6, 188.6, 193.1, 211.7, 298.6, 476.2, 506.4, 836.6, 862.1, 932.7, 964.5, 1060.6, 1107.1, 1156.1, 1294.9, 1322.4, 1379.8, 1419.3, 1480.6, 1494.8, 1718.9, 3014.0, 3057.0, 3092.0, 3151.8, 3157.5, 3779.0
Cis-CH_3CH=CHOOH	88.0, 152.0, 200.3, 236.5, 348.7, 492.6, 639.5, 769.3, 885.0, 932.3, 953.9, 1023.9, 1066.3, 1149.7, 1265.6, 1381.0, 1389.4, 1424.6, 1485.9, 1493.1, 1714.8, 3016.9, 3059.6, 3115.8, 3156.0, 3180.4, 3779.2
$CH_3(CH_3)$C=CHOOH	86.8, 161.6,172.7, 207.2, 216.8, 287.2, 350.0, 418.9, 506.9, 556.3, 813.1, 856.2, 876.6, 970.7, 1007.1, 1056.8, 1101.9, 1145.5, 1241.1, 1357.7, 1367.8, 1410.5, 1421.3, 1473.6, 1482.7, 1492.5, 1496.8, 1717.7, 3007.4, 3013.2, 3050.3, 3053.2, 3097.1, 3128.4, 3153.9, 3778.2
CH_2=C(CH_3)OOH	106.6, 182.3, 207.6, 304.4, 413.7, 472.0, 552.9, 739.8, 831.6, 872.4, 893.7, 962.0, 1021.5, 1072.5, 1268.6, 1370.7, 1399.4, 1427.3, 1471.5, 1493.2, 1710.9, 3032.2, 3096.6, 3128.0, 3160.0, 3252.6, 3781.8
CH_3CH=C(CH_3)OOH	93.3, 154.6, 173.9, 211.0, 241.6, 264.3, 297.6, 435.6, 566.1, 590.0, 788.7, 831.6, 861.1, 959.5, 1047.4, 1064.3, 1075.1, 1107.0, 1215.5, 1339.6, 1358.8920, 1408.4852, 1413.8951, 1472.6, 1484.0, 1492.1, 1500.1, 1733.8, 3014.7, 3026.9, 3054.7, 3088.6, 3121.3, 3124.5, 3154.3, 3779.8

[a]Frequencies are calculated at B3LYP/6-311 level of theory.

Table B.6 (cont): Vibrational Frequencies[a] (cm^{-1}) for CH_2=CHCH_2OOH, CH≡COOH, CH_3C≡COOH, and C_6H_5OOH

Molecules	Frequencies ν^a
CH_2=CHCH_2OOH	92.4, 110.9, 186.4, 256.1, 414.9, 465.3, 639.1, 918.7, 942.0, 966.1, 993.8, 1033.8, 1041.2, 1175.8, 1240.4, 1314.7, 1356.9, 1382.9, 1457.8, 1511.6, 1712.3, 2998.7, 3038.3, 3128.9, 3150.4, 3214.6, 3773.4
CH≡COOH	126.7270, 229.8413, 442.0211, 520.6386, 562.9493, 712.1315, 754.5012, 1067.6054, 1343.6821, 2218.3537, 3489.4161, 3757.4715
CH_3C≡COOH	27.8, 118.4, 143.7, 198.4, 324.3, 439.6, 527.7, 715.2, 779.2, 1045.9, 1059.5, 1245.7, 1337.0, 1417.3, 1473.8, 1485.4, 2337.5, 3019.8, 3076.3, 3077.5, 3761.9
C_6H_5OOH	59.8, 175.4, 226.5, 261.5, 419.7, 440.7, 511.6, 599.8, 628.3, 703.0, 764.3, 811.3, 838.9, 909.8, 958.1, 974.7, 991.8, 1011.5, 1045.1, 1101.4, 1176.4, 1180.2, 1226.0, 1332.6, 1349.0, 1396.2, 1489.4, 1516.6, 1637.8, 1638.9, 3166.0, 3174.1, 3188.4, 3195.9, 3233.4, 3767.1

[a]Frequencies are calculated at B3LYP/6-311 level of theory.

Table B.7: Vibrational Frequencies[a] (cm^{-1}) for $CH_2=CHOOCH_3$, *trans*-$CH_3CH=CHOOCH_3$, *cis*-$CH_3CH=CHOOCH_3$, and $CH_3(CH_3)C=CHOOCH_3$

Molecules	Frequencies ν[b]
$CH_2=CHOOCH_3$	**30.1, 204.8, 219.9**, 318.2, 436.5, 604.3, 710.1, 846.4, 853.5. 969, 971.7, 1029.6, 1153.9, 1174.5, 1215.1, 1322.1, 1420.4, 1451.9, 1460.4, 1511, 1699.9, 3014.1, 3084.4, 3115.1, 3167, 3182.1, 3270.3
	[b]**C=CO—OC, C=C—OOC, C=COO—C**
trans-$CH_3CH=CHOOCH_3$	**40.3. 93.3.** 156.9. **210.4. 238.2.** 290.2. 358.9. 442.8. 564.0. 813.2. 837.5. 933.8. 962.7 1033.5. 1060.6. 1111.9. 1153.0. 1174.2. 1231.6. 1298.4. 1323.3. 1418.4. 1447.9. 1455.5. 1480.6. 1495.1. 1513.4. 1717.8. 3007.5. 3012.7. 3055.4. 3073.1. 3090.7. 3110.5. 3149.9. 3156.5
	[b]**tr-CC=CO—OC. tr-CC=C—OOC. tr-C—C=COOC. tr-CC=COO—C**
cis-$CH_3CH=CHOOCH_3$	**44.6. 72.2. 143.8.** 193.3. **227.0.** 332.0. 417.3. 486.1. 637.2. 766.2. 846.0. 933.1. 952.1. 1030.9. 1033.7. 1066.7. 1146.9. 1174.1. 1221.0. 1273.6. 1387.5. 1423.7. 1448.2. 1455.4. 1485.5. 1493.1. 1513.4. 1713.7. 3008.5. 3016.1. 3058.6. 3074.9. 3111.4. 3114.8. 3155.2. 3179.2
	[b]**cis-CC=CO—OC. cis-CC=C—OOC. cis-C—C=COOC. cis-CC=COO—C**
$CH_3(CH_3)C=CHOOCH_3$	**45.1. 71.1.** 138.9. **171.7. 203.7. 228.5.** 275.8. 337.7. 373.6. 424.9. 486.0. 598.5. 780.4. 843.0. 874.3. 970.4. 1007.2. 1031.4. 1060.6. 1101.5. 1147.4. 1174.1. 1221.0. 1248.4. 1358.7. 1410.1. 1420.8. 1447.4. 1454.7. 1473.5. 1483.0. 1492.5. 1496.4. 1513.8. 1716.2. 3004.9. 3006.6. 3012.5. 3049.1. 3052.0. 3069.5. 3097.0. 3107.3. 3128.3. 3152.7
	[b]**C(C)C=CO—OC. C(C)C=C—OOC. C(—C)C=COOC. C—(C)C=COOC. C(C)C=COO—C**

[a]Frequencies are calculated at B3LYP/6-311 level of theory. [b]Frequecies in bold are the rotational frequencies. the corresponding type of rotation is shown in the table.

Table B.7 (cont): Vibrational Frequencies[a] (cm^{-1}) of $CH_2=CHCH_2OOCH_3$, $CH_2=C(CH_3)OOCH_3$, $CH_3CH=C(CH_3)OOCH_3$, $CH\equiv COOCH_3$, $CH_3C\equiv COOCH_3$, and $C_6H_5OOCH_3$

Molecules	Frequencies ν[b]
$CH_2=CHCH_2OOCH_3$	**19.2. 81.4. 105.7.** 192.5. **251.6.** 383.7. 404.0. 481.6. 646.7. 875.6. 941.0. 963.7. 995.3. 1032.2. 1043.6. 1062.3. 1173.0. 1176.8. 1220.6. 1248.8. 1315.3. 1378.8. 1450.3. 1455.8. 1457.5. 1511.1. 1516.4. 1712.3. 2996.6. 3004.2. 3038.3. 3065.9. 3108.7. 3129.7. 3149.4. 3214.8
	[b]**C=CCO—OC. C=CC—OOC. C=C—COOC. C=CCOO—C**
$CH_2=C(CH_3)OOCH_3$	**39.6. 89.8. 190.3. 207.8.** 272.4. 359.7. 457.5. 480.9. 534.3. 735.1. 817.4. 865.6. 884.7. 971.9. 1025.9. 1035.0. 1073.2. 1173.8. 1207.2. 1293.5. 1401.3. 1427.6. 1448.3. 1455.0. 1473.4. 1493.0. 1514.1. 1707.0. 3007.3. 3036.4. 3072.9. 3098.6. 3110.0. 3126.9. 3159.6. 3251.7
	[b]**C=C(C)O—OC. C=C(C)—OOC. C=C(—C)OOC. C=C(C)OO—C**
$CH_3CH=C(CH_3)OOCH_3$	**43.3. 73.4. 147.6. 173.8. 205.1.** 227.2. 245.5. 284.8. 410.9. 439.2. 556.7. 580.9. 789.5. 805.9. 861.6. 966.7. 1029.5. 1050.5. 1063.6. 1075.2. 1107.4. 1174.1. 1196.1. 1241.4. 1346.3. 1409.5. 1413.6. 1447.1. 1454.8. 1472.6. 1483.1. 1491.5. 1499.9. 1514.3. 1730.3. 3004.5. 3014.0. 3029.7. 3053.8. 3069.0. 3090.4. 3105.1. 3120.4. 3123.3. 3154.1
	[b]**CC=C(C)O—OC. CC=C(C)—OOC. C—C=C(C)OOC. CC=C—(C)OOC. CC=C(C)OO—C**
$CH\equiv COOCH_3$	15.8. 174.3. 229.5. 378.3. 433.2. 496.6. 512.3. 712.7. 716.6. 1003.3. 1097.8. 1173.4. 1220.6. 1438.0. 1443.7. 1514.5. 2208.8. 3010.3. 3084.4. 3103.3. 3490.1
$CH_3C\equiv COOCH_3$	19.7. 20.8. 107.3. 192.1. 221.5. 294.9. 362.3. 444.1. 470.8. 669.6. 808.3. 1009.1. 1044.6. 1059.6. 1173.0. 1199.5. 1288.9. 1416.5. 1436.9. 1443.3. 1472.4. 1486.2. 1514.7. 2326.8. 3004.5. 3017.5. 3073.5. 3074.1. 3075.6. 3097.0
$C_6H_5OOCH_3$	**94.0. 209.1. 256.8.** 273.8. 425.2. 446.7. 520.0. 561.4. 630.0. 703.4. 767.0. 799.3. 828.4. 895.6. 967.5. 986.7. 1010.0. 1042.5. 1068.0. 1102.4. 1170.3. 1178.1. 1194.4.1205.0. 1276.6. 1335.4. 1357.3. 1476.2. 1487.8. 1491.9. 1507.3. 1530.6.1627.1. 1646.8. 2999.7.3056.3. 3131.5. 3162.6.3170.3. 3186.4. 3194.4.3202.0
	[b]**phO—OC. ph—OOC. phOO—C**

[a]Frequencies are calculated at the B3LYP/6-311g(d.p) level of theory. [b]Frequecies in bold are the rotational frequencies. the corresponding type of rotation is shown in the table.

Table B.8: Moment of Inertia[a] of CH_2=CHOOH, trans- CH_3CH=CHOOH, cis- CH_3CH=CHOOH, $CH_3(CH_3)$C=CHOOH, CH_3CH=C(CH_3)OOH, CH_2=CHCHOOH, CH_2=C(CH_3)OOH, CH≡COOH, CH_3C≡COOH and C_6H_5-OOH

	I_a [b]	I_b	I_c
CH_2=CHOOH	41.13	4.59	4.28
Trans- CH_3CH=CHOOH	25.09	2.19	2.10
Cis- CH_3CH=CHOOH	14.77	2.75	2.46
$CH_3(CH_3)$C=CHOOH	7.32	1.98	1.66
CH_2=CHCHOOH	24.05	2.27	2.23
CH_2=C(CH_3)OOH	8.17	4.33	3.06
CH_3CH=C(CH_3)OOH	5.11	2.70	1.90
CH≡COOH	41.56	5.11	4.60
CH_3C≡COOH	27.60	2.19	2.06
C_6H_5OOH	5.07	1.55	1.19

[a]Optimized at the B3LYP/6-311g(d,p) level of theory. [b]Units in GHz

Table B. 8: Moment of Inertia[a] of CH_2=CHOOCH$_3$, *trans*-CH_3CH=CHOOCH$_3$, *cis*-CH_3CH=CHOOCH$_3$, $CH_3(CH_3)$C=CHOOCH$_3$, CH_3CH=C(CH_3)OOCH$_3$, CH_2=CHCHOOCH$_3$, CH_2=C(CH_3)OOCH$_3$, CH≡COOCH$_3$, CH_3C≡COOCH$_3$ and C_6H_5OOCH$_3$

	I_a [b]	I_b	I_c
CH_2=CHOOCH$_3$	12.97	3.32	2.90
trans-CH_3CH=CHOOCH$_3$	20.26	1.35	1.32
cis-CH_3CH=CHOOCH$_3$	11.83	1.63	1.51
$CH_3(CH_3)$C=CHOOCH$_3$	6.91	1.19	1.07
CH_2=CHCHOOCH$_3$	15.23	1.44	1.41
CH_2=C(CH_3)OOCH$_3$	7.57	2.26	1.86
CH_3CH=C(CH_3)OOCH$_3$	4.21	1.68	1.28
CH≡COOCH$_3$	32.83	2.62	2.46
CH_3C≡COOCH$_3$	23.21	1.34	1.28
C_6H_5OOCH$_3$	5.05	1.56	1.20

[a]Optimized at the B3LYP/6-311g(d.p) level of theory. [b]Units in GHz

Appendix C: Error Analysis

We estimate error limits on the target $\Delta_f H^0_{298}$ by considering the errors reported for the reference species in the work reactions plus the error due to the calculation method. For the reference species we add the absolute value of the reported errors. We choose not to take the square root of the sum of squared errors, because these errors are not considered conventional "random statistical errors".

The error resulting from the B3LYP/6-311g(d,p) and the G3MP2B3 work reaction computation methods, where the total energy include zero-point vibration energy (ZPVE), and the thermal energy, is estimated by comparing the calculated enthalpy of reaction for a series of work reactions, with $\Delta H^0_{rxn,298}$ from accepted literature data. The reactions are chosen to be appropriate to the current study; that is they involve species that are estimated to be representative of species in this study. The enthalpies ($\Delta H^0_{rxn,298}$) for a series of work reactions, where each species in the reaction has a literature value, are determined and compared to the experimental enthalpy of reaction, $\Delta H^0_{rxn,298}$. The difference $\Delta H^0_{rxn,298}$(cal) - $\Delta H^0_{rxn,298}$(exp) on these work reactions should provide a reasonable evaluation of the error of the calculation method. The mean absolute average from the 27 calculated difference values for B3LYP computation method (0.78 kcal mol^{-1}) and the mean absolute average from the 15 calculated difference values for G3MP2B3 computation method (0.31 kcal mol^{-1}) are considered as the error in the calculation method for this study.

A hypothetical standard deviation S_h determined for each method for the differences, $\Delta H^0_{rxn,298}$(cal) - $\Delta H^0_{rxn,298}$(exp), using different work reactions are given as well. We note that each reaction is different and we do not wish to imply that this S_h be used to represent an error bound; however it may be of some value for a reader to evaluate this number.

Table C1: Determination of the error resulting from the computation methods using work reactions with all species having known $\Delta_f H^0_{298}$ values.

Work Reactions (B3LYP)	$\Delta H^0_{rxn,298\,(cal)} - \Delta H^0_{rxn,298\,(exp)} = X_{cal} - X_{exp}$
R1: $CH_3CH_2OH + CH_3CH_2CH_3 \rightarrow CH_3CH_2CH_2OH + CH_3CH_3$	-0.5
R2: $CH_3CH_2OH + CH_3CH{=}CH_2 \rightarrow CH_2{=}CH_2 + CH_3CH_2CH_2OH$	0.04
R3: $CH_3C({=}O)CH_3 + CH_3CH_2CH_3 \rightarrow CH_3CH_2C({=}O)CH_3 + CH_3CH_3$	0.39
R4: $CH_3CH_2CH{=}O + CH_3CH_2OH \rightarrow CH_3CH{=}O + CH_3CH_3CH_2OH$	-0.05
R5: $CH_3CH{=}O + CH_3CH_3 \rightarrow CH_3CH_2CH{=}O + CH_4$	-0.25
R6: $C_6H_5OH + CH_3OCH_3 \rightarrow CH_3OH + C_6H_5OCH_3$	1.38
R7: $CH_3CH_2\bullet + CH_2{=}CH_2 \rightarrow CH_3CH_3 + CH_2{=}CH\bullet$	0.03
R8: $C_6H_5\bullet + CH_3CH_3 \rightarrow C_6H_6 + CH_3CH_2\bullet$	-0.27
R9: $C_6H_5\bullet + CH_2{=}CH_2 \rightarrow C_6H_6 + CH_2{=}CH\bullet$	-0.24
R10: $C_5H_5\bullet + CH_3CH_3 \rightarrow C_5H_6 + CH_3CH_2\bullet$	0.25
R11: $CH_3C\bullet{=}O + CH_3CH_3 \rightarrow CH_3CH{=}O + CH_3CH_2\bullet$	-0.23
R12: $CH_3OOCH_3 + CH_3O\bullet \rightarrow CH_3OO\bullet + CH_3OCH_3$	1.15
R13: $CH_3OOCH_3 + CH_3CH_2O\bullet \rightarrow CH_3OO\bullet + CH_3CH_2OCH_3$	0.72
R14: $CH_3CH_2OO\bullet + CH_3OH \rightarrow CH_3OO\bullet + CH_3CH_2OH$	-0.71
R15: $CH\bullet{=}O + CH_3OCH_3 \rightarrow CH_3CH{=}O + CH_3O\bullet$	-0.78
R16: $CH_2\bullet CH{=}O + CH_3OCH_3 \rightarrow CH_2\bullet OCH_3 + CH_3CH{=}O$	1.0
R17: $CH_2{=}O + CH_3O\bullet \rightarrow CH\bullet{=}O + CH_3OH$	2.0
R18: $CH_2{=}CHOOH + CH_3CH_2O\bullet \rightarrow CH_2{=}CHOO\bullet + CH_3CH_2OH$	1.75
R19: $CH_2{=}CHO\bullet + CH_2{=}CHOOH \rightarrow CH_2{=}CHOO\bullet + CH_2{=}CHOH$	1.85
R20: $CH_2{=}CHCH_2OO\bullet + CH_3OH \rightarrow CH_3OO\bullet + CH_2{=}CHCH_2OH$	-0.99
R21: $C_6H_5\bullet + CH_3OCH_3 \rightarrow C_6H_5OCH_3 + CH_3\bullet$	-0.22
R22: $C_6H_5\bullet + CH_3OH \rightarrow C_6H_5OH + CH_3\bullet$	-1.60
R23: $C_6H_5O\bullet + CH_3OCH_3 \rightarrow C_6H_5OCH_3 + CH_3O\bullet$	-0.52
R24: $C_6H_5O\bullet + CH_3CH_2OH \rightarrow C_6H_5OH + CH_3CH_2O\bullet$	-1.56
R25: $C_6H_5O\bullet + CH_2{=}CHOH \rightarrow C_6H_5OH + CH_2{=}CHO\bullet$	-0.51
R26: $C_6H_5O\bullet + CH_2{=}CHOCH_3 \rightarrow C_6H_5OCH_3 + CH_2{=}CHO\bullet$	0.35
R27: $CH_2{=}CHO\bullet + C_6H_5OOH \rightarrow C_6H_5OO\bullet + CH_2{=}CHOH$	1.7
Average Absolute deviation = (|R1| + |R2| + … + |R27|) / 27	0.78
$S_h = \sqrt{\dfrac{\sum (X_{cal} - X_{exp})^2}{(N-1)}}$	1.01

Table C2: Estimation of error from G3MP2B3 calculations using work reactions. All species have literature $\Delta_f H^0_{298}$ values.

Work Reactions	$\Delta H^0_{rxn,298\,(cal)} - \Delta H^0_{rxn,298\,(exp)} = X_{cal} - X_{exp}$ [a]
	G3MP2B3
R1: $CH_4 + CH_3CH_2OH \rightarrow CH_3OH + CH_3CH_3$	0.14
R2: $CH_3CH_2CH_3 + CH_3CH_2OH \rightarrow CH_3CH_2CH_2OH + CH_3CH_3$	-0.25
R3: $CH_3OCH_3 + CH_3CH_3 \rightarrow CH_3CH_2OCH_3 + CH_4$	0.55
R4: $CH_3OCH_3 + CH_3CH_2CH_3 \rightarrow CH_3CH_2OCH_3 + CH_3CH_3$	0.07
R5: $CH_3OCH_3 + CH_3CH_2CH_2CH_3 \rightarrow CH_3CH_2OCH_3 + CH_3CH_2CH_3$	0.61
R6: $CH_3OCH_3 + CH_3CH_2OH \rightarrow CH_3CH_2OCH_3 + CH_3OH$	0.69
R7: $CH_3OCH_3 + CH_3CH_2CH_2OH \rightarrow CH_3CH_2OCH_3 + CH_3CH_2OH$	0.37
R8: $CH_3CH_2OH + CH_2{=}CH_2 \rightarrow CH_3CH{=}CH_2 + CH_3OH$	0.11
R9: $CH_3CH_2OH + CH_3CH{=}CH_2 \rightarrow CH_2{=}CH_2 + CH_3CH_2CH_2OH$	0.2
R10: $CH_3C({=}O)CH_3 + CH_3CH_2CH_3 \rightarrow CH_3CH_2C({=}O)CH_3 + CH_3CH_3$	-0.34
R11: $CH_3CH_2CH{=}O + CH_3OH \rightarrow CH_3CH{=}O + CH_3CH_2OH$	-0.17
R12: $CH_3CH_2CH{=}O + CH_3CH_2OH \rightarrow CH_3CH{=}O + CH_3CH_3CH_2OH$	0.15
R13: $CH_3CH{=}O + CH_3CH_3 \rightarrow CH_3CH_2CH{=}O + CH_4$	0.02
R14: $CH_3CH{=}O + CH_3CH_3CH_3 \rightarrow CH_3CH_2CH{=}O + CH_3CH_3$	-0.44
R15: $CH_3CH{=}CH_2 + CH_3CH_3 \rightarrow CH_3CH_2CH_3 + CH_2{=}CH_2$	0.51
Average Absolute deviation = (|R1| + |R2| + … + |R15|) / 15	0.31
$S_h = \sqrt{\frac{\sum (X_{cal} - X_{exp})^2}{(N-1)}}$	0.38

Appendix D: Thermochemical Properties

TABLE D1: Ideal Gas-Phase Thermodynamic Properties[a], $\Delta_f H^0_{298}$, S^0_{298} and Cp(300-1500)

Species			$\Delta_f H^0_{298}$ [b]	S^0_{298} [c]	Cp (T) cal mol^{-1} K^{-1}						
					300 K	400 K	500 K	600 K	800 K	1000 K	1500 K
CH_2=C(CH=O)CH=CH_2	TVR[f]			71.08	19.10	25.25	30.61	35.07	41.84	46.66	53.89
(1)[d]	IR[g]	C_d—C_d		5.82	2.41	2.20	2.05	1.92	1.70	1.53	1.29
(1)[e]		C_d—CH=O		3.72	2.80	3.15	3.25	3.20	2.92	2.60	1.99
	Total		**-2.56**	80.62	24.31	30.6	35.91	40.19	46.46	50.79	57.17
CH_2=CHC(OH)$_2$CH=CH_2	TVR[f]			75.13	24.53	32.11	38.43	43.52	51.09	56.52	65.00
(1)[d]	IR[g]	C_d—C		5.57	2.82	2.60	2.38	2.18	1.86	1.64	1.34
(2)[e]		C_d—C		5.57	2.82	2.60	2.38	2.18	1.86	1.64	1.34
		C—OH		1.55	2.40	3.00	3.20	3.12	2.65	2.21	1.59
		C—OH		1.55	2.40	3.00	3.20	3.12	2.65	2.21	1.59
	Total		**-61.30**	89.37	34.97	43.31	49.59	54.12	60.11	64.22	70.86
CH_2=CHCH=O	TVR[f]			62.70	14.14	17.69	20.88	23.59	27.80	30.85	35.48
(2)[d] (2)[e]	IR[g]	Cd—CH=O		3.79	2.83	3.07	3.08	2.97	2.62	2.28	1.72
	Total		**-18.65**	66.49	16.97	20.76	23.96	26.56	30.42	33.13	37.2
	Nist[h]			-	17.11	20.97	24.21	26.87	30.81	33.51	37.38
(YC_5H_5)C(OH)=O	TVR[f]			76.87	24.22	32.49	39.27	44.60	52.26	57.47	65.12
(1)[d]	IR[g]	C—C=O		7.91	1.26	1.17	1.12	1.08	1.05	1.03	1.01
(1)[e]		CO—OH		0.57	1.30	1.86	2.29	2.62	2.99	3.11	2.87
	Total		**-52.81**	85.35	26.78	35.52	42.68	48.3	56.3	61.61	69.00
(YC_5H_5)OCH=O	TVR[f]			78.38	24.06	32.05	38.79	44.19	52.05	57.44	65.27
(1)[d]	IR[g]	C—O		6.64	2.84	2.47	2.17	1.93	1.61	1.42	1.20
(1)[e]		O—CH=O		3.76	3.33	3.16	2.91	2.73	2.54	2.43	2.15
	Total		**-34.67**	88.78	30.23	37.68	43.87	48.85	56.2	61.29	68.62
(YC_5H_5O)CH=O	TVR[f]			81.15	26.11	33.99	40.64	46.03	53.95	59.40	67.29
0[d](1)[e]	IR[g]	C—CH=O		4.81	2.41	2.68	2.76	2.69	2.39	2.07	1.57
	Total		**-25.27**	85.96	28.52	36.67	43.40	48.72	56.34	61.47	68.86
Y(C_6H_6)=O (1)[d](1)[e]	TVR[f]		**-4.88**	77.54	24.44	31.82	38.08	43.15	50.65	55.87	63.50
Furan (1)[d](1)[e]	TVR[f]			66.46	15.75	21.19	25.71	29.26	34.32	37.76	42.81
	Nist[h]		**-8.29**		15.74	21.22	25.77	29.34	34.44	37.90	42.94
CH_3OCH=O	TVR[f]			61.26	12.42	15.51	18.54	21.23	25.52	28.68	33.43
(1)[d]	IR[g]	C—O		5.69	1.03	1.01	1.00	1.00	0.99	0.99	0.99
(3)[e]		O—CH=O		2.80	1.62	1.91	2.28	2.63	3.12	3.29	3.04
	Total		**-83.39**	69.75	15.07	18.43	21.82	24.86	29.63	32.96	37.46
	Nist[h]		**-84.97**	-	15.44	18.54	21.58	24.27	28.56	31.71	36.44
CH_2=CHCH(CH=O)C≡CH (1)[d] (1)[e]	TVR[f]		**45.23**	75.71	23.23	29.61	34.91	39.24	45.82	50.56	57.75
CH_2=C(CH=O)C≡CH (1)[d] (1)[e]	TVR[f]		**39.18**	72.41	19.82	24.55	28.37	31.47	36.17	39.55	44.67

[a]Thermodynamic properties are referred to a standard state of an ideal gas at 1 atm. torsional frequencies are excluded in the calculations of entropies and heat capacities. Instead, a more exact contribution from hindered rotations bonds is included; [b] $\Delta_f H^0_{298}$ in kcal mol^{-1}; [c] S^0_{298} in cal mol^{-1} K^{-1}; [d]optical isomers number; [e]symmetry number; [f]The sum of contributions from translations, external rotations and vibrations; [g]Contribution from internal rotation about the corresponding bonds. [h]http://webbook.nist.gov/chemistry/

TABLE 5 D1 (con't): Ideal Gas-Phase Thermodynamic Properties[a], $\Delta_f H^0_{298}$, S^0_{298} and Cp(300-1500)

Species			$\Delta_f H^0_{298}$ [b]	S^0_{298} [c]	Cp (T) cal mol^{-1} K^{-1} 300 K	400 K	500 K	600 K	800 K	1000 K	1500 K
$CH(OH)_2CH{=}O$	TVR[f]			67.50	15.31	19.16	22.52	25.30	29.49	32.47	37.04
(1)[d]	IR[g]	C—OH		3.04	1.52	1.67	1.77	1.81	1.79	1.70	1.46
(1)[e]		C—OH		3.04	1.52	1.67	1.77	1.81	1.79	1.70	1.46
		C—CH=O		4.19	3.03	3.59	3.56	3.24	2.52	2.02	1.44
	Total		**119.08**	77.77	21.38	26.09	29.62	32.16	35.59	37.89	41.4
$CH_3C(OH)_2CH{=}O$	TVR[f]			70.87	19.76	25.09	29.72	33.55	39.38	43.60	50.14
(2)[d]	IR[g]	C—C		4.10	2.39	2.61	2.64	2.56	2.25	1.96	1.51
(3)[e]		C—OH		2.48	1.98	2.03	2.07	2.06	1.95	1.79	1.49
		C—OH		2.48	1.98	2.03	2.07	2.06	1.95	1.79	1.49
		C—CH=O		3.88	1.96	1.99	1.88	1.75	1.53	1.38	1.19
	Total		**-130.57**	83.81	28.07	33.75	38.38	41.98	47.06	50.52	55.82
$CH_2{=}C(OH)_2$	TVR[f]			64.55	14.62	18.19	21.11	23.43	26.86	29.35	33.38
(1)[d]	IR[g]	C_d—OH		2.36	2.54	2.56	2.46	2.32	2.02	1.78	1.42
(2)[e]		C_d—OH		2.36	2.54	2.56	2.46	2.32	2.02	1.78	1.42
	Total		**-73.95**	69.27	19.7	23.31	26.03	28.07	30.9	32.91	36.22
$CH_2{=}C(OH)CH{=}CH_2$	TVR[f]			68.65	18.70	24.14	28.70	32.40	37.98	42.03	48.36
(1)[d]	IR[g]	C_d—OH		2.16	3.11	3.04	2.75	2.45	1.99	1.70	1.34
(1)[e]		C_d—C_d		4.28	2.36	2.86	3.09	3.08	2.74	2.34	1.70
	Total		**-18.0**[h]	75.09	24.17	30.04	34.54	37.93	42.71	46.07	51.4
$CH_2{=}CHCH(OH)CH{=}CH_2$	TVR[f]			72.12	21.60	28.67	34.77	39.78	47.37	52.87	61.39
(1)[d]	IR[g]	C_d—C		6.00	2.94	2.45	2.10	1.85	1.54	1.36	1.17
(2)[e]	IR[g]	C_d—C		6.00	2.94	2.45	2.10	1.85	1.54	1.36	1.17
		C—OH		2.96	3.07	2.57	2.15	1.86	1.52	1.34	1.15
	Total		**-12.35**	87.08	30.55	36.14	41.12	45.34	51.97	56.93	64.88
$CH_2{=}CHOCH{=}CH_2$	TVR[f]			66.86	18.07	23.50	28.15	31.97	37.77	41.99	48.49
(1)[d]	IR[g]	C_d—O		5.84	2.00	2.03	2.00	1.92	1.73	1.57	1.31
(2)[e]		C_d—O		5.84	2.00	2.03	2.00	1.92	1.73	1.57	1.31
	Total		**-6.15**	78.54	22.07	27.56	32.15	35.81	41.23	45.13	51.11
$CH_2{=}CHCH_2CH{=}O$	TVR[f]			68.61	17.40	22.60	27.26	31.21	37.33	41.79	48.53
(1)[d]	IR[g]	C_d—C		5.47	2.11	1.76	1.54	1.40	1.24	1.15	1.07
(1)[e]		C—CH=O		5.73	1.91	1.63	1.46	1.35	1.22	1.15	1.07
	Total		**-20.62**	79.81	21.42	25.99	30.26	33.96	39.79	44.09	50.67
$CH_2{=}CHCH(CH_3)CH{=}CH_2$	TVR[f]			73.27	23.25	31.14	38.20	44.17	53.45	60.27	70.82
(1)[d]	IR[g]	C_d—C		6.54	2.14	1.91	1.71	1.55	1.35	1.24	1.11
(3)[e]		C_d—C		6.54	2.14	1.91	1.71	1.55	1.35	1.24	1.11
		C—C		4.28	2.09	2.14	2.05	1.92	1.67	1.49	1.25
	Total		**17.98**	90.63	29.62	37.1	43.67	49.19	57.82	64.24	74.29
$CH_2{=}CHCH(CH{=}O)CH{=}CH_2$	TVR[f]			76.77	24.23	31.68	38.22	43.66	52.01	58.03	67.13
(1)[d]	IR[g]	C_d—C		6.88	1.99	1.68	1.48	1.36	1.21	1.14	1.06
(2)[e]		C_d—C		6.88	1.99	1.68	1.48	1.36	1.21	1.14	1.06
		C—CH=O		6.49	2.09	1.84	1.64	1.50	1.32	1.22	1.10
	Total		**-0.83**	97.02	30.3	36.88	42.82	47.88	55.75	61.53	70.35
$CH_2{=}CHCH(CH{=}O)CH{=}O$	TVR[f]			75.82	22.12	28.38	33.87	38.46	45.49	50.49	57.89
(1)[d]	IR[g]	C_d—C		6.56	2.39	1.99	1.74	1.56	1.35	1.24	1.11
(1)[e]		C—CH=O		6.29	1.90	1.88	1.78	1.67	1.47	1.34	1.17
		C—CH=O		6.29	1.90	1.88	1.78	1.67	1.47	1.34	1.17
	Total		**-39.56**	94.96	28.31	34.13	39.17	43.36	49.78	54.41	61.34

[a]Thermodynamic properties are referred to a standard state of an ideal gas at 1 atm; [b] $\Delta_f H^0_{298}$ in kcal mol^{-1}; [c] S^0_{298} in cal mol^{-1} K^{-1}; [d]optical isomers number; [e]symmetry number; [f]The sum of contributions from translations, external rotations and vibrations; [g]Contribution from internal rotation about the corresponding bonds. [h]http://webbook.nist.gov/chemistry/

TABLE D2: Ideal Gas-Phase Thermodynamic Properties[a], $\Delta_f H^0_{298}$, S^0_{298} and Cp(300-1500)

Species					Cp (T) cal mol^{-1} K^{-1}						
			$\Delta_f H^0_{298}$ [b]	S^0_{298} [c]	300 K	400 K	500 K	600 K	800 K	1000 K	1500 K
$C_6H_5C(CH_3)_2CH=O$	TVR[f]			83.07	35.78	48.35	59.29	68.33	82.00	91.68	106.05
(1)[d]	IR[g]	C_b—C		7.57	2.33	2.03	1.81	1.64	1.43	1.30	1.15
(9)[e]		C—C		4.52	2.09	2.06	1.93	1.78	1.55	1.39	1.19
		C—C		4.52	2.09	2.06	1.93	1.78	1.55	1.39	1.19
		C—CH=O		5.64	3.23	2.77	2.35	2.03	1.64	1.42	01.19
	Total		**-21.96**	105.32	45.52	57.27	67.31	75.56	88.17	97.18	110.77
$C_6H_5C(=CH_2)CH=O$	TVR[f]			85.24	31.41	41.74	50.48	57.54	67.95	75.15	85.62
(1)[d]	IR[g]	C_b—C_d		7.24	2.30	2.07	1.84	1.66	1.42	1.28	1.13
(1)[e]		C_d—CH=O		4.08	3.02	3.05	2.97	2.88	2.66	2.42	1.92
	Total		**11.98**	96.56	36.73	46.86	55.29	62.08	72.03	78.85	88.67
$C_6H_5C(=CH_2)CH=CH_2$	TVR[f]			84.37	33.47	45.13	54.95	62.88	74.60	82.78	94.93
(1)[d]	IR[g]	C_b—C_d		6.68	3.11	2.77	2.37	2.05	1.64	1.42	1.18
(2)[e]		C_d—C_d		4.76	3.44	3.38	3.12	2.82	2.31	1.95	1.48
	Total		**52.84**	95.81	40.02	51.28	60.44	67.75	78.55	86.15	97.59
$C_6H_5C(=CH_2)CH_3$	TVR[f]			77.99	30.48	41.33	50.58	58.14	69.43	77.40	89.30
(1)[d]	IR[g]	C_b—C_d		7.45	1.63	1.56	1.47	1.39	1.27	1.19	1.09
(6)[e]		C_d—C		4.86	2.07	1.88	1.68	1.53	1.33	1.22	1.10
	Total		**29.05**	90.3	34.18	44.77	53.73	61.06	72.03	79.81	91.49
$C_6H_5C(=CH_2)OH$	TVR[f]			79.21	29.09	39.04	47.28	53.85	63.43	70.05	79.88
(1)[d]	IR[g]	C_b—C_d		6.72	2.23	2.11	1.96	1.81	1.58	1.42	1.21
(2)[e]		C_d—OH		2.64	2.68	2.45	2.25	2.08	1.81	1.61	1.33
	Total		**-8.14**	88.57	34	43.6	51.49	57.74	66.82	73.08	82.42
$C_6H_5C(OH)_2CH_3$	TVR[f]			79.49	32.03	43.26	52.69	60.28	71.43	79.22	90.92
(1)[d]	IR[g]	C_b—C		6.57	2.29	2.34	2.29	2.18	1.94	1.72	1.40
(6)[e]		C—OH		2.06	3.11	3.19	2.92	2.59	2.06	1.73	1.34
		C—OH		2.06	3.11	3.19	2.92	2.59	2.06	1.73	1.34
		C—C		4.21	2.04	2.15	2.11	2.01	1.78	1.59	1.31
	Total		**-79.52**	94.39	42.58	54.13	62.93	69.65	79.27	85.99	96.31
$C_6H_5CH=O$	TVR[f]			75.04	24.58	32.61	39.48	45.05	53.24	58.87	66.99
(1)[d] (2)[e]	IR[g]	C_b—CH=O		4.67	2.00	2.04	2.06	2.09	2.15	2.17	2.03
	Total		**-7.46**	79.71	26.58	34.65	41.54	47.13	55.39	61.04	69.02
	Nist[h]		**-8.90**	-	26.84	35.01	41.92	47.47	55.50	60.95	68.67
$C_6H_5OCH=O$	TVR[f]			80.19	26.17	34.80	42.08	47.94	56.49	62.31	70.68
(1)[d]	IR[g]	C_b—O		7.43	1.48	1.46	1.42	1.37	1.27	1.20	1.11
(1)[e]		O—CH=O		5.02	2.47	2.27	2.12	2.05	2.02	1.99	1.82
	Total		**-51.49**	92.64	30.12	38.53	45.62	51.36	59.78	65.5	73.61
$CH_3C_6H_4CH=O$	TVR[f]			80.25	28.75	38.03	46.10	52.77	62.79	69.79	80.03
(1)[d]	IR[g]	C—C_b		5.73	0.99	0.99	0.99	0.99	0.99	0.99	0.99
(3)[e]		C_b—CH=O		4.64	2.00	2.03	2.06	2.08	2.14	2.17	2.04
	Total		**-16.37**	90.62	31.74	41.05	49.15	55.84	65.92	72.95	83.06
$C_6H_5CH=CH_2$	TVR[f]			77.43	26.90	36.21	44.11	50.50	59.94	66.54	76.30
(1)[d](1)[e]	IR[g]	C_b—C_d		5.95	1.81	1.87	1.89	1.86	1.75	1.61	1.37
	Total			83.38	28.71	38.08	46.00	52.36	61.69	68.15	77.67
	Nist[h]		**35.11**	82.48	28.90	38.19	46.03	52.34	61.66	68.16	77.72
$C_6H_5CH_2OH$	TVR[f]			75.64	25.72	34.70	42.45	48.76	58.10	64.62	74.28
(1)[d]	IR[g]	C_b—C		6.74	2.20	1.98	1.78	1.61	1.40	1.27	1.13
(2)[e]		C—OH		3.28	2.31	2.16	1.92	1.72	1.45	1.29	1.13
	Total		**-22.60[h]**	85.66	30.23	38.84	46.15	52.09	60.95	67.18	76.54

[a]Thermodynamic properties are referred to a standard state of an ideal gas at 1 atm.; [b] $\Delta_f H^0_{298}$ in kcal mol^{-1}; [c] S^0_{298} in cal mol^{-1} K^{-1}; [d]optical isomers number; [e]symmetry number; [f]The sum of contributions from translations, external rotations and vibrations; [g]Contribution from internal rotation about the corresponding bonds; [h]http://webbook.nist.gov/chemistry/

TABLE D3: Ideal Gas-Phase Thermodynamic Properties[a]

Species			$\Delta_f H^0_{298}$ [b]	S^0_{298} [c]	Cp (T) cal mol^{-1} K^{-1} 300 K	400 K	500 K	600 K	800 K	1000 K	1500 K
Furanj1 $(1)^d(1)^e$	TVR[f]		**61.67**	66.42	15.40	20.15	24.04	27.08	31.36	34.20	38.29
Furanj2 $(1)^d(1)^e$	TVR[f]		**61.80**	66.25	15.23	19.98	23.89	26.95	31.26	34.13	38.25
$CH_3OC{\bullet}{=}O$	TVR[f]			62.43	12.09	14.59	17.01	19.15	22.54	25.03	28.77
$(1)^d(3)^e$	IR[g]	C—O		5.06	0.99	0.99	0.99	0.99	0.99	0.99	0.99
		O—C•=O		4.14	2.11	2.12	2.16	2.21	2.29	2.29	2.07
	Total		**-42.02**	71.43	15.17	17.69	20.16	22.35	25.81	28.29	31.73
$CH_2{=}C{\bullet}CH{=}O$	TVR[f]			68.62	16.38	19.21	21.70	23.82	27.12	29.51	33.08
$(1)^d(2)^e$	IR[g]	Cd•—CH=O		No rotor							
	Total		**41.97**	68.62	16.38	19.21	21.70	23.82	27.12	29.51	33.08
$CH_2{=}CHC{\bullet}{=}O$	TVR[f]			66.46	15.70	18.68	21.27	23.42	26.74	29.15	32.79
$(1)^d(2)^e$	IR[g]	Cd—C•=O		No rotor							
	Total		**20.22**	66.46	15.70	18.68	21.27	23.42	26.74	29.15	32.79
$CH_2{=}CHC{\bullet}(CH{=}O)CH{=}O$	TVR[f]			80.50	25.00	30.81	35.84	40.02	46.32	50.74	57.20
$(1)^d(1)^e$	IR[g]	C_d—C•		4.48	2.00	2.05	2.07	2.09	2.13	2.16	2.18
		C•—CH=O		No rotor							
	Total		**-18.99**	84.98	27.00	32.86	37.91	42.11	48.45	52.90	59.38
$C{\bullet}(OH)_2CH{=}O$	TVR[f]			67.80	14.19	17.59	20.50	22.85	26.33	28.75	32.42
$(1)^d$	IR[g]	C•—OH		1.69	2.50	2.82	2.88	2.81	2.53	2.23	1.70
$(1)^e$		C•—OH		1.69	2.50	2.82	2.88	2.81	2.53	2.23	1.70
		C•—CH=O		2.08	1.96	2.09	2.18	2.28	2.52	2.77	3.11
	Total		**-95.78**	73.26	21.15	25.32	28.44	30.75	33.91	35.98	38.93
$(YC_5H_5)C(O{\bullet}){=}O$	TVR[f]			80.42	25.88	33.49	39.70	44.55	51.46	56.10	62.78
$(1)^d(1)^e$	IR[g]	C—C=O		6.70	2.21	2.11	1.98	1.86	1.64	1.49	1.27
	Total		**5.10**[h]	87.12	28.09	35.60	41.68	46.41	53.10	57.59	64.05
$(YC_5H_5)OC{\bullet}{=}O$	TVR[f]			77.66	23.90	31.72	38.35	43.62	51.01	55.74	62.06
$(1)^d$	IR[g]	C—O		7.66	1.88	1.57	1.39	1.28	1.16	1.10	1.04
$(1)^e$		O—C•=O		4.27	2.37	2.29	2.27	2.29	2.34	2.34	2.11
	Total		**9.54**	89.59	28.15	35.58	42.01	47.19	54.51	59.18	65.21
$(YC_5H_5O)C{\bullet}{=}O$	TVR[f]			80.47	26.06	33.35	39.34	44.10	51.02	55.75	62.60
$(1)^d(1)^e$	IR[g]	C—C•=O		5.22	2.86	2.74	2.54	2.32	1.94	1.67	1.33
	Total		**10.56**	85.69	28.92	36.09	41.88	46.42	52.96	57.42	63.93
$Y(C_6H_5{\bullet}){=}O$ $(1)^d(1)^e$	TVR[f]		**12.77**	75.26	23.06	30.25	36.19	40.91	47.68	52.27	58.89
$C_6H_5C{\bullet}{=}O$	TVR[f]			77.82	24.27	31.70	37.93	42.93	50.22	55.18	62.31
$(1)^d(2)^e$	IR[g]	C_b—C•=O		5.22	2.13	2.22	2.24	2.18	1.98	1.77	1.43
	Total		**30.75**	83.04	26.4	33.92	40.17	45.11	52.20	56.95	63.74
$CH_2{=}CHC{\bullet}(CH{=}O)C{\equiv}CH$ $(1)^d(1)^e$	TVR[f]		**62.80**	84.41	28.31	34.23	39.05	42.94	48.76	52.90	59.10

[a]Thermodynamic properties are referred to a standard state of an ideal gas at 1 atm. [b] $\Delta_f H^0_{298}$ in kcal mol^{-1}; [c] S^0_{298} in cal mol^{-1} K^{-1}; [d]optical isomers number; [e]symmetry number; [f]The sum of contributions from translations. external rotations and vibrations; [g]Contribution from internal rotation about the corresponding bonds; [h]Ref. 126

Appendix E: Enthalpy of Formation Calculation

Table E1: Calculated $\Delta_f H^0_{298}$ in kcal mol^{-1} for species at G3MP2B3 level versus work reaction

Reactions Series – G3MP2B3	$\Delta_f H^0_{298}$	Error limit[a]
$\boldsymbol{CH_2{=}CHCH(CH{=}O)CH{=}O} + CH_3CH_3 \rightarrow CO + CH_2{=}O + CH_2{=}CHCH_2CH_2CH_3$	-40.07	± 1.61
$\boldsymbol{CH_2{=}CHCH(CH{=}O)CH{=}O} + CH_4 \rightarrow CO + CH_2{=}O + CH_2{=}CHCH_2CH_3$	-41.45	± 1.75
$\boldsymbol{CH_2{=}CHCH(CH{=}O)CH{=}O} + CH_4 \rightarrow CO + CH_3CH{=}O + CH_2{=}CHCH_3$	-39.96	± 2.07
Average	**-40.76 ± 0.90**	
$\boldsymbol{CH_2{=}CHCH(CH{=}O)C{\equiv}CH} + CH_3CH_3 \rightarrow CO + CH{\equiv}CH + CH_2{=}CHCH_2CH_2CH_3$	43.68	± 1.01
$\boldsymbol{CH_2{=}CHCH(CH{=}O)C{\equiv}CH} + CH_4 \rightarrow CO + CH_3C{\equiv}CH + CH_2{=}CHCH_3$	43.46	± 1.14
Average	**43.57 ± 0.11**	
$\boldsymbol{CH(OH)_2CH{=}O} + CH_4 \rightarrow 2CH_3OH + CO$	-118.48	± 0.85
$\boldsymbol{CH(OH)_2CH{=}O} + CH_3CH_3 \rightarrow CH_3CH_2OH + CH_3OH + CO$	-118.62	± 0.97
Average	**-118.55 ± 0.09**	
$\boldsymbol{CH_3C(OH)_2CH{=}O} + 2CH_4 \rightarrow CH_3CH_2CH{=}O + 2CH_3OH$	-132.47	± 2.77
$\boldsymbol{CH_3C(OH)_2CH{=}O} + CH_4 \rightarrow CH_3CH_2CH{=}O + CH_2(OH)_2$	-129.69	± 2.38
$\boldsymbol{CH_3C(OH)_2CH{=}O} + 2CH_3CH_3 \rightarrow CH_3CH_2CH{=}O + 2CH_3CH_2OH$	-132.76	± 2.01
Average	**-131.64 ± 1.69**	
$\boldsymbol{CH_2{=}CHCH(CH{=}O)CH{=}CH_2} + 2CH_4 \rightarrow CH_3CH{=}O + 2CH_2{=}CHCH_3$	0.69	± 3.56
$\boldsymbol{CH_2{=}CHCH(CH{=}O)CH{=}CH_2} + CH_4 \rightarrow CH_3CH{=}O + CH_2{=}CHCH_2CH{=}CH_2$	-2.27	± 3.27
Average	**-0.79 ± 2.09**	
$\boldsymbol{CH_2{=}CHC(OH)_2CH{=}CH_2} + CH_4 \rightarrow CH_2(OH)_2 + CH_2{=}CHCH_2CH{=}CH_2$	-61.33	± 1.73
$\boldsymbol{CH_2{=}CHC(OH)_2CH{=}CH_2} + CH_4 \rightarrow 2CH_2{=}CHCH_2OH$	-60.04	± 2.12
Average	**-60.68 ± 0.91**	
$\boldsymbol{CH_2{=}C(OH)_2} + CH_4 \rightarrow CH_2{=}CHOH + CH_3OH$	-73.67	± 2.83
$\boldsymbol{CH_2{=}C(OH)_2} + CH_3CH_3 \rightarrow CH_2{=}CHOH + CH_3CH_2OH$	-73.81	± 2.95
$\boldsymbol{CH_2{=}C(OH)_2} + CH_2CH_2 \rightarrow 2CH_2{=}CHOH$	-74.19	± 4.81
Average	**-73.89 ± 0.27**	
$\boldsymbol{CH_2{=}CHCH_2CH{=}O} + CH_4 \rightarrow CH_2{=}CHCH_3 + CH_3CH{=}O$	-19.12	± 1.10
$\boldsymbol{CH_2{=}CHCH_2CH{=}O} + CH_4 \rightarrow CH_2{=}CHCH_2CH_3 + CH_2{=}O$	-20.61	± 1.78
$\boldsymbol{CH_2{=}CHCH_2CH{=}O} + CH_2{=}CH_2 \rightarrow CH_2{=}CHCH_3 + CH_2{=}CHCH{=}O$	-21.23	± 1.19
Average	**-20.32 ± 1.08**	
$\boldsymbol{CH_2{=}CHCH(CH_3)CH{=}CH_2} + CH_2CH_2 \rightarrow CH_2{=}CHCH_3 + CH_2{=}CHCH_2CH{=}CH_2$	16.12	± 1.29
$\boldsymbol{CH_2{=}CHCH(CH_3)CH{=}CH_2} + CH_3CH_3 \rightarrow CH_3CH_2CH_3 + CH_2{=}CHCH_2CH{=}CH_2$	16.63	± 1.29
$\boldsymbol{CH_2{=}CHCH(CH_3)CH{=}CH_2} + CH_4 \rightarrow CH_3CH_3 + CH_2{=}CHCH_2CH{=}CH_2$	16.15	± 1.22
Average	**16.3 ± 0.28**	
$\boldsymbol{CH_2{=}C(CH{=}O)CH{=}CH_2} + CH_2CH_2 \rightarrow CH_2{=}CHCH{=}O + CH_2{=}CHCH{=}CH_2$	-1.70	± 1.55
$\boldsymbol{CH_2{=}C(CH{=}O)CH{=}CH_2} + CH_3CH_3 \rightarrow CH_3CH_2CH{=}CH_2 + CH_2{=}CHCH{=}O$	-2.29	± 1.57
Average	**-2.00 ± 0.41**	
$\boldsymbol{CH_2{=}C(CH{=}O)C{\equiv}CH} + CH_3CH_3 \rightarrow CH_3CH_2C{\equiv}CH + CH_2{=}CHCH{=}O$	39.31	± 1.59
$\boldsymbol{CH_2{=}C(CH{=}O)C{\equiv}CH} + H_2 \rightarrow CH_2{=}CHCH{=}O + CH{\equiv}CH$	38.72	± 1.45
Average	**39.01 ± 0.41**	
$\boldsymbol{CH_2{=}CHCH(OH)CH{=}CH_2} + CH_4 \rightarrow CH_2{=}CHCH_2OH + CH_3CH{=}CH_2$	-11.76	± 2.13
$\boldsymbol{CH_2{=}CHCH(OH)CH{=}CH_2} + CH_2{=}O \rightarrow CH_2{=}CHCH_2OH + CH_2{=}CHCH{=}O$	-13.91	± 1.31
$\boldsymbol{CH_2{=}CHCH(OH)CH{=}CH_2} + CH_3CH_3 \rightarrow CH_2{=}CHCH_2OH + CH_2{=}CHCH_2CH_3$	-13.32	± 2.21
Average	**-12.99 ± 1.11**	
$\boldsymbol{CH_2{=}CHOCH{=}CH_2} + CH_2{=}O \rightarrow CH_2{=}CHOH + CH_2{=}CHCH{=}O$	-5.53	± 3.14
$\boldsymbol{CH_2{=}CHOCH{=}CH_2} + CH_3CH_3 \rightarrow CH_2{=}CHOH + CH_2{=}CHCH_2CH_3$	-4.94	± 3.04
Average	**-5.23 ± 0.29**	
$\boldsymbol{CH_2{=}CHCH{=}O} + CH_3CH_3 \rightarrow CH_2{=}CHCH_2CH_3 + CH_2{=}O$	-18.06	± 1.16
$\boldsymbol{CH_2{=}CHCH{=}O} + CH_2{=}CH_2 \rightarrow CH_2{=}CHCH{=}CH_2 + CH_2{=}O$	-17.47	± 1.14
Average	**-17.76 ± 0.41**	
$\boldsymbol{CH_3OCH{=}O} + CH_3CH_3 \rightarrow CH_3OCH_2CH_3 + CH_2{=}O$	-81.72	± 1.14
$\boldsymbol{CH_3OCH{=}O} + CH_4 \rightarrow CH_3OCH_3 + CH_2{=}O$	-82.27	± 1.05
$\boldsymbol{CH_3OCH{=}O} + CH_4 \rightarrow CH_3OH + CH_3CH{=}O$	-82.56	± 1.33
Average	**-82.18 ± 0.42**	

[a]Reported errors for each of the standard species (when available) + estimated error due to the method (see Appendix C)

Table E2: Calculated $\Delta_f H^0_{298}$ in kcal mol^{-1} for species at G3MP2B3 level versus work reaction

Reactions Series – G3MP2B3	$\Delta_f H^0_{298}$	Error limit[a]
C6H5C(OH)2CH3 + 2CH3CH3 → C6H5CH2CH3 + 2CH3CH2OH	-82.07	± 3.14
C6H5C(OH)2CH3 + CH4 → C6H5CH2CH3 + CH2(OH)2	-79.00	± 2.51
C6H5C(OH)2CH3 + 2CH4 → C6H5CH3 + CH2(OH)2 + CH3CH3	-78.88	± 2.87
Average	**-79.98 ± 1.8**	
C6H5C(CH2)CH=CH2 + CH2CH2 → C6H5CH=CH2 + CH2=CHCH=CH2	50.82	± 1.53
C6H5C(CH2)CH=CH2 + CH3CH3 → C6H5 CH2CH3 + CH2=CHCH=CH2	51.74	± 1.51
C6H5C(CH2)CH=CH2 + CH4 → C6H5CH3 + CH2=CHCH=CH2	51.86	± 1.41
Average	**51.47 ± 0.56**	
C6H5CH=O → C6H6 + CO	-8.09	± 1.48
C6H5CH=O + CH4 → C6H5CH3 + CH2=O	-9.50	± 1.65
Average	**-8.79 ± 0.99**	
CH3C6H4CH=O → C6H5CH3 + CO	-15.84	± 0.89
CH3C6H4CH=O + CH3CH3 → C6H6 + CH3CH2CH3 + CO	-15.93	± 1.34
CH3C6H4CH=O + CH4 → C6H6 + CH3CH3 + CO	-16.41	± 1.27
CH3C6H4CH=O + CH3CH3 → C6H5CH=O + CH3CH2CH3	-16.62	± 2.06
Average	**-16.2 ± 0.37**	
C6H5C(=CH2)CH=O + CH4 → C6H5CH3 + CO + CH2=CH2	11.16	± 1.43
C6H5C(=CH2)CH=O + CH4 → C6H6 + CO + CH2=CHCH3	10.55	± 1.46
C6H5C(=CH2)CH=O → C6H5CH=CH2 + CO	10.12	± 1.15
Average	**10.61 ± 0.52**	
C6H5OCH=O + CH3CH3 → C6H5OH + CH3CH2CH=O	-53.55	± 2.30
C6H5OCH=O → C6H5OH + CO	-51.53	± 1.93
Average	**-52.54 ± 1.42**	
C6H5C(=CH2)CH3 + 2CH3CH3 → C6H6 + CH3CH2CH2CH=CH2	27.61	± 1.34
C6H5C(=CH2)CH3 + CH4 → C6H5CH=CH2 + CH3CH3	27.37	± 1.30
C6H5C(=CH2)CH3 + CH4 → C6H5CH3 + CH3CH=CH2	28.38	± 1.25
C6H5C(=CH2)CH3 + CH3CH3 → C6H5CH=CH2 + CH3CH2CH3	27.85	± 1.37
Average	**27.80 ± 0.43**	
C6H5C(=CH2)OH + CH3CH3 → C6H5CH2CH3 + CH2=CHOH	-7.47	± 2.88
C6H5C(=CH2)OH + CH3CH3 → C6H5CH=CH2 + CH3CH2OH	-7.22	± 1.04
C6H5C(=CH2)OH + CH4 → C6H6CH3 + CH2=CHOH	-7.35	± 2.78
Average	**-7.35 ± 0.12**	
Y(C6H6)=O + 2CH4 → CH2=CHCH=O + CH2=CH2 + CH3CH=CH2	-4.32	± 2.22
Y(C6H6)=O + CH2CH2 → C6H6 + CH3CH=O	-5.43	± 1.56
Average	**-4.88 ± 0.78**	
Y(C5H5)C(OH)=O + CH2CH2 → Y(C5H6) + CO + CH2=CHOH	-53.46	± 3.69
Y(C5H5)C(OH)=O + CH3CH3 → Y(C5H6) + CO + CH3CH2OH	-53.09	± 1.83
Y(C5H5)C(OH)=O + 2CH4 → Y(C5H6) + CH3OH + CH3CH=O	-54.99	± 2.20
Average	**-53.84 ± 1.00**	
Y(C5H5O)CH=O + CH2=CH2 → CH2=CHCH=CHCH=CH2 + CH2=O + CO	-25.23	± 2.44
Y(C5H5O)CH=O → Y(C5H6) + CO2	-24.55	± 0.87
Y(C5H5O)CH=O → Y(C5H5)OCH=O	-25.47	± 1.65
Y(C5H5O)CH=O → Y(C5H5)C(OH)=O	-25.76	± 1.73
Average	**-25.25 ± 0.51**	
Y(C5H5)OCH=O + CH4 → Y(C5H6) + CH3OCH=O	-35.55	± 1.43
Y(C5H5)OCH=O + CH3CH3 → Y(C5H6) + CO + CH3CH2OH	-34.03	± 1.75
Y(C5H5)OCH=O + CH4 → Y(C5H6) + CO + CH3OH	-33.89	± 1.63
Average	**-34.49 ± 0.92**	

[a]Reported errors for each of the standard species (when available) + estimated error due to the method (see Appendix C)

TABLE E3: Calculated $\Delta_f H^0_{298}$ in kcal mol^{-1} for Radicals at G3MP2B3 level versus work reaction

Reactions Series – G3MP2B3			$\Delta_f H^0_{298}$	Error
$CH_3OC{\bullet}{=}O$ + CH_3CH_3	→ $CH_3OCH{=}O$ + $CH_3CH_2{\bullet}$		-41.94	± 1.63
$CH_3OC{\bullet}{=}O$ + CH_4	→ $CH_3OCH{=}O$ + $CH_3{\bullet}$		-41.13	± 1.28
		Average	**-41.53 ± 0.57**	
$CH_2{=}C{\bullet}CH{=}O$ + CH_3CH_3	→ $CH_2{=}CHCH{=}O$ + $CH_3CH_2{\bullet}$		41.93	± 1.90
$CH_2{=}C{\bullet}CH{=}O$ + CH_2CH_2	→ $CH_2{=}CHCH{=}O$ + $CH_2CH{\bullet}$		42.34	± 1.38
$CH_2{=}C{\bullet}CH{=}O$ + CH_4	→ $CH_2{=}CHCH{=}O$ + $CH_3{\bullet}$		42.75	± 1.55
$CH_2{=}C{\bullet}CH{=}O$ + $CH_3CH{=}O$	→ $CH_2{=}CHCH{=}O$ + $CH_3C{\bullet}{=}O$		42.86	± 2.33
		Average	**42.47 ± 0.42**	
$CH_2{=}CHC{\bullet}(CH{=}O)C{\equiv}CH$ + CH_3CH_3	→ $CH_3CH_2{\bullet}$ + $CH_2{=}CHCH(CH{=}O)C{\equiv}CH$		63.27	± 3.02
$CH_2{=}CHC{\bullet}(CH{=}O)C{\equiv}CH$ + CH_4	→ CO + $CH{\equiv}C{\bullet}$ + $CH_2{=}CHCH_2CH_3$		60.58	± 4.53
$CH_2{=}CHC{\bullet}(CH{=}O)C{\equiv}CH$ + CH_4	→ $CH_2{=}O$ + $CH_2{=}CH{\bullet}$ + $CH{\equiv}CCH_2CH_3$		63.74	± 3.42
		Average	**62.53 ± 1.70**	
$C{\bullet}(OH)_2CH{=}O$ + CH_3CH_3	→ CH_3CH_2OH + $CH_2{=}O$ + $CH{\bullet}{=}O$		-95.49	± 1.05
$C{\bullet}(OH)_2CH{=}O$ + CH_4	→ CH_3OH + $CH_2{=}O$ + $CH{\bullet}{=}O$		-95.35	± 0.93
$C{\bullet}(OH)_2CH{=}O$ + $CH_2{=}CH_2$	→ $CH_2{=}CHOH$ + $CH_2{=}O$ + $CH{\bullet}{=}O$		-95.86	± 3.91
		Average	**-95.56 ± 0.26**	
$CH_2{=}CHC{\bullet}(CH{=}O)CH{=}O$ + CH_3CH_3	→ $CH{\bullet}{=}O$ + $CH_2{=}O$ + $CH_2CHCH_2CHCH_2$		-19.13	± 1.42
$CH_2{=}CHC{\bullet}(CH{=}O)CH{=}O$ + CH_3CH_3	→ $CH_3CH_2{\bullet}$ + $CH_2CHCH(CHO)CHO$		-18.25	± 2.45
$CH_2{=}CHC{\bullet}(CH{=}O)CH{=}O$ + CH_2CH_2	→ CO + $CH{\bullet}{=}O$ + $CH_2CHCH_2CHCH_2$		-18.72	± 1.44
		Average	**-18.70 ± 0.44**	

[a]Reported errors for each of the standard species (when available) + estimated error due to the method (see Appendix C)

Table E4: Calculated $\Delta_f H^0_{298}$ in kcal mol^{-1} for Radicals at G3MP2B3 level versus work reaction

	Reactions Series – G3MP2B3	$\Delta_f H^0_{298}$	Error Limit[a]
FURANJ1 	***Furanj1*** + CH_3CH_3 → Furan + $CH_3CH_2\bullet$	59.95	± 2.33
	Furanj1 + CH_2CH_2 → Furan + $CH_2CH\bullet$	60.36	± 2.81
	Furanj1 + $CH_3CH{=}O$ → Furan-CH=O + $CH_3\bullet$	62.34	± 2.26
	Average	**60.88 ± 1.27**	
FURANJ2 	***Furanj2*** + CH_3CH_3 → Furan + $CH_3CH_2\bullet$	60.02	± 2.33
	Furanj2 + CH_2CH_2 → Furan + $CH_2CH\bullet$	60.44	± 2.81
	Furanj2 + $CH_3CH{=}O$ → Furan-CH=O + $CH_3\bullet$	62.41	± 2.26
	Average	**60.96 ± 1.27**	
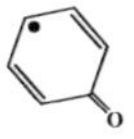	***Y(C₆H₅•)=O*** + $2CH_4$ → $CH_2{=}CHCH{=}O + CH_2CH_2 + CH_2{=}CHCH_2\bullet$	15.41	± 2.54
	Y(C₆H₅•)=O + CH_2CH_2 → $Y(C_6H_6){=}O + CH_2CH\bullet$	15.44	± 3.47
	Y(C₆H₅•)=O + CH_3CH_3 → $Y(C_6H_6){=}O + CH_3CH_2\bullet$	15.03	± 2.89
	Y(C₆H₅•)=O + CH_4 → $C_6H_6 + CH_3O\bullet$	15.52	± 1.85
	Average	**15.35 ± 0.22**	
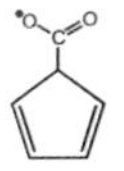	***Y(C₅H₅)C(O•)=O*** + CH_2CH_2 → $Y(C_5H_5\bullet) + CO + CH_2{=}CHOH$	5.12	± 3.29
	Y(C₅H₅)C(O•)=O + CH_3CH_3 → $Y(C_5H_5\bullet) + CO + CH_3CH_2OH$	6.29	± 1.39
	Y(C₅H₅)C(O•)=O + $2CH_4$ → $Y(C_5H_6) + CH_3OH + CH_3C\bullet{=}O$	5.94	± 2.15
	Average	**5.78 ± 0.60**	
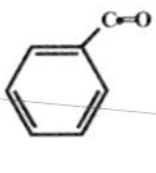	***C₆H₅C•=O*** + CH_4 → $C_6H_5CH{=}O + CH_3\bullet$	30.82	± 2.56
	C₆H₅C•=O + CH_4 → $C_6H_6 + CH_3C\bullet{=}O$	29.57	± 2.50
	C₆H₅C•=O + CH_4 → $C_6H_6 + CO + CH_3\bullet$	31.50	± 1.53
	C₆H₅C•=O + CH_3CH_3 → $C_6H_5CH{=}O + CH_3CH_2\bullet$	30.00	± 2.91
	Average	**30.47 ± 0.86**	
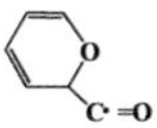	***Y(C₅H₅O)C•=O*** + CH_3CH_3 → $CH_3CH{=}CHCH_3 + CH_2{=}C\bullet CH{=}O + CO$	10.64	± 1.66
	Y(C₅H₅O)C•=O + CH_3CH_3 → $CH_2{=}CHCH{=}CH_2 + CH_3CHCHO + CH\bullet{=}O$	10.69	± 1.56
	Y(C₅H₅O)C•=O + CH_3CH_3 → $2CH_2{=}CH_2 + CO + CH_2{=}C\bullet CH{=}O$	11.05	± 1.73
	Y(C₅H₅O)C•=O + CH_4 → $CH_2{=}CHCH{=}CH_2 + CO + CH_3C\bullet{=}O$	11.59	± 1.84
	Y(C₅H₅O)C•=O + CH_3CH_3 → $Y(C_5H_5O)CH{=}O + CH_3CH_2\bullet$	11.19	± 2.32
	Average	**11.02 ± 0.38**	
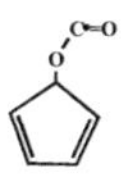	***Y(C₅H₅)OC•=O*** + CH_4 → $Y(C_5H_5\bullet) + CO + CH_3OH$	9.70	± 0.90
	Y(C₅H₅)OC•=O + CH_4 → $Y(C_5H_6) + CH\bullet{=}O + CH_2{=}OH$	9.82	± 0.81
	Y(C₅H₅)OC•=O + CH_3CH_3 → $Y(C_5H_5\bullet) + CO + CH_3CH_2OH$	9.56	± 1.02
	Y(C₅H₅)OC•=O → $Y(C_5H_5\bullet) + CO_2$	10.02	± 0.55
	Average	**9.78 ± 0.19**	

[a]Reported errors for each of the standard species (when available) + estimated error due to the method (see Appendix C);

Appendix F: Reaction of Phenyl + O_2 System

Scheme F1: Input file for QRRK analysis with master equation analysis for Fall-off

PH + O_2 => Products

```
Temp
7  600.  900.  1200.  1500. 1800. 2100. 2500.

Pres
4        0.01   0.1   1.  10.

Chemact
Dissoc

INPUT (A, n ,alpha, E kcal    k=AT^nexp(-alpha*T)exp(-E/RT) )
 3.0e13   0.0  0.0  0.0
# hadad+sebbar  6.5068E+05   2.70856   0.0  -2.3998E+03
#  PHJ + O2 => TS2-1

param (mass), sigma (A), e/k (K)
     109.  5.82    617.
collider (mass), sigma(A), e/k(K)  air here
       29.1  3.55   103.
# COLLIDER (take user's data if collider name preceded !  )
#  !N2
#  1. 40.0 3.621 97.5 500.
# EXP (temperature exponent to alpha, alpha=alpha(300)*(T/300)^N )
# 0.03
delta (E)
 745.  0.
Kfit  (uncomment if want modified Arrhenius fits)
INT   (integration interval in kcal)
1.0
EHEAD integration stops at Ehead + highest barrier, def. 75 kcal
 75.
--------------------------------------------------------------------------------------------------------------------
WELL 1
 PHOOJ
 FREQ
 3   453.6   10.723
    1172.9   16.614
    3055.2    5.663
 REACTANT
    PHJ + O2
    1.0784E+11    1.63768   0.0    47.8
# hadad+sebbar PHOOJ => TS2*1
 ISOMER
    Y(C6J)Y(CO2)
    1.7523E+06  1.86486   0.0   28
#  sebbar+hadad+lin  PHOOJ  => TS1*8LIN
 PRODUCT
    Y(C6JO)DO
     1.1940E+07  1.45802   0.0   47.15
#  sebbar+hada PHOOJ => TS1*19 via  Y(C6J)Y(C2O2)
 PRODUCT
```

```
    PHJOOH
    4.3679E+05  1.94114   0.0   40.45
#  PHOOJ => TSPHJOOH (with H-tunneling effect with average G3MP2B3/B3LYP 40.4=nadia )
    PRODUCT
    PHOJ + O
    2.0243E+08    1.74234   0.0    41.78
# sebbar+lin  PHOOJ => TS1*3LIN
 END
------------------------------------------------------------------------------------------------------------------------
WELL 2
 Y(C6J)Y(CO2)
 FREQ
 3   528.0  14.081
     1254.4  13.787
     3125.1   5.132
 ISOMER
    PHOOJ
    4.0030E+11  .50706    0.0   3.49
# sebbar+lin Y(C6J)Y(CO2) => TS1*8LIN
 ISOMER
    Y(C6JO)DO
     6.9944E+09    .85638    0.0    10.58
#  sebbar+lin YC6JYCO2 => TS8*10LIN
END
------------------------------------------------------------------------------------------------------------------------
WELL 3
 Y(C6JO)DO
 FREQ
 3   423.4  10.784
    1177.1  16.403
    3020.2   5.814
 ISOMER
    Y(C6J)Y(CO2)
    2.4361E+08    1.16303    0.0    79.10
# sebbar+lin Y(C6JO)DO => TS8*10LIN
 ISOMER
    Y(C5J)Y(C3O)DO
    3.3844E+11       .36456  0.0    39.4
# sebbar+hadad Y(C6JO)DO => TS10*11
 ISOMER
    ODC6JDO
    1.5182E+11   .51609    0.0    26.21
# sebbar+hadad Y(C6JO)DO => TS21*10
 PRODUCT
    Y(C5JO) + CO
    7.4451E+09   .52332    0.0  50.01
# sebbar+hadad  Y(C6JO)DO => TS10*17 via Y(C5O)CJDO
END
------------------------------------------------------------------------------------------------------------------------
WELL 4
 Y(C5J)Y(C3O)DO
 FREQ
 3   523.8   12.063
    1209.3   15.729
    3140.7    5.208
  ISOMER
    Y(C6JO)DO
    2.2169E+12  0.36728   0.0  29.57
```

```
# sebbar+hadad Y(C5J)Y(C3O)DO => TS10*11
  PRODUCT
    Y(C5J) + CO2
    2.7487E+12   0.31959  0.0  8.76
# sebbar+lin  Y(C5J)Y(C3O)DO  => TS11*12
  ISOMER
    Y(C5)OCJDO
    2.7359E+12    0.36514  0.0   28.97
# sebbar+hadad Y(C5J)Y(C3O)DO => TS 11*13
  END
-----------------------------------------------------------------------------------------------------------------
WELL 5
  Y(C5)OCJDO
  FREQ
  3   445.8   11.817
     1208.4   15.583
     3046.5    5.601
  ISOMER
   Y(C5J)Y(C3O)DO
   1.1625E+10  .35831    0.0   17.20
# sebbar+hadad Y(C5)OCJDO => TS11*13
  PRODUCT
    Y(C5J) + CO2
    2.7191E+10    .49892     0.0   23.13
# sebbar+hadad Y(C5)OCJDO => TS13*14
END
-----------------------------------------------------------------------------------------------------------------
WELL 6
  ODC6JDO
  FREQ
  3   250.0    9.152
    1082.1   16.891
     3128.4   6.957
  ISOMER
   Y(C6JO)DO
   3.4654E+10  .67656  0.0   8.56
# sebbar-hadad from ODC6JDO = TS21*10
  ISOMER
   O-ODY(C6)OJ
   5.2789E+08   .65990  0.0  13.33
# sebbar from ODC6JDO    = TSODC5*COJ
  ISOMER
   ODY(C5J)CDO
   1.7498E+09   .63879  0.0  7.408
# sebbar from ODC6JDO = TODYC5JCDO
  PRODUCT
   CJC4DO+CO
   2.1524E+11  .85136  0.0   34.97
# sebbar from ODC6JDO    = TSOCJ*C5DO
END
-----------------------------------------------------------------------------------------------------------------
WELL 7
 O-ODY(C6)OJ
  FREQ
  3  434.0  11.900
    1189.8  15.511
     2979.1   5.589
  ISOMER
```

```
    ODC6JDO
    1.0190E+10  .24319  0.0  2.78
# sebbar from O*ODYC6OJ  = TSODC5*COJ
  PRODUCT
    o-quinone+H
    8.8450E+08  .46827  0.0  9.83
# sebbar from O*ODYC6OJ  = TSODYC6OJ
END
--------------------------------------------------------------------------------
WELL 8
 ODY(C5J)CDO
  Freq
  3  545.7  13.924
     1414.2  14.941
     3922.6  3.635
  ISOMER
    ODC6JDO
    3.2539E+11   .46009   0.0   25.93
# sebbar from ODY(C5J)CDO = TSODYC5JCDO
  PRODUCT
    ODY(C5)+CJDO
    1.1801E+11  1.20059  0.0   34.14
# sebbar from ODY(C5J)CDO = TODYC5J*CO
END
```

Table F2: Rate constants from chemical activation QRRK – Master Equation analysis

$Ph + O_2 = [PhOO\bullet]^* = Products$

Pressure = 0.01, 0.1, 1 and 10 atm, $k = A(T/K)^n \exp(-E_a/RT)$ ($600 \leq T/K \leq 2500K$)

Reactions	Reactions[a]	*A*	*n*	*Ea* *(Kcal mol^{-1})*	*P* *(atm)*
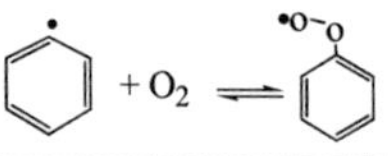	$PHJ + O_2 = PHOOJ$	2.34+129	-36.22	42101.	0.01
		1.74+134	-37.16	48119.	0.1
		1.60+137	-37.57	53144.	1
		2.27+136	-36.92	55977.	10
Reverse reaction	$PHJ + O_2 = PHJ + O_2$	8.03E+21	-2.31	7252.	0.01
		3.87E+28	-4.14	13056.	0.1
		8.89E+36	-6.41	20742.	1
		2.05E+44	-8.38	28536.	10
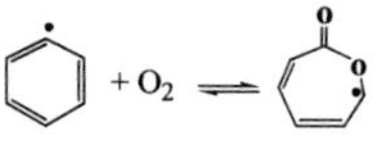	$PHJ + O_2$ = Y(C6JO)DO	1.22E+18	-2.52	6909.	0.01
		7.67E+24	-4.38	12834.	0.1
		1.75E+33	-6.65	20555.	1
		3.01E+40	-8.58	28293.	10
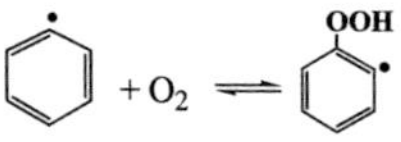	$PHJ + O_2$ = PHJOOH	2.74E+18	-2.45	5115.	0.01
		6.69E+25	-4.46	11717.	0.1
		9.47E+33	-6.67	19548.	1
		1.75E+40	-8.32	26822.	10
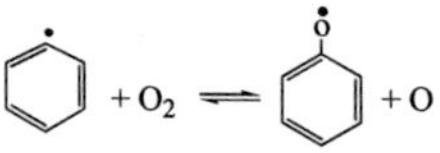	$PHJ + O_2$ = PHOJ + O	3.05E+21	-2.74	4927.	0.01
		8.50E+28	-4.77	11608.	0.1
		1.06E+37	-6.96	19437.	1
		1.36E+43	-8.56	26632.	10
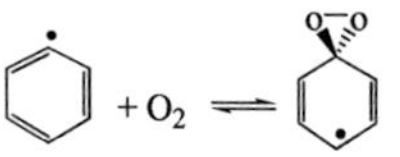	$PHJ + O_2$ = Y(C6J)Y(CO2)	6.58+117	-35.53	42376.	0.01
		4.57+122	-36.47	48365.	0.1
		2.34+125	-36.81	53138.	1
		2.60+123	-35.86	54813.	10
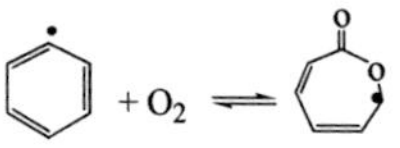	$PHJ + O_2$ = Y(C6JO)DO	1.22E+18	-2.52	6909.	0.01
		7.67E+24	-4.38	12834.	0.1
		1.75E+33	-6.65	20555.	1
		3.01E+40	-8.58	28293.	10
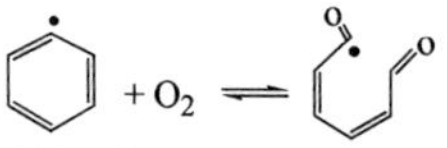	$PHJ + O_2$ = ODC6JDO	7.72E+13	-7.31	9509.	0.01
		2.07E+17	-7.94	12619.	0.1
		-	-	-	1
		-	-	-	10
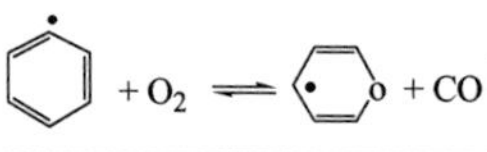	$PHJ + O_2$ = Y(C5JO) + CO	6.35E+12	-1.52	3630.	0.01
		3.57E+20	-3.64	10603.	0.1
		9.01E+28	-5.91	18689.	1
		4.51E+36	-7.95	27237.	10
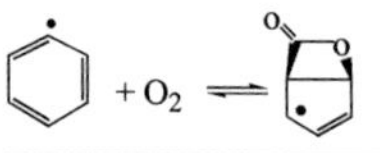	$PHJ + O_2$ = Y(C5J)Y(C3O)DO[a]	6.91E-16	0.35	-11983.	0.01
		3.60E-07	-1.76	-5097.	0.1
		5.14E+08	-5.62	7740.	1
		6.17E+01	-3.41	1892.	10
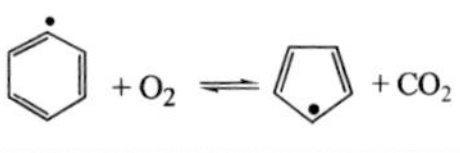	$PHJ + O_2$ = Y(C5J) + CO_2	4.16E+18	-2.65	3730.	0.01
		2.42E+26	-4.76	10855.	0.1
		2.05E+34	-6.90	18754.	1
		5.92E+40	-8.59	26510.	10
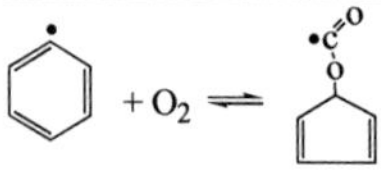	$PHJ + O_2$ = Y(C5)OCJDO[a]	-	-	-	0.01
		-	-	-	0.1
		6.47+174	-53.08	65903.	1
		5.49+204	-59.69	95374.	10
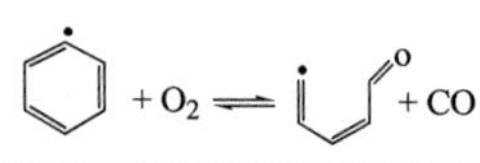	$PHJ + O_2$ = CJC4DO+CO	1.34E+15	-1.80	3821.	0.01
		6.43E+22	-3.90	10675.	0.1
		2.09E+31	-6.20	18762.	1
		1.09E+39	-8.26	27189.	10

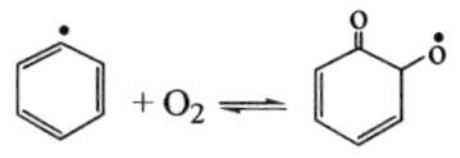	PHJ + O_2 = O-ODY(C6)OJ[a]	1.98E-03	-2.57	1023.	0.01
		7.45E+01	-3.49	5659.	0.1
					1
					10
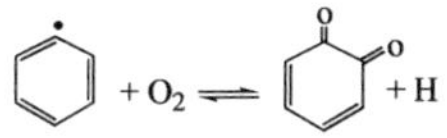	PHJ + O_2 = o-quinone + H	1.53E+14	-2.02	3600.	0.01
		8.85E+21	-4.13	10679.	0.1
		1.05E+30	-6.30	18636.	1
		5.35E+36	-8.07	26536.	10
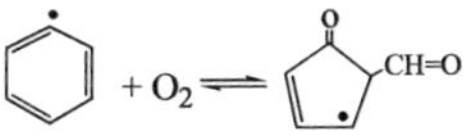	PHJ + O_2 = ODY(C5J)CDO[a]	8.44E-26	2.77	-30414.	0.01
		1.88E-91	20.83	-94810.	0.1
		3.57+209	-63.38	86717.	1
		6.83+242	-70.22	123785.	10
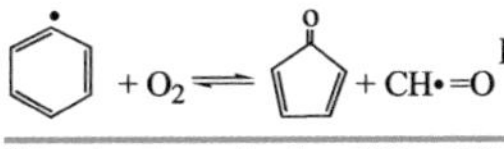	PHJ + O_2 = ODY(C5)+CJDO	9.89E+20	-3.40	4407.	0.01
		4.93E+28	-5.49	11579.	0.1
		1.65E+36	-7.51	19288.	1
		3.94E+41	-8.90	26296.	10

[a]kinetics determined at (900≤ T/K ≤ 2100K)

Table F3: Resulting Rate Constants in QRRK calculations for Dissociation Reactions of PhOOJ (dissociation of stabilized adducts)

Calculated Reaction Parameters at P = 0.01, 0.1, 1 and 10 atm, $k = A(T/\mathrm{K})^n \exp(-E_a/RT)$ (600≤ T/K ≤ 2500K)

Reactions	Reactions	*A*	*n*	*Ea* *(Kcal mol⁻¹)*	*P* *(atm)*
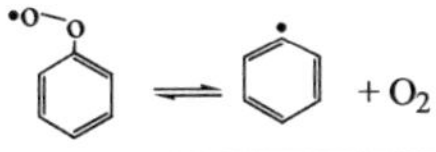	PHOOJ = PHJ + O_2	1.94E+80	-19.80	81.82	0.01
		4.31E+82	-20.15	83.87	0.1
		2.08E+84	-20.32	86.41	1
		8.05E+71	-16.43	81.68	10
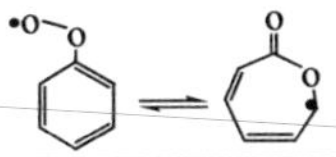	PHOOJ = Y(C6JO)DO	3.23E+76	-19.99	81.83	0.01
		6.28E+78	-20.32	83.86	0.1
		2.24E+80	-20.46	86.31	1
		1.06E+68	-16.59	81.60	10
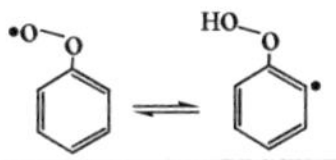	PHOOJ = PHJOOH	1.17E+76	-19.51	81.84	0.01
		1.97E+78	-19.82	83.86	0.1
		3.02E+79	-19.86	86.09	1
		2.34E+67	-16.08	81.35	10
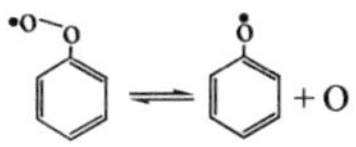	PHOOJ = PHOJ + O	7.71E+78	-19.71	81.84	0.01
		1.26E+81	-20.02	83.86	0.1
		1.57E+82	-20.03	86.04	1
		1.41E+70	-16.28	81.31	10
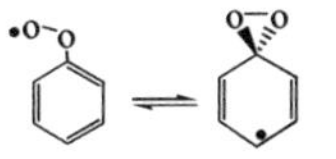	PHOOJ = Y(C6J)Y(CO2)	2.20E+79	-19.60	81.86	0.01
		2.62E+81	-19.87	83.85	0.1
		7.31E+81	-19.71	85.64	1
		3.01E+70	-16.18	81.00	10

Table F4: Resulting Rate Constants in QRRK calculations for Dissociation Reactions of ODC6JDO

Calculated Reaction Parameters at P = 0.01, 0.1, 1 and 10 atm, $k = A(T/K)^n \exp(-E_a/RT)$ (600≤ T/K ≤ 2500K)

Reactions	Reactions	*A*	*n*	*Ea* (*Kcal mol⁻¹*)	*P* (*atm*)
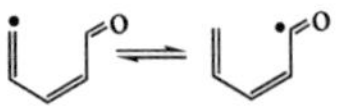	CJC4DO ↔ C4CJDO	8.29E+27	-5.92	19.45	0.01
		2.21E+36	-7.96	26.25	0.1
		3.79E+32	-6.53	26.39	1
		4.62E+33	-6.54	29.28	10
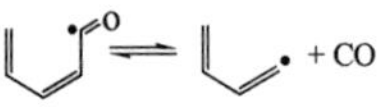	C4CJDO ↔ CDCCDCJ + CO	1.99E+29	-6.17	34.47	0.01
		4.51E+30	-6.26	34.89	0.1
		2.45E+32	-6.46	36.08	1
		2.17E+34	-6.71	37.92	10
	CDCCDCJ ↔ CDCJ + CTC	1.98E+26	-5.17	46.08	0.01
		1.99E+27	-5.17	46.08	0.1
		2.09E+28	-5.18	46.10	1
		3.16E+29	-5.23	46.30	10
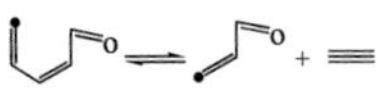	CJC4DO ↔ CJDCCDO + CTC	2.28E+64	-15.66	62.97	0.01
		1.36E+70	-16.98	68.59	0.1
		2.17E+61	-14.11	66.20	1
		3.57E+58	-13.03	66.89	10
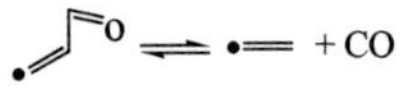	CJDCCDO ↔ CDCJ + CO	1.39E+10	-0.99	3.20	0.01
		1.39E+11	-0.99	3.20	0.1
		1.39E+12	-0.99	3.20	1
		1.39E+13	-0.99	3.21	10
	CJDCCDO ↔ CTC + CJDO	2.00E+24	-4.78	33.26	0.01
		2.00E+25	-4.78	33.27	0.1
		2.05E+26	-4.78	33.28	1
		2.59E+27	-4.81	33.41	10
	CJDCCDO ↔ CDCCJDO	3.54E+21	-4.10	25.06	0.01
		4.00E+22	-4.12	25.13	0.1
		9.10E+23	-4.21	25.63	1
		6.45E+25	-4.44	27.25	10
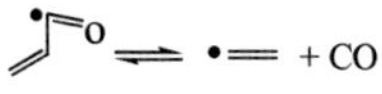	CDCCJDO ↔ CDCJ + CO	7.67E+22	-4.44	28.63	0.01
		7.75E+23	-4.44	28.64	0.1
		8.57E+24	-4.45	28.69	1
		1.88E+26	-4.54	29.10	10
	CDCJ ↔ CTC + H	5.37E+28	-5.91	40.19	0.01
		1.01E+30	-5.98	40.47	0.1
		1.40E+32	-6.29	41.80	1
		4.03E+34	-6.66	44.30	10
CH•=O ⇌ CO + H	CJDO ↔ CO + H	1.71E+18	-3.21	15.86	0.01
		1.78E+19	-3.22	15.88	0.1
		2.56E+20	-3.26	16.04	1
		1.75E+22	-3.48	16.98	10

Appendix G: Enthalpy of Formation Calculations

Table G1: Calculated $\Delta_f H^0_{298}$ in kcal mol^{-1} for Radicals versus work reaction

	Reactions Series		$\Delta_f H^0_{298}$	
			B3LYP	GA[b]
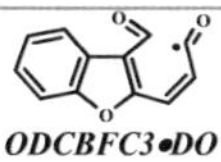	**ODCBFC3•DO** + CH_4 → Benzofuran + CO + CH_2=C=O + CH_2=CH•		-8.31	
	ODCBFC3•DO + CH_4 → Benzofuran + CO + CH•=C=O + CH_2=CH_2		-8.58	
ODCBFC3•DO		**Average**	**-8.44 ± 0.19**	**-8.19**
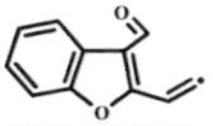	***ODCBFCDC•*** + CH_3CH_3 → Benzofuran + CH•=O + CH_2=CHCH=CH_2		51.77	
	ODCBFCDC• + CH_3CH_3 → Benzofuran + CH_2=O + CH•=CHCH=CH_2		49.81	
	ODCBFCDC• + CH_2=CH_2 → Benzofuran + CO + CH•=CHCH=CH_2		54.11	
	ODCBFCDC• + CH_4 → Benzofuran + CO + CH_2•CH=CH_2		49.2	
ODCBFCDC•		**Average**	**51.22 ± 2.21**	**52.32**
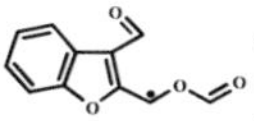	***ODCBFC•OCDO*** + CH_4 → Benzofuran + CO + CH_3CH_2• + CO_2		-63.46	
	ODCBFC•OCDO → Benzofuran + 2CO + CH•=O		-61.34	
ODCBFC•OCDO	***ODCBFC•OCDO*** + CH_2CH_2 → Benzofuran + CO + CH_2CHCH_2• + CO_2		-61.71	
		Average	**-62.17 ± 1.13**	**-63.54**
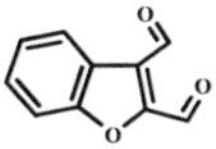	***ODCBFCDO*** + CH_4 → Benzofuran + CO + CH_3CH=O		-51.55	
	ODCBFCDO + 2CH_4 → Benzofuran + CH_3CH=O		-55.46	
	ODCBFCDO + CH_2=CH_2 → C_6H_6 + furan-CH=O + CH_2=O		-55.63	
	ODCBFCDO + CH_2=CH_2 + H_2 → C_6H_6 + furan + CH_2=O + CO		-54.99	
ODCBFCDO	***ODCBFCDO*** + CH_4 → Benzofuran + CO + CH_2=O		-54.61	
		Average	**-54.45 ± 1.66**	**-52.38**
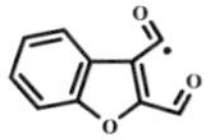	***ODC•BFCDO*** + CH_4 → Benzofuran + CO + CH_3C•=O		-15.32	
	ODC•BFCDO + CH_2=CH_2 → Benzofuran + CO + CH•=CHCH=O		-15.59	
	ODC•BFCDO + CH_3CH_3 → Benzofuran + CO + CH_3CH_2C•=O		-15.40	
ODC•BFCDO		**Average**	**-15.43 ± 0.14**	**-14.01**
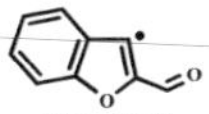	***BF•CDO*** + CH_2=CH_2 + H_2 → C_6H_6 + Furan + CH•=O		40.82	
	BF•CDO + H_2 → C_6H_6 + Benzofuran• + CH_2=O		40.48	
	BF•CDO + CH_2=CH_2 → Benzofuran• + CH_2=CHCH=O		41.46	
BF•CDO		**Average**	**40.92 ± 0.49**	**40.34**
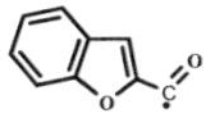	***BFC•DO*** + CH_2=CH_2 → Benzofuran + CH_2=CHC•=O		10.78	
	BFC•DO + 2CH_2=CH_2 → C_6H_6 + Furan + CH_2=CHC•=O		10.33	
BFC•DO	***BFC•DO*** + CH_2=CH_2 + H_2 → C_6H_6 + Furan + CH•=O		10.54	
		Average	**10.55 ± 0.22**	**9.12**
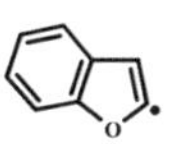	***BFURANJ1*** + CH_2=CH_2 → Benzofuran + CH_2=CH•		72.15	
	BFURANJ1 + C_6H_6 → Benzofuran + C_6H_5•		71.90	
	BFURANJ1 + CH_4 → C_6H_6 + CH_2=CHC•=O		72.02	
BFURANJ1	***BFURANJ1*** + CH_2=CH_2 → C_6H_6 + Furanj1		72.25	
		Average	**72.08 ± 0.15**	**72.91**
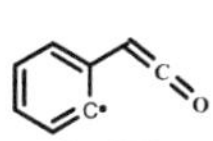	***B•KE*** + CH_4 + H_2 → C_6H_6 + CH•=O + CH_2=CH_2		69.71	
	B•KE → Benzofuranj1		69.58	
B•-KE	***B•KE*** + CH_2=CH_2 → C_6H_5• + CO + CH_2=C=CH_2		69.4	
	B•KE + CH_3CH_3 + H_2 → C_6H_6 + CH•=O + CH_3CH=CH_2		69.37	
		Average	**69.51 ± 0.16**	**68.90**
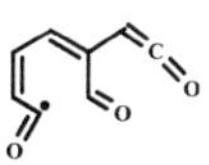	***ODC•C5DO-KE*** + CH_2CH_2 → CH_2=CHCH=CHCH=CH• + CH_2=C=O + 2CO		-6.95	
	ODC•C5DO-KE + CH_3CH_3 → O=C•CH=CH_2 + CH_2=CHCH=CHCH=O + CH_2=C=O		-6.99	
	ODC•C5DO-KE + CH_3CH_3 → CH_2=CHCH=O + CH_2=CHCH=CHCH=O + CH•=C=O		-7.52	
ODC•C5DO-KE		**Average**	**-7.15 ± 0.32**	**-8.40**

[a]Unit: kcal mol^{-1}; [b]values from group additivity method; [d]Ref. 71; [e]Ref. 30; [f]WS=wrong structure; [g]ref. 28

Table G1 (con't): Calculated $\Delta_f H^0_{298}$ in kcal mol^{-1} for Radicals versus work reaction

	Reactions Series	$\Delta_f H^0_{298}$ B3LYP	$\Delta_f H^0_{298}$ GA[b]
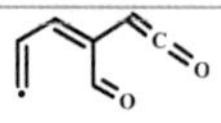	*C•C4DO-KE* + H_2 → CH•=CHCH=CHCH=O + CH_2=C=O	103.19	Verificati
	C•C4DO-KE + $2H_2$ → CH•=CHCH=O + CH_2=C=O + CH_2=CH_2	100.25	
	C•C4DO-KE + H_2 + CH_2=CH_2 → CH•=CHCH=O + CH_2=C=O + CH_2=CHCH=CH_2	102.75	
C•C4DO-KE	**Average**	**102.07 ± 1.58**	**36.91**
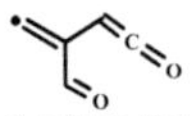	*C•C2DO-KE* + H_2 → CH•=CHCH=O + CH_2=C=O	28.29	
	C•C2DO-KE + CH_2=CH_2 → CH•=CHCH=CHCH=O + CH_2=C=O	31.22	
	C•C2DO-KE + CH_2=CH_2 → CH_2=CHCH=CHC•=O + CH_2=C=O	31.19	
C•C2DO-KE	**Average**	**30.23 ± 1.68**	**27.32**
HC≡CCH=O **CTCCDO**	*CTCCDO* + H_2 → CH≡CH + CH_2=O	26.90	
	CTCCDO + CH_2CH_2 → CH_2=CHCH=O + CH≡CH	28.98	
	CTCCDO + CH_4 → CH≡CCH_3 + CH_2=O	28.75	
	Average	**28.21 ± 1.14**	**25.9**
HC≡CCH=C=O **CTCCDCDO**	*CTCCDCDO* + CH_4 → CH≡CCH_3 + CH_2=C=O	38.81	
	CTCCDCDO + CH_3CH_3 → CH_2=CHCH=CH_2 + CH_2=C=O	40.33	
	CTCCDCDO + CH_2=O → CH≡CCH=O + CH_2=C=O	38.56	
	Average	**39.24 ± 0.95**	**41.06**
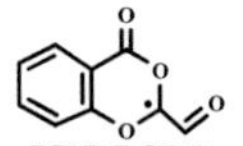	*BY(RJ)CDO* + CH_2CH_2 → C_6H_5OH + 3CO + CH_2=CH•	-73.60	
	BY(RJ)CDO → C_6H_5(CH=O)O•	-73.57	
BY(RJ)CDO	**BY(RJ)CDO** → C_6H_5O• + 3CO	-71.26	
	Average	**-72.81 ± 1.34**	**-72.84**
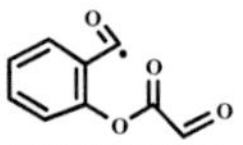	*OCJBOCOCO* + CH_2CH_2 → C_6H_5OH + 2CO + CH_2=CHC•=O	-61.76	
	OCJBOCOCO + CH_3CH_3 → C_6H_5OH + 2CO + CH_3CH_2C•=O	-59.58	
OCJBOCOCO	*OCJBOCOCO* + H_2 → C_6H_5OH + 2CO + HC•=O	-61.54	
	Average	**-60.96 ± 1.2**	**-64.19**
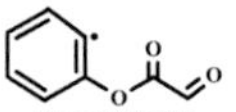	**BJOCOCO** + CH_4 → $C_6H_5OCH_3$ + CO + HC•=O	-12.14	
	BJOCOCO + $2CH_4$ → C_6H_6 + CH_3OCH_3 + CO + HC•=O	-12.97	
BJOCOCO	*BJOCOCO* + CH_4 → C_6H_5• + CH_3OH + 2CO	-12.75	
	Average	**-12.62 ± 0.43**	**-13.32**
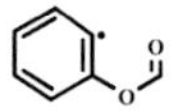	*BJOCO* + CH_4 → C_6H_5• + CH_3OCH=O	6.81	
	BJOCO + CH_4 → $C_6H_5OCH_3$ + HC•=O	8.72	
BJOCO	*BJOCO* + H_2 → C_6H_5O• + CH_2=O	5.93	
	Average	**7.16 ± 1.42**	**9.18**
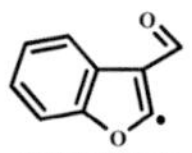	**BFJ1CDO** + CH_3CH_3 → BENZOFURANJ1+ CH_3CH_2CH=O	46.97	
	BFJ1CDO + CH_2=CH_2 → C_6H_5• + Furan-CH=O	46.14	
BFJ1CDO	**BFJ1CDO** → BENZOFURANJ1 + CO	47.35	
	Average	**46.82 ± 0.62**	**49.78**
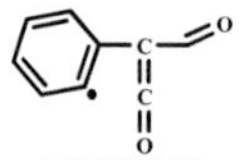	**BJKE(CO)** + CH_3CH_3 → C_6H_5• + CH_2=C=O + CH_2=CHCH=O	40.46	
	BJKE(CO) + CH_3CH_3 → C_6H_6 + CH_2=C=O + CH_2=CHC•=O	40.35	
BJKE(CO)	**Average**	**40.46 ±**	**45.77**
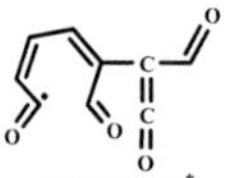	**OCJC5OR**[*] + $2H_2$ → CH_2=C=O + CH_2=O + 2CO + CH•=CHCH=CH_2	-36.80	
	OCJC5OR[*] + $2H_2$ → CH_2=C=O + CH_2=O + CO + CH_2=CHCH=CHC•=O	-40.82	
	OCJC5OR[*] + $2CH_4$ → CH_2=C=O + CH_2=O + CH_2=CHCH=O +CH•=CHCH=CHCH=O	-35.96	
OCJC5OR[*]	**Average**	**-37.86 ± 2.59**	**-37.13**
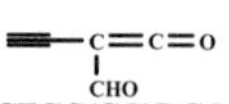	**CTCC(CO)DCO** + H_2 → CH_2=C=O + CH≡CH + CO	9.96	
	CTCC(CO)DCO + CH_4 → CH_2=C=O + CH_3C≡CH + CO	11.80	
CTCC(CO)DCO	**CTCC(CO)DCO** → + HC≡CCH=C=O + CO	12.15	
	Average	**11.30 ± 1.17**	**12.07**

Table G1 (con't): Calculated $\Delta_f H^0_{298}$ in kcal mol[-1] for Radicals versus work reaction

	Reactions Series	$\Delta_f H^0_{298}$ B3LYP	$\Delta_f H^0_{298}$ GA[b]
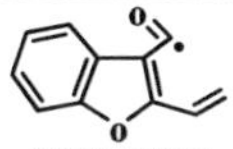 OCJBFCDC	**OC•BFCDC** + CH_2CH_2 → C_6H_5• + Furan-CH=CH_2 + CO	**27.78**	**30.12**
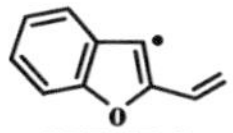 BFJ2CDC	**BFJ2CDC** + CH_2=CH_2 → C_6H_5• + Furan-CH=CH_2	86.26	
	BFJ2CDC + 2CH_2=CH_2 → C_6H_6 + Furanj2 + CH_2=CHCH=CH_2	83.59	
	BFJ2CDC + CH_2=CH_2 → C_6H_6 + Furan + CH•=CHCH=CH_2	82.83	
	Average	**84.23 ± 1.8**	**86.31**
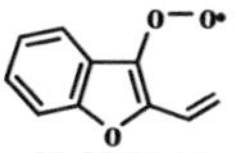 CDCBFOOJ	**CDCBFOOJ** + H_2 → Benzofuran + CH_2=CHOO•	27.24	
	CDCBFOOJ + H_2 → Furan + C_6H_5OO•	26.45	
	Average	**26.84 ± 0.56**	**26.00**
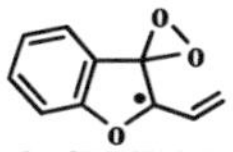 CDCBFJYOO	CDCBFOO• + H_2 → **CDCBFJYOO•**	36.34	
	CDCBFJYOO• + H_2 → FuranOO• + C_6H_6	36.12	
	Average	**36.23 ± 0.15**	**38.50**
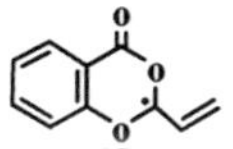 BYJCDC	**BYJCDC** + H_2 → C_6H_5CH=O + •OC(=O)CH=CH_2	-46.27	
	BYJCDC + H_2 → C_6H_5OH + CH_2=CHC•=O + CO	-44.23	
	Average	**-45.25 ± 1.44**	**-45.47**
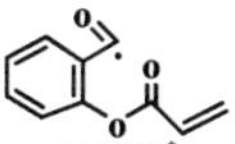 OCJBR*	**OCJBR*** + H_2 → C_6H_5• + CO + HOC(=O)CH=CH_2	-34.33	
	OCJBR*+ H_2 → C_6H_6 + CO + •OC(=O)CH=CH_2	-33.98	
	Average	**-34.16 ± 0.25**	**-32.42**
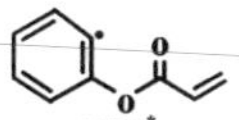 BJR*	**BJR*** + H_2 → Y(C_6H_5•)=O + CH_2=CHCH=O	15.67	
	BJR* + H_2 → Y(C_6H_6)=O + CH_2=CHC•=O	14.90	
	Average	**15.28 ± 0.54**	**19.76**
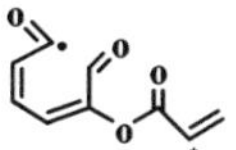 OCJC5OR*	**OCJC5OR*** + H_2 → CO + OHC(=O)CH=CH_2 + CH•=CHCH=CHCH=O	-60.73	
	OCJC5OR* + H_2 → CO + •OC(=O)CH=CH_2 + CH_2=CHCH=CHCH=O	-60.36	
	Average	**-60.54 ± 0.26**	**-60.37**
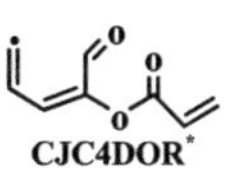 CJC4DOR*	**CJC4DOR*** + H_2 → OHC(=O)CH=CH_2 + CH•=CHCH=CHCH=O	-12.10	
	CJC4DOR*+ H_2 → •OC(=O)CH=CH_2 + CH_2=CHCH=CHCH=O	-11.79	
	CJC4DOR*+ 2H_2 → HC(=O)OH + CH_2=CH_2 + CH•=CHCH=CHCH=O	-11.82	
	Average	**-11.90 ± 0.17**	**-11.17**
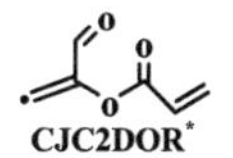 CJC2DOR*	**CJC2DOR*** → •OC(=O)CH=CH_2 + CH≡CCH=O	-14.87	
	CJC2DOR* + H_2 → CO_2 + CH•=CHCH=O + CH_2 CH_2	-10.18	
	CJC2DOR* + H_2 → HOCH=O + CH•=CH_2 + CH≡CCH=O	-14.69	
	Average	**-13.24 ± 2.66**	**-13.06**
 JO(CO)CDC	**JO(CO)CDC** + H_2 → HOCH=O + CH_2=CH•	-26.29	
	JO(CO)CDC + CH_2CH_2 → HOC(=O)CH=CH_2 + CH_2=CH•	-26.58	
	JO(CO)CDC + CH_3CH_3 → HOC(=O)CH=CH_2 + CH_3CH_2•	-26.54	
	Average	**-26.47 ± 0.16**	**-27.66**

*R = OC(=O)CH=CH_2

Appendix H: Reaction of DBF Radical + O_2 System

Scheme H1: Input file for QRRK analysis with master equation analysis for Fall-off DBFJ + O2 => Products with chirico value

```
Temp
5    1200.  1500. 1800. 2100. 2500.

Pres
4    0.01   0.1    1.  10. 50

chemact
dissoc

INPUT (A, n ,alpha, E kcal    k=AT^nexp(-alpha*T)exp(-E/RT) )
6.5068E+05   2.70856    0.0   0.0
# 3.0e13   0.0  0.0  0.0
#  ph + o2 => TS2-1

param (mass), sigma (A), e/k (K)
     199     5.82    617.0

collider (mass), sigma(A), e/k(K)  air here
        29.1   3.55   103.

#COLLIDER (take user's data if collider name preceded !  )
#   !N2
#   1. 40.0 3.621 97.5 400.
# EXP (temperature exponent to alpha, alpha=alpha(300)*(T/300)^N )
#  0.03
# Alternate collider input method
# COLLIDER (take default look up table data)
# N2
# 1.0

delta (E)
 745.  0.

Kfit   (uncomment if want modified Arrhenius fits)
INT   (integration interval in kcal)
  1.0

EHEAD integration stops at Ehead + highest barrier, def. 75 kcal
 75.
------------------------------------------------------------------------------------------------------------
WELL 1
 DBFOOJ
 FREQ
 3     483.2   25.403
     1322.7   26.829
     3003.7   7.769
 REACTANT
    DBFJ + O2
    1.0784E+11    1.63768   0.0    47.8
# from PHOOJ => TS2*1
 ISOMER
```

```
    BFYC6JYCO2
    1.75E+06    1.864    0.0   24.56
# Derived from PHOOJ => TS1*8LIN
  PRODUCT
    BFY(C6JO)DO
     1.19E+07  1.458  0.0   62.94
# Derived from phenyl PHOOJ => TS1*19 via Y(C6J)Y(C2O2)
   PRODUCT
    DBFJOOH
    4.33E+05   1.941   0.0    45.85
# PHOOJ => TSPHJOOH (with H-tunneling effect with G3MP2B3 42.5=chiungchu)
# PHOOJ => TSPHJOOH (with H-tunneling effect with G3MP2B3 45.8=nadia )
   PRODUCT
    DBZOFOJ + O
    2.02E+08    1.742   0.0    38.54
# Derived from PHOOJ =>  TS1*3LIN
  END
--------------------------------------------------------------------------------
WELL 2
  BFYC6JYCO2
  FREQ
  3   457.0     24.399
     1254.2    27.579
     2981.3     8.022
  ISOMER
    DBFOOJ
    4.00E+11   0.507     0.0           0.08
# Derived von phenyl Y(C6J)Y(CO2) => TS1*8LIN
  ISOMER
    BFY(C6JO)DO
    6.99E+09   0.856   0.0   10.58
# Derived from phenyl YC6JYCO2 => TS8*10LIN
END
--------------------------------------------------------------------------------
WELL 3
  BFY(C6JO)DO
  FREQ
  3  481.5   25.988
     1343.1  26.788
     3105.3   7.224
  ISOMER
    BFYC6JYCO2
    2.43E+08  1.163  0.0   84.45
# Derived from phenyl Y(C6JO)DO => TS8*10LIN
  ISOMER
    BFYC5JYC3ODO
    3.25E+11    0.364   0.0   63.67
# Derived from phenyl Y(C6JO)DO => TS10*11
  ISOMER
    BFODCJC3DO
    1.52E+11   0.516   0.0   38.01
# Derived from phenyl Y(C6JO)DO => TS21*10
  PRODUCT
    BFY(C5JO) + CO
    7.44E+09  0.523   0.0   69.70
# Derived from Y(C6JO)DO => TS10*17 via Y(C5O)CJDO
   END
--------------------------------------------------------------------------------
```

```
WELL 4
 BFYC5JYC3ODO
 FREQ
 3   452.9  22.023
     1215.0 29.299
     2950.0  8.678
 ISOMER
    BFY(C6JO)DO
    2.2169E+12  .36728  0.0  25.58
# Derived from phenyl Y(C5J)Y(C3O)DO => TS10*11
 PRODUCT
   BFYC5J + CO2
   2.7487E+12    0.31959   0.0   8.76
# Derived from phenyl Y(C5J)Y(C3O)DO  => TS11*12
 ISOMER
   BFY(C5)OCJDO
   2.7359E+12     0.36514   0.0    21.02
# Derived from phenyl Y(C5J)Y(C3O)DO => TS 11*13
  END
--------------------------------------------------------------------------------
WELL 5
 BFY(C5)OCJDO
 FREQ
 3   465.4    25.417
     1307.2   26.741
     3012.8    7.842
 ISOMER
   BFYC5JYC3ODO
   1.1625E+10  .35831    0.0    17.20
# Derived from phenyl Y(C5)OCJDO => TS11*13
 PRODUCT
   BFY(C5J) + CO2
   2.7191E+10    .49892     0.0    23.13
# sebbar+hadad Y(C5)OCJDO => TS13*14
 END
--------------------------------------------------------------------------------
WELL 6
  BFODCJC3DO
  FREQ
  3   459.4   25.830
      1260.0  26.932
      3097.1   7.238
  ISOMER
   BFY(C6JO)DO
   3.4654E+10   .67656   0.0    4.9603
# Derived from phenyl ODC6JDO = TS21*10
  PRODUCT
   ODCBFC3JDO
   2.6356E+12   .30166  0.0   0.0
# sebbar from BFODCJC3DO => TODBFC6JDO abstraction
END
```

Appendix I: Basis Sets

A basis sets is the mathematical description of the orbitals within a system (which in turn combine to approximate the total electronic wavefunction) used to perform the theoretical calculation. Larger basis sets more accurately approximate the orbitals by imposing fewer restrictions on the locations of the electrons in space. In the true quantum mechanical picture, electrons have a finite probability of existing anywhere in space; this limit corresponds to the infinite basis set expansion in the chart we looked at previously.

There are two types of basis functions (also called Atomic Orbitals, AO) commonly used in electronic structure calculations: Slater Type Orbitals (STO) and Gaussian Type Orbitals (GTO). In term of computational efficiency, GTOs are preferred, and used almost universally as basis functions in electronic structures calculations. Having decided on the type of function, the most important factor is the number of functions to be used.

Basis sets assign a group of basis functions to each atom within a molecule to approximate its orbitals. These functions themselves are composed of a linear combination of gaussian functions; such basis functions are referred to as *contracted* functions, and the component Gaussian functions are referred as *primitives*.

Contracted basis sets

The contracted basis sets is developed from "core electrons". Considering that the important chemical region is the outer valence and the "inner" part changes very little and that the computational cost increases as the fourth power of the number of basis functions, this is very inefficient.

For example: 6-31G is a split valence basis, where the core orbitals are a contraction of 6 primary Gaussian Type Orbitals (PGTO), the inner part of the valence orbitals is a contraction of three PGTOs and the outer part of the valence represented by one PGTO.

6-311G is a triple valence basis consisting of 6 contractions of PGTO and of three split functions represented by three, one and one PGTOs respectively.

Split valence basis sets

The first way that a basis set can be made larger is to increase the number of basis functions per atom. Split valence basis sets, such as 3-21G and 6-31, have two (or more) sizes of basis function for each valence orbital. For example, hydrogen and carbon are represented as:

H: 1s, 1s'

C: 1s, 2s, 2s', $2p_x$, $2p_y$, $2p_z$, $2p_{x'}$, $2p_{y'}$, $2p_{z'}$

Where the prime and unprimed orbitals orbitals differ in size.

Polarised basis sets

Split valence basis sets allow orbitals to change size, but not to change shape. Polarized basis sets remove this limitation; for example, polarized sets add **d** functions to carbon atom and **p** functions for hydrogen atom.

6-31G(d) or also known as 6-31G* indicates that it is the 6-31G basis set with d functions added to heavy atoms. Another popular polarized basis sets is 6-31G(d,p), also known as 6-31G**, which adds p functions to hydrogen atoms in addition to the d functions on heavy atoms.

Diffuse functions

Diffuse functions are large-size versions of s- and p-type functions. They allow orbitals to occupy a larger region of space. Diffuse functions are important for systems where electrons are relatively far from the nucleus. The 6-31+G(d) basis set is the 6-31G(d) basis set with diffuse functions added to heavy atoms. The double plus, 6-31++G(d), adds diffuse functions to the hydrogen atoms as well, which make a significant difference in accuracy.

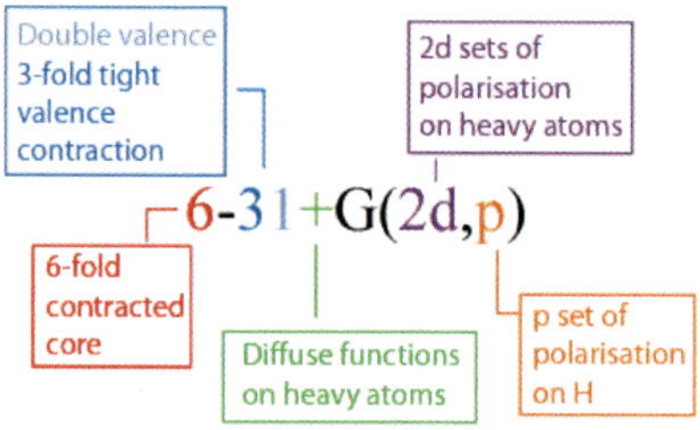

Figure I1: Basis Sets definition

10. References

[1] Word Health Organisation, http://www.who.int/en/

[2] Benson, S. W. *J. Phys. Chem.* **1996**, *100*, 13544-13547.

[3] Rauk, A.; Yu, D.; Armstrong, D. A. *J. Am. Chem. Soc.* **1994**, *116*, No. 18.

[4] Niki, F. F.; Marker, P. D.; Savage, C. M.; Breitenbach, L. P. *J. Chem. Phys.* **1982**, *86*, 3925.

[5] (a) Wallington, T. J.; Gierzak, C. A.; Ball, J. C.; Japar, S. M. *Int. J. Chem. Phys.* **1989**, *21*, 1077.

(b) Wallington, T. J.; Japar, S. M. *Int. J. Chem. Phys. Lett.* **1990**, *166*, 495.

(c) Wallington, T. J.; Japar, S. M. *Int. J. Chem. Phys. Lett.* **1990**, *167*, 513.

[6] Wallington, T. J.; Dagaut, P. *Chem. Rev.* **1992**, 92, 1967.

[7] Wu, F.; Carr, R. W. *J. Phys. Chem.* **1992**, *66*, 1743.

[8] Griffith, J. F. *Combust. Flame* **1993,** *93*, 202-206

[9] Koert, D. N.; Pitz, W. J.; Bozzelli, J. W.; Cernansky, N. P. *Twenty-Sixth Symposium (International) on Combustion. The Combustion Institute,* Pittsburgh PA, **1996**, 633.

[10] Sheng, C; Bozzelli, J. W. *J. Phys. Chem. A* **2002**, *106*, 1113-1121.

[11] Gutman, D. *J. Chem. Phys.* **1987**, *84*, 409-414.

[12] Wagner, A. F.; Slagle, I. R.; Sarzynski, D.; Gutman, D. *J. Chem. Phys.* **1990**, *94*, 1853-1868.

[13] Morgan, C. A.; Pilling, M. J.; Tulloch, M. J.; Ruiz R. P.; Bayes, K. D. *J. Chem. Soc., Faraday Trans. 2*, **1982**, 78, 1323-1330.

[14] Knyazev V. D.; Slagle I. R.; *J. Phys. Chem. A*. **1998**, *102*; 1770-1778.

[15] Clifford, E. P.; Wenthold, P. G.; Gareyev, R.; Lineberger, W. C.; DePuy, C. H.; Bierbaum, V. M.; Ellison, G. B. *J. Chem. Phys.* **1998**, *109* (23), 10293.

[16] Blanksby, S. J.; Ramond, T. M.; Davico; G. E.; Nimlos, M. R.; Kato, S.; Bierbaum, V. M.;

Lineberger, W. C.; Ellison, G. B.; Okumura, M. *J. Am. Chem. Soc.* **2001**, *123*, 9585-9596.

[17] Yu, T.; Lin, M.C. *J. Am. Chem. Soc* **1994**, *116*, 9571.

[18] Curtis, L. A.; Raghavachari, K.; Redfern, P. C.; Pople, J.A. *J. Chem. Phys. A* **2000**, *112*, 7374.

[19] Baboul, A. G.; Curtiss, L. A.; Redfern, P. C.; Raghavachari, K. *J. Chem. Phys.*, **1999**, *110* (16), 7650-7657.

[20] Benson, S. W. *Thermochemical Kinetics*, 2nd ed.; Wiley Interscience: New York, **1976**.

[21] Joback, Kevin G., "A Unified Approach to Physical Property Estimation Using Multivariate Statistical Techniques", MS Thesis, MIT, June **1982**.

[22] Reid, R. C.; Prausnitz, J. M.; Poling, B. E. *The properties of gases and liquids*, 4th Ed. Mc. Graw-Hill Book Company **1987**.

[23] Cohen, N. *J. Phys. Chem.ref. Data* **1996**, *25*, 1411.

[24] Cohen, N.; Benson, S. W. *Chem. Rev.*(Washington, D.C) **1993**, *93*, 2419.

[25] Lay, T. H.; Bozzelli, J. W.; Dean, A. M..; Ritter, E. R. *J. Phys. Chem.* **1995**, *99*, 14514-14527.

[26] Sebbar, N.; Bockhorn, H.; Bozzelli, J. W. *Phys. Chem. Chem. Phys.* **2002**, *4*, 3691-3703.

[27] Sebbar, N.; Bockhorn, H.; Bozzelli, J. W. *Phys. Chem. Chem. Phys.* **2003**, *5*, 300-307.

[28] Sebbar, N.; Bockhorn, H.; Bozzelli, J. W. *J. Phys. Chem. A* **2004,** *108,* 8353-8366.

[29] Sebbar, N; Bockhorn, H.; Bozzelli, J. W. *J. Phys. Chem. A* **2005,** *109,* 2233-2253.

[30] Sebbar, N; Bockhorn, H.; Bozzelli, J. W. *Int. J. Chem. Kinet.* **2005**, *37*, 633-648,.

[31] Sebbar, N.; Bockhorn, H.; Bozzelli, submitted to *31th International Symposium on Combustion.*

[32] Fadden, M.J.; Barckholtz, C.; Hadad, C. M. *J. Phys. Chem A* **2000**, *104*, 3004-3011.

[33] Mebel, A.M.; Diau, E.W.G.; Lin, M.C.; Morokuma, K.J. *J. Am. Chem. Soc.* **1996**, *118*, 9759.

[34] Sebbar, N.; Bockhorn, H.; Bozzelli, J. W. *in preparation.*

[35] Jensen, Frank "*Introduction to Computational Chemistry*" John Wiley & Sons Ltd, Baffins Lanes, Chichester, West Sussex PO19 1UD, England

[36] Foresman, James B. "*Exploring chemistry with electronic structure methods: a guide to using Gaussian*" Æleen Frisch. - 1. ed., Pittsburg, Pa.: Gaussian, 1993.

[37] Hehre, W. J.; Radom, L.; Schleyer, P. v. R.; Pople, J. A. "*Ab initio Molecular Orbital Theory*" (New York: Wiley, 1986), 135-226.

[38] Curtiss, L. A.; Raghavachari, K.; Redfern, P. C.; Rassolov, V.; Pople, J. A.; *J. Chem. Phys.A* **1998**, *109*, 7764.

[39] Baboul, A. G.; Curtiss, L. A.; Redfern, P. C.; Raghavachari, K. *J. Chem. Phys.A* **1999**, *110*, 7650.

[40] D. R. Salahub and M. C. Zerner, Eds., The Challenge of d and f Electrons (ACS, Washington, D.C., 1989).

[41] R. G. Parr and W. Yang, Density-Functional Theory of atoms and molecules (Oxford Univ. Press, Oxford, 1989.

[42] Becke, A. D. *J. Chem. Phys.*, **1993**, *98*, 1372.

[43] Lee, C.; Yang, W.; Parr, R. G. *Phys. Rev.* **1988**, B41, 785.

[44] http://gems.mines.edu/~reactionxml/hindered-rotor-methods.doc.

[45] Shokhirev, N. V. and Krasnoperov, L. N. ROTATOR-A Computer Code to Calculate Energy Levels of a One-Dimensional Rotor with Arbitrary Potential94.

[46] Sumathi, R.; Carstensen, H. H.; Green, W. H. Jr. *J. Phys. Chem. A* **2001**, *105*, 6910-6925.

[47] Lay, T. H.; Krasnoperov, L. N.; Venanzi, C. A.; Bozzelli, J. W.; Shokhirev, N. V. *J. Phys. Chem.* **1996**, *100*, 8240-8249.

[48] Redfern, P. C.; Zapol, P.; Curtiss, L. A.; Raghavachari, K.; *J. Phys. Chem.* **2000**, *104*, 5850-5854.

[49] Curtiss, L. A.; Raghavachari, K.; Redfern, P. C.; Stefanov, B. B.; *J. Chem. Phys.* **1998**, *108* (2), 692.

[50] Benson, S. W.; Buss J. H. *J. Chem. Phys.* **1958**, *29*, 546.

[51] Furuyama, S.; Golden, D.M.; Benson, S.W., *J. Chem. Thermodyn.*, **1969**, *1*, 363-375.

[52] Lindemann, F. A. *Trans. Faraday Soc.* **1922**, *17*, 598.

[53] Steinfeld, J. I.; Francisco, J. S.; Hase, W. L. Chemical Kinetics and Dynamics Prentice Hall: NJ, **1989**.

[54] Robinson, P. J.; Holbrook, K. A. Unimolecular Reactions; John Wiley and Sons: New York, **1971**.

[55] Hinshelwood, C. N. *Proc. Roy. Soc. A* **1927**, *113*, 230.

[56] Slater, N. B. *Proc. Comb. Phil. Soc.* **1939**, *56*, 35.

[57] Rice, O. K.; Ramsperger, H. C. *J. Am. Chem. Soc.* **1927**, *49*, 1617.

[58] Kassel, L. S. *J. Phys. Chem.* **1928**, *32*, 225.

[59] Kassel, L. S. *J. Phys. Chem.* **1928**, *32*, 1065.

[60] Kassel, L. S. Kinetics of Homogenous Gas Reaction; Chemical Catalog Co.: New York, **1932**.

[61] Marcus, R. A. *J. Chem. Phys.*, **1952**, *20*, 359.

[62] Bozzelli, J. W.; Dean, A. M.; Chang, A. *Int. J. Chem. Kinet.* **1997**, *29*, 161-170.

[63] Chang, A. Y.; Bozzelli, J. W.; Dean, A. M. *Zeit. Phys. Ch.* **2000**, 1533-1568.

[64] Gaussian 98 (Revision A.11), M. J. Frisch, G. W. Trucks, H. B. Schlegel, G. E. Scuseria, M. A. Robb, J. R. Cheeseman, V. G. Zakrzewski, J. A. Montgomery, Jr., R. E. Stratmann, J. C. Burant, S. Dapprich, J. M. Millam, A. D. Daniels, K. N. Kudin, M. C. Strain, O. Farkas, J. Tomasi, V. Barone, M. Cossi, R. Cammi, B. Mennucci, C. Pomelli, C. Adamo, S. Clifford, J. Ochterski, G. A. Petersson, P. Y. Ayala, Q. Cui, K. Morokuma, P. Salvador, J. J. Dannenberg, D. K. Malick, A. D. Rabuck, K. Raghavachari, J. B. Foresman, J. Cioslowski, J. V. Ortiz, A. G. Baboul, B. B. Stefanov, G. Liu, A. Liashenko, P. Piskorz, I. Komaromi, R. Gomperts, R. L. Martin, D. J. Fox, T. Keith, M. A. Al-Laham, C. Y. Peng, A. Nanayakkara, M. Challacombe, P. M. W. Gill, B. Johnson, W. Chen, M. W. Wong, J. L. Andres, C. Gonzalez, M. Head-Gordon, E. S. Replogle, and J. A. Pople, Gaussian, Inc., Pittsburgh PA, 2001.

[65] http://www.gaussian.com/index.htm.

[66] Healy, E. F.; Holder, A. *J. Mol. Struct.* **1993**, *281*, 141.

[67] Pople, J. A.; Schlegel, H. B.; Krishnan, R.; DeFrees, D. J.; Binkley, J. S.; Frisch, M. J.; Whiteside, R. A.; Hout, R. F.; Hehre, W. J. *Int. J. Quantum Chem., Quantum Chem. Symp.* **1981**, *15*, 269.

[68] Peterson, P. E.; Abu-Omar, M.; Johnson, T.W.; Parham, R.; Goldin, D.; Henry, C.; Cook, A.; Dunn, K. M. *J. Phys. Chem.* **1995**, *99*, 5927.

[69] Pople, J. A.; Scott, A. P.; Wong, M. W.; Radom, L. *Int. J. Chem.* **1993**, *33*, 345.

[74] Sheng, C; PhD Dissertation **2002** Department of Chemical Engineering, Chemistry and Environmental Science, New Jersey Institute of Technology, Newark NJ 07102 USA.

[75] NIST Standard Reference Database; Number 69 ed.; http://webbook.nist.gov/chemistry/: February **2000**.

[76] TRC.; Thermodynamic Propertie s of the Substances in Ideal Gas State; 1.0M ed.; Thermodynamic Research Center of the Texas Engineering Experiment Station: Texas A&M University, **1994**.

[77] Pitzer, K. S.; Gwinn, W. D., *J. Chem. Phys.* **1942**, *10*, 428.

[78] McClurg, R. B.; Flagan, R. C.; Goddard III, W. A. *J. Chem. Phys.* **1997**, *106*, 6675.

[79] McClurg, R. B. *J. Chem. Phys.* **1999**, *111*, 7165.

[80] Knyazev, V. D. *J. Chem. Phys.* **1999**, *111*, 7161.

[81] Ritter, E. R. *J. Chem. Inf. Comput. Sci.* **1991**, *31*, 400 (THERM is distributed free by writing to the authors).

[82] Ritter, E.; Bozzelli, J. W. *Int. J. Chem. Kinet.* **1991**, *23*, 767. (THERM is distributed free by writing to the authors).

[83] Bozzelli, J. W.; Dean, A. M.; Chang, A. *Int. J. Chem. Kinet.*, **1997**, *29*, 161-170.

[84] Olzmann, M.; Kraka, E.; Cremer, D.; Gutbrod, R.; Andersson, S. *J. Phys. Chem. A* **1997**, *101*, 9421-9429.

[85] Baldwin, R. W.; Dean, C. E.; Walker, R. W. *J. Chem. Soc., Faraday Trans* 2 **1986**, *82*, 1445.

[86] Chen, C., Bozzelli, J., W.; *J. Phys. Chem. A* **2000**, 104, 4997-5012.

[87] Chen, C., Bozzelli, J., W.; *J. Phys. Chem. A* **2000**, 104, 9715-9732.

[88] Yamada, T.; Lay, T. H.; Bozzelli, J. W.; *J. Chem. Phys.A.* **1998**, *102*, 7286.

[89] Yamada, T.; Lay, T. H.; Bozzelli, J. W.; *J. Chem. Phys. A.* **1999**, *103*, 5602.

[90] M.; Jonsson, M.; *J. Phys. Chem.* **1996,** *100,* 6814-6818.

[91] Brinck, T.; Lee, H.-N.; Jonsson, M. *J. Phys. Chem. A* **1999**, *103*, 7094-7104.

[92] Sheng, C.; Dean, A. M.; Bozzelli, J. W.; Chang, A. Y. *J. Phys. Chem. A* **2002**, *106*, 7276-7293.

[93] Ignatyev, I. S.; Xie, Y.; Allen, W. D.; Schaefer, H. F. *J. Chem. Phys. A* **1997**, *103*, 141.

[94] Quelch, G. E.; Gallo, M.; Schaefer III, H. F. *J. Am. Chem. Soc* **1992**, *114*, 8239.

[95] Sumathi R. and Green, Jr. W. H. *Phys. Chem. Chem. Phys.*, **2003**, *5*, 3402-3417.

[96] Lee, C.; Yang, W.; Parr, R. G. *Phys. Rev.* **1988**, *B41*, 785.

[97] Becke, A. D. *J. Chem. Phys.* **1993**, *98*, 1372.

[98] Montgomery, J. A.; Ocherski, J. W.; Peterson, G. A. *J. Chem. Phys.* **1994**, *101*, 5900.

[99] Durant, J. L. *Chem. Phys. Lett.* **1996**, 256, 595.

[100] E. F. Byrd, D. Sherril, M. H. Gordon *J. Phys. Chem.* **2001,** 105.

[101] Curtiss, L. A.; Raghavachari, K.; Redferm, P. C.; Pople, J. A. *J. Chem. Phys.* **1997**, 106, 1063.

[102] Durant, J. L. Rohlfing, C. M. *J. Chem. Phys.* **1993**, 98, 8031.

[103] Petersson, G. A.; Malick, D. K.; Wilson, W. G. *J. Chem. Phys.* **1998**, *109*, 10570.

[104] Lay, T. H., Bozzelli, J., W.; *J. Phys. Chem. A* **1997**, 101, 9505-9510.

[105] Benassi, R.; Folli, U.; Sbardellati, S.; Taddei, F. *J. Comput. Chem.* **1993**, 14, 379

[106] Mordaunt, D. H.; Ashfold, M. N. R. *J. Phys. Chem.* **1994**, 101, 2630-2631.

[107] Sumathi R. and Green, Jr. W. H. *Phys. Chem. Chem. Phys.*, **2003**, *5*, 3402-3417.

[108] Slagle, I. R.; Gutman, D. J. *J. Am. Chem. Soc.* **1985,** *107*, 5342

[109] Khachatryan, L. A.; Niazyan, O. M.; Mantashyan, A. A.; Vedeneev, V. I.; Teitelboim, M. A. *Int. J. Chem. Kinet.* **1982**, *14*, 1231

[110] Knyazev V. D.; Slagle I. R.; *J. Phys. Chem. A*. **1998**, *102*; 1770-1778.

[111] Lee, J.; Bozzelli, J. W. *Inc. Int. J. Chem. Kinet.*, **2003**, *35*, 20–44.

[112] J. A. Montgomery, Jr., M. J. Frisch, J. W. Ochterski, and G. A. Petersson *J. Chem. Phys.*, **2000**, *112*, 6532.

[113] Bach, R. D.; Ayala, P. Y.; Schlegel, H. B., *J. Am. Chem. Soc* **1996**, *118*, 12578-12765.

[114] Chen, C.-C.; Bozzelli, J. W. *J. Phys. Chem. A* **2003**, *107*, 4531-4546.

[115] Lee, J.; Bozzelli, J. W. *J. Phys. Chem. A* **2003**, *107*, 3778-3791.

[116] Lee, J.; Chen, C.-J.; Bozzelli, J. W. *J. Phys. Chem. A* **2002**, *106*, 7155-7170.

[117] Turecek, F., Cramer, C. J. *J. Am. Chem. Soc.* **1995**, *117*, 12243.

[118] Turecek, F., Brabec, L.; Korvola, J. *J. Am. Chem. Soc*. **1988**, *110*, 7984.

[119] Carballeira, L.; Mosquera, R. A.; Rios, M. A.; *J. Comput. Chem.* **1988**, *9*, 851.

[120] Scott, A. P.; Radom, L. *J. Phys. Chem.* **1996**, *100*, 16502-16513.

[121] Stocker, D. W.; Pilling, M. J., manuscript in preparation. Preliminary results presented at the *14th International Symposium on Gas Kinetics*, Sept **1996**, Leeds, UK.

[122] Sun, H.; Bozzelli, J. W. *J. Phys. Chem. A* **2000**, *104*, 8270-8282.

[123] Reints, W.; Pratt, D. A.; Korth, H.-G.; Mulder, P. *J. Phys. Chem. A* **2000**, *104*, 10713-10720.

[124] Seiser, R.; Pitsch, H.; Seshadri, K.; Pitz, W. J.; Curran, H. J.; *28th International Symposium on Combustion, 2000, University of Edinburgh, Scotland.*

[125] Ribaucour, M.; Minetti, R.; Sochet, L. R.; Curran, H. J.; Pitz, W. J.; Westbrook, C. K.; *28th International Symposium on Combustion, 2000, University of Edinburgh, Scotland*

[126] Fadden, M. J.; Hadad, C. M. *J. Phys. Chem. A* **2000**, *104*, 8121.

[127] Chen, C; Wong, D. K.; Bozzelli, J. W. *J. Phys. Chem.* **1998**, *102*, 4551.

[128] Turecek, F.; Havlas, Z., *J. Org. Chem.*, **1986**, *51*, 4066.

[129] Cohen, N. *J. Phys. Chem. Ref. Data,* **1996**, *25*, 1411.

[130] Chen, C.; Bozzelli, J. W. *J. Phys. Chem. A* **1998**, *102*, 4551-4558.

[131] Stein; S., E. and Fahr, A, *J. Phys. Chem.*, **1985**, *89* (17), 3714.

[132] Sumathi, R; Green, W. H. Jr. *J. Phys. Chem. A* **2002**, 106, 11141.

[133] Sun, H.; Bozzelli, J. W. *J. Phys. Chem. A* **2001**, 105, 4504-4516.

[134] Gutman *J. Phys. Chem.*, 96 5889, **1992**.

[135] Ervin, K. M.; Gronert, S.; Barlow, S. E.; Gilles, M. K.; Harrison A. G.; Bierbaum, V. M.; DePuy, C. H.; Lineberger, W. C.; Ellison, G. B. *J. Am. Chem. Soc.* **1990,** *112,* 5750-5759.

[136] Melius, C. F.; Allendorf, M. D. *J. Phys. Chem. A* **2000**, *104,* 2168-2177

[137] Blanksby, S. J.; Ellison, B. G. *Acc. Chem. Res.* **2003**, *36*, 255-263.

[138] May, K.; Dapprich, S.; Furche, F.; Unterreiner, B. V.; Ahlrichs R. *Phys. Chem. Chem. Phys.* **2000**, *2*, 5084-5088.

[139] May, K.; Unterreiner, B. V.; Dapprich, S.; Ahlrichs, R. *Phys. Chem. Chem. Phys.* **2000**, *2*, 5089-5092.

[140] Janoschek, R.; Rossi M. J. *Int. J. Chem. Kinet.* **2002**, *34*: 550-560.

[141] Raghavachari, K.; Stefanov B. B.; Curtiss, L. A. *Molecular Physics*, **1997**, *91*(3), 555-559.

[142] Jursic, B. S. *Int J Quant Chem* **1997**, *64*: 263-269.

[143] Swart, M.; Snijders, J.G. *Theor Chem Acc*, **2003**, *110*:34–41.

[144] Byrd, E. F. C.; Sherrill, C. D.; Head-Gordon, M. *J. Phys. Chem. A*, **2001**, *105*, 9736-9747.

[145] Janoschek, R. *Pure Appl Chem*, **2001**, *73*, 9, 1521–1553.

[146] Sumathi, R.; Green, Jr. W. H. *Phys. Chem. Chem. Phys.* **2003**, *5*(16), 3402-3417.

[147] Hine, J.; Klueppet, A.W. *J. Am. Chem. Soc.* **1974**, *96*, 2924-2929.

[148] Hall, H.K., Jr.; Baldt, J.H. *J. Am. Chem. Soc.* **1971**, *93*, 140-145.

[149] Curtiss, L. A.; Raghavachari, K.; Redfern, P. C.; Rassolov, V.; Pople, J. A. *J. Chem. Phys.* **1998**, *109*, 7764-7776.

[150] Curtiss, L. A.; Raghavachari, K.; Trucks, G. W.; Pople, J. A. *J. Chem. Phys.* **1991**, *94*, 7221.

[151] Barckholtz C.; Fadden, M. J.; Hadad, C. M. *J. Phys. Chem. A* **2000**, *104*, 3001-3011.

[152] Blanksby, S. J.; Ellison, B. G. *Acc Chem Res* **2003**, *36*, 255-263.

[153] Lovell, A. B.; Brezinsky, K.; Glassman, I. *Twenty-Second Symposium (International) on Combustion. The Combustion Institute,* Pittsburgh, PA **1988**, 1063.

[154] Shandross, R. A.; Longwell, J. P.; and Howard, J. B. *Twenty-Sixth Symposium (International) on Combustion. The Combustion Institute,* Pittsburgh, PA **1996**, 711.

[155] Lindstedt, R. P.; Skevis, G. *Combust. Flame* **1994**, *99*, 551.

[156] Zhang, H.Y.; and Mckinnon, J.T., *The Combust. Sci. and Tech.*, **1995**, *107*, 261.

(157) Fujii, N.; and Asabi, T., *Fourth-teenth Symposium (International) on Combustion. The Combustion Institute,* Pittsburgh, PA **1972**, 433.

[158] Bittner, J.D.; Howard, J.B. *Eighteenth Symposium (International) on Combustion. The Combustion Institute,* Pittsburgh, PA **1980**, 1105.

[159] Brezinsky, K.; Lovell, A. B.; and I. Glassman; *Combust. Sci. and Tech.*, **1990**, *70*, 33.

[160] Brezinsky, K.; Litzinger, T. A. ; and Glassman, I.; *Int. J. Chem. Kinet.* **1984**, *16*, 1053.

[161] Manion, J. A.; and Louw, R. *J. Phys. Chem.* **1989**, *93*, 3564.

[162] Zhong, X.; Bozzelli, J. W., *Int. J. Chem. Kinet.* **1997**, *29*, 893.

[163] Chen, C.-C.; Bozzelli, J. W.; Farrell, J. T. *J. Phys. Chem. A* **2004,** *108*, 4632-4652.

[164] Colussi, A.J.; Zabel, F.; Benson, S.W., *Int. J. Chem. Kinet*. **1977**, *9*, 161.

[165] Lin, C-Y.; Lin, M. C. *Int. J. Chem. Kinet.* **1985,** *17*, 1025.

[166] Olivella, S.; Solé, A.; Garcia, Raso, A. *J. Phys. Chem*. **1995**, *99*, 10549.

[167] Zhu, L., Bozzelli J. W. *J. Phys. Chem. A* **2003,** *107*, 3696-3703.

[168] Frank, P.; Herzler, J.; Just, Th.; Wahl, C. *Twenty-Fifth Symposium (International) on Combustion. The Combustion Institute,* Pittsburgh, PA **1994**, 833.

[169] Mebel, A. M.; Lin, M. C. *J. Am. Chem. Soc.* **1994**, *116*, 9577.

[170] Bittker, D.A., *Combut. Sci. Technol*, **1991**, *79*, 47.

[171] Davis, S. G.; Wang, H.; Brezinsky, K.; and Law, C. K. *Twenty-Sixth Symposium (International) on Combustion. The Combustion Institute,* Pittsburgh, PA **1996**, 1025.

[172] Tan, Y.; Frank P.; *Twenty-Sixth Symposium (International) on Combustion. The Combustion Institute,* Pittsburgh, PA **1996**, 677.

[173] Emdee, J. I.; Brezinsky,K.; and I. Glassman; *J. Phys. Chem.* **1992**, *96*, 2151.

[174] Venkat, C.; Brezinsky, K; and Glassman, I., *Ninth-tenth Symposium (Internatinal) on Combustion. The Combustion Institute*, Pittsburgh, PA **1982**, 143-152.

[175] Brezinsky, K. *Prog. Energy Combust. Sci.* **1986**, *12*, 1.

[176] Glarborg, P.; Alzueta, M. U.; Dam-Johansen, K. *International Journal of Chemical Kinetics* **2000**, *32* (8), 498.

[177] DiNaro, J. L.; Howard, J. B.; Green, W. H.; Tester, J. W.; Bozzelli, J. W. *J. Phys. Chem. A.;* **2000**, *104*, 10576-10586.

[178] Barckholtz, C.; Fadden, M. J.; Hadad, C. M. *J. Phys. Chem. A* **1999,** *103*, 8108-8117.
Fadden, M. J.; Haddad, C. M. *J. Phys. Chem. A* **2000,** *104*, 8121.

[175] Carpenter, B. K. . *J. Am. Chem.* **1993**, 115, 9806-9807.
Carpenter, B. K. . *J. Am. Chem.* **1995**, 99, 9801-9810.

[180] Tokmakov, I. V.; Kim, G-S; Kislov, V. V.; Mebel, A. M.; M. C. Lin *J. Phys. Chem. A.;* **2005**; *109*(27); 6114-6127

[181] Mebel, A. M.; Morokuma, K.; Lin, M. C. *J. Chem. Phys.* **1995**, 103, 7414.

[182] Curtiss, L. A.; Raghavachari, K.; Trucks, G. W.; Pople, J. A. *J. Chem. Phys.* **1991**, 94, 7221.

[183] Curtiss, L. A.; Ragavachari, K.; Pople; J. A. *J. Chem. Phys.* **1993**, 98, 1293.

[184] Curtiss, L. A.; Carpenter, J. E.; Ragavachari, K., Pople; J. A. *J. Chem. Phys.* **1992**, 96, 9030.

[185] Prosen, E. J.; Maron, F. W.; Rossini, F. D. *J. Res. NBS*, **1951**, 46, 106–112.

[186] Warnatz, J., Combustion Chemistry, ed. W.C. Gardiner,Jr., pub. Springer-Verlag,NY, 1984

[187] Knyazev, V.D.; Slagle, I.R., *J. Phys. Chem.*, 100, **1996**, 16899 – 16911.

[188] Friedrichs, G.; Herbon, J.T.; Davidson, D.F.; Hanson, R.K. *Phys. Chem. Chem. Phys*., 4, **2002**, 5778 – 5788.

[189] Mebel, A. M.; Kislov, V. V. *J. Phys. Chem. A.;* (Letter**), 2005**, *109*(32); 6993-6997.

[190] Sebbar, N.; Bockhorn, H.; Bozzelli. J. W. *5th International Conference on Chemical Kinetics*, 16 – 20 July **2001**, National Institute of standards and Technologies, Gaithersburg, MD USA

[191] Sebbar, N.; Bockhorn, H.; Chen, C-J.; Bozzelli, J. W. *17th International Symposium on Gas Kinetics*, Essen, August 24-29, **2002.**

[192] Sebbar, N.; Bockhorn, H.; Bozzelli. J. W. *The 8th International Congress on Combustion by-Products: Origin, Fate and Health Impacts*, Umeå, Sweden, June 17-19, **2003**.

[193] Sebbar, N.; Bockhorn, H.; Bozzelli. J. W. *Proceeding of the European Combustion Meeting 2003*, Orléans, France, October 25-28, **2003.**

[194] Chase, M. W. *NIST-JANAF Themochemical Tables; Fourth Edition*, *J. Phys. Chem. Ref. Data*, Monograph 9, New York, **1998,** 1-1951.

[195] Frenkel, M.; Kabo, G. J.; Marsh, K. N.; Roganov, G. N.; Wilhoit, R. C. *Thermodynamics of Organic Compounds in the Gas State;* volume 1, Thermodynamic Research Center, Texas A&M University System, College Station, Texas, **1994**.

[196] Stull, D. R.; Westrum, E. F. Jr.; Sinke, G. C. *The Chemical Thermodynamics of Organic Compounds;* Robert E. Krieger, Publishing, Malabar, Florida, **1987**.

[197] Steinfeld, J. I.; Francisco, J. S.; Hase, W. L. "*Chemical Kinetics And Dynamics*", 1989 Prentice-Hall, Inc. A Simon & Schuster Company Englewoods, New Jersey 07632.

[198] Wigner E., Trans. Faraday Soc. **1938**, 34, 29.

[199] Westmoreland, P. R., Howard, J. B., Longwell, J. P. and Dean, A. M., *AICHE J.* **1986**, 32, 1971.

[200] Bozzelli, J. W., Dean, A. M., Ritter, E. R., *Combustion Science and Technology* **1991**, 80, 69

[201] Chen, C-J; Bozzelli, J., W. Personal Communication.

[202] Tsang, W.; Hampson, R.F., *J. Phys. Chem. Ref. Data*, 15, 1986

[203] Zaslonko, I.S.; Mukoseev, Yu.K.; Tyurin, A.N., *Kinet. Catal.*, 29, **1988**.

[204] Dean, A.M.; *J. Phys. Chem.*, 89, 1985.

[205] Krasnoperov, L. N.; Chesnokov, E. N.; Stark, H.; Ravishankara, A. R. *J. Phys. Chem. A* **2004,** *108,* 11526-11536.

[206] Hippler, H.; Striebel, F.; Viskolcz, B. *Phys. Chem. Chem. Phys.*, 3, **2001**, 2450 - 2458

[207] H.Y. Zhang, J.T. Mckinnon, The Combust. Sci. and Tech. 107 (1995) 261-300.

[208] R. A. Shandross, J. P. Longwell, J. B. Howard, *Twenty-Sixth Symp (Int'l) on Combustion. The Combustion Institute*, Pgh, Pa **(1997)** 711-719

[209] P. Glarborg, M.U. Alzueta, K. Dam-Johansen, *International Journal of Chemical Kinetics* 32 (8) **(2000)** 498.

[210] K, Brezinsky, *Prog. Energy Combust. Sci.* **1986**, *12*, 1-24.

[211] Becke, A. D. *J. Chem. Phys.* **1993**, *98*, 5648.

[212] Lee, C.; Yang, W.; Parr, R. G. *Phys. Rev. B* **1988**, *37*, 785.

[213] Montgomery, J. A.; Ocherski, J. W.; Petersson, G. A., *J. Chem. Phys.* **1994**, *101,* 5900.

[214] R. D. Chirico; B. E. Gammon; S. E. Knipmeyer; A. Nguyen; M. M. Strube; C. Tsonopoulos; W. V. Steele *The J. Chem. Thermodyn.* **1990**, *22* (11), 1075-1096.

[215] Sabbah, R.; *Bull. Soc. Chim. Fr.* **1991**, *128*, 350.

[216] Kline, J.M.; Penner, S.S.; *Proc. Int. Symp. Shock Tubes Waves*, 1982, 13

[217] Prosen, E.J.; Rossini, F.D. *J. Res. NBS*, **1945**, 263-267.

[218] Pittam, D.A.; Pilcher, G., *J. Chem. Soc. Faraday Trans. 1*, **1972**, *68*, 2224-2229

[219] Chase, M.W., Jr., NIST-JANAF Themochemical Tables, Fourth Edition, *J. Phys. Chem. Ref. Data, Monograph 9*, **1998**, 1-1951.

[220] Pedley, J. B.; Rylance, J. Sussex-N.P.L., *Computer Analyred Thermochemical Data: Organic and Organometallic Compounds* (University of Sussex, England, 1977)

[221] Furuyama, S.; Golden, D.M.; Benson, S.W., *J. Chem. Thermodyn.*, **1969**, *1*, 363-375.

[222] Prosen, E.J.; Maron, F.W.; Rossini, F.D., *J. Res. NBS*, **1951**, *46*, 106-112.

[223] Cox, J. D.; Pilcher, G. *Thermochemistry of Organic and Organometallic Compounds*; Academic Press: New York, 1970.

[224] Fraser, F.M.; Prosen, E.J., *J. Res. NBS*, **1955**, *54*, 143-148.

[225] Fang, W.; Rogers, D.W., *J. Org. Chem.*, **1992**, *57*, 2294-2297.

[226] Wagman, D.D.; Kilpatrick, J.E.; Pitzer, K.S.; Rossini, F.D. *J. Res. NBS*, **1945**, *35*, 467-496

[227] Prosen, E.J.; Gilmont, R.; Rossini, F.D *J. Res. NBS*, **1945**, 34, 65-70.

[228] Prosen, E.J.; Rossini, F.D., *J. Res. NBS*, **1945**, 34, 59-63.

[229] Furuyama, S.; Golden, D.M.; Benson, S.W., *J. Chem. Thermodyn.*, **1970**, *2*, 161-169.

[230] Roth, W.R.; Adamczak, O.; Breuckmann, R.; Lennartz, H.-W.; Boese, R., *Chem. Ber.*, **1991**, *124*, 2499-2521.

[231] Cox, J.D.; Wagman, D.D.; Medvedev, V.A., CODATA Key Values for Thermodynamics, *Hemisphere Publishing Corp.*, New York, **1984**, 1

[232] Stull, D. R.; Prophet, H.; *JANAF Thermochemical Tables*, 2nd Ed.; U.S. Government Printing Office: Washington DC, 1970.

[233] Tsang, W., Martinho Simoes, J.A.; Greenberg, A.; Liebman, J.F., eds., Blackie Academic and Professional, London, 1996, 22-58

[234] Feller, D.; Dixon, D. A.; Francisco, J. S. *J. Phys. Chem. A* **2003,** *107,* 1604-1617.

[235] Knyazev V. D.; Slagle I. R.; *J. Phys. Chem. A*. **1998**, *102*; 1770-1778.

[236] Sun, H., Bozzelli, J., W.; *J. Chem. Phys. A* **2001**, *105*, 4504.

[237] Mebel, A. M.; Diau, E. W. G.; Lin, M. C.; Morokuma, K. *J. Am. Chem. Soc*. **1996**, *118*, 9759.

[238] Cox, J.D.; Wagman, D.D.; Medvedev, V.A., CODATA Key Values for Thermodynamics, *Hemisphere Publishing Corp.*, New York, **1984**, 1

[239] Ruscic, B.; Feller, D.; Dixon, D. A., Peterson, K. A.; Harding, L. B.; Asher, R. L and Wagner, A. F. *J. Phys. Chem. A* **2001**, *105*, 1-4.

[240] Ribaucour, M.; Minetti, R.; Sochet, L. R.; Curran, H. J.; Pitz, W. J.; Westbrook, C. K.; *28th International Symposium on Combustion, 2000, University of Edinburgh, Scotland*

[241] Chen, C-C; Bozzelli,J. W. Conference

[242] Da Silva, G.; Bozzelli, J. W.; Sebbar, N.; Bockhorn, H. *ChemPhysChem* **2006**, *7*, 1-9.

[243] Wiberg, K.B.; Crocker, L.S.; Morgan, K.M., *J. Am. Chem. Soc.*, **1991**, *113*, 3447-3450.

[244] Pedley, J. B.; Naylor, R. D.; Kirby, S. P. *thermochemical Data of Organic Compounds,* 2nd ed.; Chapman and Hall: London, 1986.

[245] Green, J.H.S., *Chem. Ind. (London)*, **1960**, 1215-1216.

[246] Dolliver, M.A.; Gresham, T.L.; Kistiakowsky, G.B.; Smith, E.A.; Vaughan, W.E., *J. Am. Chem. Soc.*, **1938**, *60*, 440-450

[247] Turecek, F.; Havlas, Z. **J**. *Org. Chem.*, **1986**, *51*, 4066-4067

[248] Turecek, F. *J. Chem. Soc. Chem. Commun.*, **1984**, 1374-1375.

[249] Turecek, F.; Brabec, L.; Korvola, J. *J. Am. Chem. Soc.*, **1988**, *110*, 7984-7990

[250] Guthrie, J.P., *Can. J. Chem.*, **1978**, *56*, 962-973.

[251] Guthrie, J.P., *J. Am. Chem. Soc.*, **1974**, *96*, 3608-3615

[252] Pilcher, G.; Pell, A.S.; Coleman, D.J., *Trans. Faraday Soc.*, **1964**, *60*, 499-505.

[249] Hall, H.K., Jr.; Baldt, J.H., *J. Am. Chem. Soc.*, **1971**, *93*, 140-145.

[254] Kim, C.-H.; Bozzelli, J. W., Masters Thesis, Department of Chemical Engineering, Chemistry and Environmental Science, New Jersey Institut of Technology, January 2002.

[255] Dolliver, M.A.; Gresham, T.L.; Kistiakowsky, G.B.; Smith, E.A.; Vaughan, W.E., *J. Am. Chem. Soc.*, **1938**, *60*, 440-450.

[262] Turecek, F.; Havlas, Z., *J. Org. Chem.*, **1986**, *51*, 4066-4067.

[257] Turecek, F. *Tetrahedron Lett.*, 1984, *25*, 5133-5134.

[258] Holmes, J.L.; Lossing, F.P., *J. Am. Chem. Soc.*, **1982**, *104*, 2648-2649.

[259] Cox, J.D., *Pure Appl. Chem.*, **1961**, *2*, 125-128.

[260] Papina, T.S.; Pimenova, S.M.; Luk'yanova, V.A.; Kolesov, V.P., *Russ. J. Phys. Chem. (Engl. Transl.)*, **1995**, *69*, 1951-1953, In original 2148.

[261] Anthoney, M.E.; Carson, A.S.; Laye, P.G., *J. Chem. Soc. Perkin Trans. 2*, **1976**, 1032-1036.

[262] Guthrie, G.B., Jr.; Scott, D.W.; Hubbard, W.N.; Katz, C.; McCullough, J.P.; Gross, M.E.; Williamson, K.D.; Waddington, G., *J. Am. Chem. Soc.*, **1952**, *74*, 4662-46.

[263] Kudchadker, S.A.; Kudchadker, A.P., *Ber. Bunsenges. Phys. Chem.*, **1975**, *12*, 432-437.

[264] Yamada, T.; Lay, T. H.; Bozzelli, J. W. *J. Phys. Chem. A,* **1999**, *103*, 5602.

[265] Fletcher, R.A.; Pilcher, G., *Trans. Faraday Soc.*, **1970**, *66*, 794-799.

[266] Sun, H., Bozzelli, J., W.; *J. Chem. Phys. A* **2001**, *105*, 4504.

[267] Orlov, V.M.; Krivoruchko, A.A.; Misharev, A.D.; Takhistov, V.V., *Bull. Acad. Sci. USSR, Div. Chem. Sci.*, **1986**, 2404-2405.

List of Publications

1. "Experimental and Numerical Investigation of the Degradation of Chlorinated Hydrocarbons in Incineration Systems"
   Kraft, M.; Fey, H.; Procaccini, C.; Smith, K. A.; Longwell, J. P.; Sarofim, A. F.; Bonni, P.; Rutz, L.; Sebbar, N.; Bockhorn, H. *VDI-Bericht 18. Deutsch-Niederländischen Flammentag*, **1997.**

2. "Structures, Thermochemical Properties (Enthalpy, Entropy and Heat Capacity), Rotation Barriers, Bond Energies of Vinyl, Allyl, Ethynyl and Phenyl hydroperoxides"
   Sebbar, N.; Bockhorn, H.; Bozzelli, J. W. *Phys. Chem. Chem. Phys.* **2002**, 4, 3691-3703.

3. "Thermodynamic Properties (*S(298), Cp(T)*), Internal Rotations and Group Additivity Parameters in Vinyl and Phenyl Hydroperoxides"
   Sebbar, N.; Bockhorn, H.; Bozzelli, J. W., *Phys. Chem. Chem. Phys.* **2003**, 5, 300-307.

4. "Reaction of Dibenzofuran radical with O_2"
   Nadia Sebbar and Henning Bockhorn and Joseph W. Bozzelli *Proceeding of the European Combustion Meeting 2003*, Orléans, France, October 25-28, **2003.**

5. "Structures, Thermochemical Properties (Enthalpy, Entropy and Heat Capacity), Rotation Barriers, Bond Energies of Vinyl, Allyl, Ethynyl and Phenyl peroxides"
   Sebbar, N.; Bockhorn, H.; Bozzelli, J. W. *J. Phys. Chem. A* **2004**, *108*(40), 8353-8366.

6. "Thermochemical Properties, Rotation Barriers, and Group Additivity for Unsaturated Oxygenated Hydrocarbons and Radicals Resulting from Reaction of Vinyl and Phenyl Radical Systems with O_2"
   Sebbar, N.; Bockhorn, H.; Bozzelli, J. W. *J. Phys. Chem. A*; **2005**, *109*, 2233-2253.

7. "Enthalpy of Formation and Bond Energies on Unsaturated Oxygenated Hydrocarbons using G3MP2B3 Calculation Methods"
   Sebbar, N.; Bockhorn, H.; Bozzelli, J. W. *Int. J. Chem. Kinet.* **2005**, *37*, 633–648.

8. "Thermodynamic and Ab Initio Analysis of the Controversial Enthalpy of Formation of Formaldehyde"
   daSilva, G.; Bozzelli, J. W.; Sebbar, N.; Bockhorn, H. *ChemPhysChem*, **2006**, *7*, 1119 – 1126.

Presentations

1. "Reaction of Phenyl Radical with O_2: Thermodynamic Properties, Important Reaction Paths and Kinetics"
   Joseph Bozzelli, Nadia Sebbar, William Pitz and Henning Bockhorn
   The 2nd Joint Meeting of the US Sections of the Combustions Institute, Oakland, California, March 25-28, **2001**

2. "Reaction of Phenyl Radical and Dibenzofuran with O_2: Thermodynamic Properties, Reaction Pathways, Kinetics and Initial Steps for Dibenzofuran Oxidation"
   Nadia Sebbar, Henning Bockhorn and Joseph W. Bozzelli
   Seventh International Congress on Toxic Combustion By-Products. June 4 – 6, **2001**, Research Triangle Park, North Carolina USA

3. "Reaction of Phenyl Radical and Dibenzofuran with O_2"
   Nadia Sebbar, Henning Bockhorn and Joseph W. Bozzelli
   Joint Meeting of the Belgian and Dutch Sections of the Combustion Institute, Brussels Belgium. April **2002**.

Posters

1. “Numerical and Experimental Investigation of Pyrolysis and Oxidation of Dioxin Precursors and other Toxic By Products”
   Petra Bonni, Leonhard Rutz, Nadia Sebbar and Henning Bockhorn
   WIP. The combustion Institute, 27th International Symposium on Combustion, **1998** University of Colorado, Boulder, USA, August 2- 7.

2. “Atomized Validation of a Detailed Mechanism for the Description of Dioxins Formation and other by products” Nadia Sebbar, Jörg Appel and Henning Bockhorn
   The Combustion Institute, Joint Meeting of the British, German and French Sections. 18 – 21 May, Nancy – France, **1999**

3. “Atomized Validation of a Detailed Mechanism for the Description of Dioxins Formation and other by products” Nadia Sebbar, Jörg Appel and Henning Bockhorn
   The Sixth International Congress on Toxic Combustion By products. June 27 – 30, **1999**, University of Karlsruhe, Germany.

4. “Reaction of Phenyl and Dibenzofuran Radicals with O_2: Thermodynamic Properties, Reaction Pathways and kinetics”
   Nadia Sebbar, Henning Bockhorn and Joseph, W. Bozzelli.
   The combustion Institute, 28th International Symposium on Combustion, 30 July – 4 August, **2000**, Edinburgh, Scotland

5. “Thermodynamic properties (Enthalpies, Entropies and Heat Capacities) and Reactions of Vinyl hydroperoxides, peroxy radicals and phenyl hydroperoxides”
   Nadia Sebbar, Henning Bockhorn and Joseph, W. Bozzelli.
   Seventh International Congress on Toxic Combustion By-Products. June 4 – 6, **2001**, Research Triangle Park, North Carolina USA

6. “Thermodynamic properties (Enthalpies, Entropies and Heat Capacities) and Reactions of Vinyl hydroperoxides, peroxy radicals and phenyl hydroperoxides”
   Nadia Sebbar, Henning Bockhorn and Joseph, W. Bozzelli.
   5th International Conference on Chemical Kinetics, 16 – 20 July **2001**, National Institute of standards and Technologies, Gaithersburg, MD USA

7. “Thermodynamic properties and reactions of vinyl hydroperoxides and phenyl hydroperoxides”
   Nadia Sebbar, Henning Bockhorn and Joseph, W. Bozzelli.
   Deutsche Bunsen-Gesellschaft für Physikalische Chemie. 77th International Discussion Meeting „Formation and Degradation of Hydrocarbons in High-Temperature Reactions, October **2001**

8. “Reaction of Phenyl Radical and Dibenzofuran with O_2: Thermodynamic Properties, Reactions Pathways,
   Kinetics and Initial Steps for Dibenzofuran”.
   Nadia Sebbar, Henning Bockhorn and Joseph, W. Bozzelli.
   Deutsche Bunsen-Gesellschaft für Physikalische Chemie. 77th International Discussion Meeting „Formation and Degradation of Hydrocarbons in High-Temperature Reactions, October **2001**.

9. “Reaction of Phenyl Radical with O_2: Thermodynamic Properties, Reaction Pathways and Kinetics”
   Nadia Sebbar and Henning Bockhorn
   Chiung-ju Chen and Joseph, W. Bozzelli.
   17th International Symposium on Gas Kinetics, Essen, August 24-29, **2002.**

10. “Reaction of Dibenzofuran radical with O_2: Thermodynamic Properties, Reaction Pathways and Kinetics”
    Nadia Sebbar, Henning Bockhorn and Joseph, W. Bozzelli.
    The 8th International Congress on Combustion by-Products: Origin, Fate and Health Impacts, Umeå, Sweden, June 17-19, **2003**.

Printed in Dunstable, United Kingdom

91343235R00136